Elements of Real Analysis

PURE AND APPLIED MATHEMATICS

A Program of Monographs, Textbooks, and Lecture Notes

EXECUTIVE EDITORS

Earl J. Taft
Rutgers University
Piscataway, New Jersey

Zuhair Nashed
University of Central Florida
Orlando, Florida

EDITORIAL BOARD

M. S. Baouendi
University of California,
San Diego

Jane Cronin
Rutgers University

Jack K. Hale
Georgia Institute of Technology

S. Kobayashi
University of California,
Berkeley

Marvin Marcus
University of California,
Santa Barbara

W. S. Massey
Yale University

Anil Nerode
Cornell University

Freddy van Oystaeyen
University of Antwerp,
Belgium

Donald Passman
University of Wisconsin,
Madison

Fred S. Roberts
Rutgers University

David L. Russell
Virginia Polytechnic Institute
and State University

Walter Schempp
Universität Siegen

Mark Teply
University of Wisconsin,
Milwaukee

MONOGRAPHS AND TEXTBOOKS IN PURE AND APPLIED MATHEMATICS

Recent Titles

J. Haluska, The Mathematical Theory of Tone Systems (2004)

C. Menini and F. Van Oystaeyen, Abstract Algebra: A Comprehensive Treatment (2004)

E. Hansen and G. W. Walster, Global Optimization Using Interval Analysis, Second Edition, Revised and Expanded (2004)

M. M. Rao, Measure Theory and Integration, Second Edition, Revised and Expanded (2004)

W. J. Wickless, A First Graduate Course in Abstract Algebra (2004)

R. P. Agarwal, M. Bohner, and W-T Li, Nonoscillation and Oscillation Theory for Functional Differential Equations (2004)

J. Galambos and I. Simonelli, Products of Random Variables: Applications to Problems of Physics and to Arithmetical Functions (2004)

Walter Ferrer and Alvaro Rittatore, Actions and Invariants of Algebraic Groups (2005)

Christof Eck, Jiri Jarusek, and Miroslav Krbec, Unilateral Contact Problems: Variational Methods and Existence Theorems (2005)

M. M. Rao, Conditional Measures and Applications, Second Edition (2005)

A. B. Kharazishvili, Strange Functions in Real Analysis, Second Edition (2006)

Vincenzo Ancona and Bernard Gaveau, Differential Forms on Singular Varieties: De Rham and Hodge Theory Simplified (2005)

Santiago Alves Tavares, Generation of Multivariate Hermite Interpolating Polynomials (2005)

Sergio Macías, Topics on Continua (2005)

Mircea Sofonea, Weimin Han, and Meir Shillor Analysis and Approximation of Contact Problems with Adhesion or Damage (2006)

Marwan Moubachir and Jean-Paul Zolésio, Moving Shape Analysis and Control: Applications to Fluid Structure Interactions (2006)

Alfred Geroldinger and Franz Halter-Koch, Non-Unique Factorizations: Algebraic, Combinatorial and Analytic Theory (2006)

Kevin J. Hastings, Introduction to the Mathematics of Operations Research with *Mathematica*®, Second Edition (2006)

Robert Carlson, A Concrete Introduction to Real Analysis (2006)

John Dauns and Yiqiang Zhou, Classes of Modules (2006)

N. K. Govil, H. N. Mhaskar, Ram N. Mohapatra, Zuhair Nashed, and J. Szabados, Frontiers in Interpolation and Approximation (2006)

Luca Lorenzi and Marcello Bertoldi, Analytical Methods for Markov Semigroups (2006)

M. A. Al-Gwaiz and S. A. Elsanousi, Elements of Real Analysis (2006)

Elements of Real Analysis

M. A. Al-Gwaiz
King Saud University
Riyadh, Saudi Arabia

S. A. Elsanousi
King Saud University
Riyadh, Saudi Arabia

Chapman & Hall/CRC
Taylor & Francis Group
Boca Raton   London   New York

Chapman & Hall/CRC is an imprint of the
Taylor & Francis Group, an informa business

Chapman & Hall/CRC
Taylor & Francis Group
6000 Broken Sound Parkway NW, Suite 300
Boca Raton, FL 33487-2742

© 2007 by Taylor & Francis Group, LLC
Chapman & Hall/CRC is an imprint of Taylor & Francis Group, an Informa business

No claim to original U.S. Government works
Printed in the United States of America on acid-free paper
10 9 8 7 6 5 4 3 2 1

International Standard Book Number-10: 1-58488-661-7 (Hardcover)
International Standard Book Number-13: 978-1-58488-661-7 (Hardcover)

This book contains information obtained from authentic and highly regarded sources. Reprinted material is quoted with permission, and sources are indicated. A wide variety of references are listed. Reasonable efforts have been made to publish reliable data and information, but the author and the publisher cannot assume responsibility for the validity of all materials or for the consequences of their use.

No part of this book may be reprinted, reproduced, transmitted, or utilized in any form by any electronic, mechanical, or other means, now known or hereafter invented, including photocopying, microfilming, and recording, or in any information storage or retrieval system, without written permission from the publishers.

For permission to photocopy or use material electronically from this work, please access www.copyright.com (http://www.copyright.com/) or contact the Copyright Clearance Center, Inc. (CCC) 222 Rosewood Drive, Danvers, MA 01923, 978-750-8400. CCC is a not-for-profit organization that provides licenses and registration for a variety of users. For organizations that have been granted a photocopy license by the CCC, a separate system of payment has been arranged.

Trademark Notice: Product or corporate names may be trademarks or registered trademarks, and are used only for identification and explanation without intent to infringe.

Visit the Taylor & Francis Web site at
http://www.taylorandfrancis.com

and the CRC Press Web site at
http://www.crcpress.com

Preface

This book presents a course on real analysis at the senior undergraduate level. It is based on lecture notes prepared (in Arabic) by the authors, and used over the last fifteen years to teach two semester courses on the subject to mathematics students, usually in their third or fourth year, at King Saud University (KSU). The students who enroll in the first course would have completed, among other things, an introductory algebra course on foundations and the usual calculus sequence. Chapter 1 and much of chapter 2 (excluding completeness) are designed to present the foundations part needed, while the following seven chapters basically cover the theory of calculus on the real line. The last two chapters, on the Lebesgue theory of measure and integration, have been somewhat expanded and could serve as an introduction to this theory for senior undergraduates or beginning graduates. The book is therefore, to a large extent, self-contained.

Analysis is one of the main pillars of mathematics, and a solid course in analysis in $\mathbb{R}$ at the undergraduate level is important for students of mathematics on two counts. On the one hand, it develops the analytical skills and structures needed for handling the basic notions of limits and continuity (and, by extension, differentiability, integrability, etc.) in a simple and concrete setting. On the other hand, it prepares the student for conducting and appreciating analysis in higher dimensions and more abstract spaces, for the same notions and ideas developed in $\mathbb{R}$ are then generalized, usually in a purely formal and predictable procedure, to fit the new setting. Thus the seeds of analysis are planted in this first encounter with real analysis.

Chapter 1 includes the basic preliminary concepts of sets, relations, and functions needed for formulating mathematical ideas. In Chapter 2 the real number system is defined axiomatically, based on its field, order, and completeness properties. Its important subsets, including the natural numbers, the integers, and the rational numbers, are then derived by imposing additional conditions. This would seem to run against the natural tendency of starting with simple structures to build more complex ones, but it has the advantage of reaching $\mathbb{R}$ quickly and precisely. A concrete method for constructing the real numbers by decimal expansions is then outlined in Section 2.5, and this is used in the following section to show that the real numbers are uncountable. Countability as such does not play a significant role in the chapters to follow, but is included here more for the beauty of its ideas than for anything else.

Sequences and their convergence properties are the subject of Chapter 3, and its importance derives from several considerations: First of all, its intimate connection with the completeness property, and hence the structure of $\mathbb{R}$, whereby every real number is seen to be the limit of a rational sequence. Secondly, its pivotal role in the treatment of infinite series and their convergence tests in Chapter 4. Thirdly, sequences provide a convenient approach to the concept of the limit of a function (through Theorem 5.1), which is the fundamental concept of real analysis. Thus many results in Chapter 5 on the limit of a function are proved either by the ε-δ method or by using the convergence properties of a sequence, whichever is more convenient.

Chapter 6 is on continuity, with particular attention to uniform continuity and continuity on an interval and on a compact set. We end the chapter with a brief word on the more general characterization of continuity of a function, namely, that the inverse image of an open set is open in the domain. Differentiation and Riemann integration are taken up in Chapters 7 and 8. Here we go over the same grounds as calculus, but with more care, depth, and rigour.

Chapter 9 is on sequences and series of functions and their modes of convergence, which naturally builds on the properties of numerical sequence and series. It allows us to define the exponential, logarithmic, and trigonometric functions analytically through power series, as it provides the tools for proving the basic theorems of Lebesgue integration in the following chapters.

The last two chapters are on the Lebesgue theory of measure and integration. This, of course, involves a qualitative jump in the level of abstraction, which we tried to minimize without sacrificing rigour. We followed the standard approach of defining outer measure on the subsets of $\mathbb{R}$, and then restricting it through Carathéodory's condition to form Lebesgue measurable sets. In defining the Lebesgue measurability of a function f, we could not resist the temptation to base it on the measurability of $f^{-1}(B)$ for every Borel set B. Such a definition is better motivated, as it corresponds to the (generalized) characterization of continuity; and the choice of Borel sets guarantees the measurability of continuous functions. The more common definition, based on the measurability of the inverse image of any semi-infinite interval, which is shown in Theorem 10.6 to be an equivalent definition, only works because the σ-algebra of Borel sets is generated by such intervals.

The Lebesgue integral is defined in stages, starting with simple functions and ending with measurable functions, as is usually done. The fruits of this theory are now available in a sequence of powerful results which include the monotone convergence theorem, Fatou's lemma, the dominated convergence theorem, and the bounded convergence theorem. They all point to the superior behaviour of the Lebesgue integral, compared to the Riemann integral, under (pointwise) limiting operations. In a certain sense, these results show how the Lebesgue integrable functions constitute a sort of "completion" of the Riemann integrable functions. In the last section, we finally provide a necessary and sufficient condition for a bounded function on a bounded interval to be Riemann integrable, a condition we could not have derived within the theory of Riemann integration as it relies on the notion of Lebesgue measure.

In treating a topic as rich and sophisticated as measure and integration, we tried to make it as readable and self-contained as we could within the space and time available. Hence the essential tools are developed at a moderate pace, and the build-up of the theory is, hopefully, gentle and motivated. The result is a "mini-course" which we realize will still be challenging to the average undergraduate, but one which we hope will be accessible none-the-less. Not all universities here offer this material at the undergraduate level, and we have had to resist attempts to remove it completely from the syllabus. Any coherent treatment of the Sturm-Liouville theory, Fourier series, and orthogonal functions, even at the undergraduate level, requires some knowledge of $\mathcal{L}^2$ space. $\mathcal{L}^1$, on the other hand, is the natural domain of definition of the Fourier transformation, and the continuity of the transform function is then a direct consequence of the dominated convergence theorem. Furthermore, with the increasing role of random effects in many disciplines, it is hoped that these last two chapters will also provide a good introduction to advanced courses in the theory of probability and stochastic processes.

For the last few years, an Arabic version of this book has been available in two volumes, tailored to cover the two-course sequence mentioned earlier and offered by our mathematics department. Each course takes up 15 weeks, delivered at the rate of three lectures and one problem session per week. In the first course, Chapter 1 and most of Chapter 2 are skimmed over, as they are covered in a previous course; only completeness is discussed at length. The rest of the time is spent on the following five chapters, up to and including Chapter 7 on differentiation. The second course covers the last four chapters of the book. We often run out of time in the final chapters, and we have to do with less than the full treatment of Lebesgue's theory as presented here.

The list of references at the end includes some classical titles which are referred to in the text, as well as some more recent titles which, more or less, address the same readership and can therefore be used for collateral reading. It is a short list, and we are almost certain that we have missed some relevant titles of which we are not aware. It is a pleasure to acknowledge the many corrections and useful comments which we have received from our colleagues at KSU. We are also grateful to the reviewer of Taylor and Francis for a comprehensive critique of the original manuscript.

Authors
Riyadh, March 2006.

Contents

Preface vii

1 Preliminaries 1
1.1 Sets 1
1.2 Functions 7

2 Real Numbers 19
2.1 Field Axioms 19
2.2 Order Axioms 23
2.3 Natural Numbers, Integers, Rational Numbers 30
2.4 Completeness Axiom 38
2.5 Decimal Representation of Real Numbers 49
2.6 Countable Sets 54

3 Sequences 65
3.1 Sequences and Convergence 65
3.2 Properties of Convergent Sequences 73
3.3 Monotonic Sequences 82
3.4 The Cauchy Criterion 88
3.5 Subsequences 97
3.6 Upper and Lower Limits 101
3.7 Open and Closed Sets 105

4 Infinite Series 113
4.1 Basic Properties 113
4.2 Convergence Tests 121

5 Limit of a Function 131
5.1 Limit of a Function 131
5.2 Basic Theorems 141
5.3 Some Extensions of The Limit 149
5.4 Monotonic Functions 155

6 Continuity — 159
- 6.1 Continuous Functions … 159
- 6.2 Combinations of Continuous Functions … 169
- 6.3 Continuity on an Interval … 173
- 6.4 Uniform Continuity … 186
- 6.5 Compact Sets and Continuity … 194

7 Differentiation — 207
- 7.1 The Derivative … 207
- 7.2 The Mean Value Theorem … 223
- 7.3 L'Hôpital's Rule … 237
- 7.4 Taylor's Theorem … 247

8 The Riemann Integral — 257
- 8.1 Riemann Integrability … 257
- 8.2 Darboux's Theorem and Riemann Sums … 272
- 8.3 Properties of the Integral … 276
- 8.4 The Fundamental Theorem of Calculus … 286
- 8.5 Improper Integrals … 295
  - 8.5.1 **Unbounded Integrand** … 295
  - 8.5.2 **Unbounded Interval** … 297

9 Sequences and Series of Functions — 303
- 9.1 Sequences of Functions … 303
- 9.2 Properties of Uniform Convergence … 312
- 9.3 Series of Functions … 323
- 9.4 Power Series … 332

10 Lebesgue Measure — 349
- 10.1 Classes of Subsets of $\mathbb{R}$ … 351
- 10.2 Lebesgue Outer Measure … 358
- 10.3 Lebesgue Measure … 365
- 10.4 Measurable Functions … 380

11 Lebesgue Integration — 393
- 11.1 Definition of The Lebesgue Integral … 394
- 11.2 Properties of the Lebesgue Integral … 403
- 11.3 Lebesgue Integral and Pointwise Convergence … 407
- 11.4 Lebesgue and Riemann Integrals … 418

References	**429**
Notation	**431**
Index	**433**

1

Preliminaries

Here we present some of the basic concepts and terminology that will be used in the following chapters of this book.

1.1 Sets

The introduction of the concept of a "set" by the nineteenth century German mathematician *Georg Cantor* may have been the most significant development in mathematics since calculus was invented in the seventeenth century. While calculus was used as a tool for solving problems involving motion, set theory was first used to study the properties of the real numbers, then it was developed, together with logic, to investigate the foundations of mathematics (see [CAN], [KAM], and [TAR], for example). Here we are mainly interested in set theory as a symbolic, or supplementary, language for expressing mathematical ideas more precisely and concisely.

According to Cantor, a *set* is a "collection into a whole of definite and separate objects (called the *elements* of the set) of our intuition or our thought", such as the odd numbers from 1 to 9, or the points on a given line segment. We shall not attempt to define what a set is, but rather accept it as a *primitive concept* which can then be used to define other concepts in the larger mathematical structure. But it is helpful, for practical purposes, to keep in mind this intuitive notion of a set as a *collection of objects*, whether physical or abstract. Though we shall have occasion to talk about sets of such entities as points or lines or intervals, we shall mainly be interested here in sets of numbers.

A set may be *finite*, such as the odd integers from 1 to 9, which is made up of five elements, and denoted by

$$\{1, 3, 5, 7, 9\} \tag{1.1}$$

in conventional set notation; or it may be *infinite*, such as the set of *natural numbers*

$$\mathbb{N} = \{1, 2, 3, 4, \cdots\}. \tag{1.2}$$

When a is an element in the set A we write $a \in A$, otherwise we write $a \notin A$ to indicate that a does not belong to A. Thus $105 \in \mathbb{N}$, while $-1 \notin \mathbb{N}$.

The essential feature of a (well-defined) set A is that, for any element a, either $a \in A$ or $a \notin A$. A set is defined either by explicitly writing down its elements, as in (1.1) and (1.2), or by giving a rule for obtaining them. For example,

$$\{x : x \in \mathbb{N}, \ x^2 - 4x + 3 = 0 \} \tag{1.3}$$

is the set of natural numbers x such that $x^2 - 4x + 3 = 0$. This can also be expressed more briefly as $\{x \in \mathbb{N} : x^2 - 4x + 3 = 0\}$.

We say that the set A is a *subset* of the set B, or that B *contains* A, if every element in A belongs to B, and we express this relation between the two sets symbolically by $A \subseteq B$. When $A \subseteq B$ and $B \subseteq A$ the sets A and B have the same elements and are said to be *equal*, and we write $A = B$. But when $A \subseteq B$ and $A \neq B$, A is called a *proper subset* of B, and the more specific relation $A \subset B$ is used. Thus $\{1, 3, 5, 7, 9\} \subset \mathbb{N}$ and $\{x \in \mathbb{N} : x^2 - 4x + 3 = 0\} = \{1, 3\}$.

In order to avoid some logical contradictions which can arise in the definition of a set, as in Exercise 1.1.9, we shall assume the existence of a fixed set U, sometimes referred to as the *universal set,* which contains all the sets under discussion. In real analysis, for example, this is usually taken to be the set of *real numbers* $\mathbb{R}$, which will be the subject of chapter 2. We shall also find it necessary to define an *empty set*, denoted by $\varnothing$, as one which contains no elements, such as the set $\{x \in \mathbb{N} : 2x = 1\}$. Hence $\varnothing \subseteq A$ for any set A.

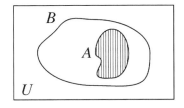

Figure 1.1

The *union* of the sets A and B, denoted by $A \cup B$, is the set of elements in A or B (or both), while their *intersection* $A \cap B$ is the set composed of those elements which belong to both A *and* B. When the two sets A and B have no elements in common, that is, when $A \cap B = \varnothing$, A and B are said to be *disjoint*. The *complement of A in B* is defined as the set $\{x : x \in B, x \notin A\}$. This is denoted by $B \setminus A$, and when B is the universal set U it is usually abbreviated to A^c, the *complement* of A.

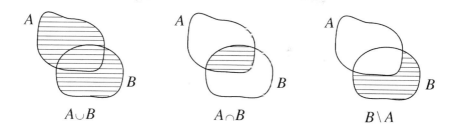

Figure 1.2

For any two sets A and B we have *De Morgan's laws*

$$(A \cup B)^c = A^c \cap B^c \qquad (1.4)$$

$$(A \cap B)^c = A^c \cup B^c. \qquad (1.5)$$

To prove the first equality, let $x \in (A \cup B)^c$, which means $x \notin A \cup B$. It then follows that $x \notin A$ and $x \notin B$, that is, $x \in A^c$ and $x \in B^c$. Hence $x \in A^c \cap B^c$. Since x was an arbitrary element in $(A \cup B)^c$, we conclude that $(A \cup B)^c \subseteq A^c \cap B^c$. By a similar argument we can show that, if x is any element in $A^c \cap B^c$, then $x \in (A \cup B)^c$, hence $A^c \cap B^c \subseteq (A \cup B)^c$. This implies that the two sets $(A \cup B)^c$ and $A^c \cap B^c$ are equal.

More generally, we shall define the union and the intersection of a finite number of sets $A_1, A_2, A_3, \cdots, A_n$, respectively, by

$$\bigcup_{i=1}^{n} A_i = \{x : x \in A_i \text{ for some } i, 1 \leq i \leq n\}$$

$$\bigcap_{i=1}^{n} A_i = \{x : x \in A_i \text{ for all } i, 1 \leq i \leq n\}.$$

These definitions have a natural extension to *any* collection of sets $\{A_\lambda : \lambda \in \Lambda\}$ indexed by the set Λ, which may be finite or infinite, given by

$$\bigcup_{\lambda \in \Lambda} A_\lambda = \{x : x \in A_\lambda \text{ for some } \lambda \in \Lambda\} \qquad (1.6)$$

$$\bigcap_{\lambda \in \Lambda} A_\lambda = \{x : x \in A_\lambda \text{ for all } \lambda \in \Lambda\}. \qquad (1.7)$$

In particular, with $\Lambda = \mathbb{N}$, the union

$$\bigcup_{i \in \mathbb{N}} A_i = \bigcup_{i=1}^{\infty} A_i$$

and the intersection
$$\bigcap_{i \in \mathbb{N}} A_i = \bigcap_{i=1}^{\infty} A_i$$
of any infinite sequence of sets $A_1, A_2, A_3, \cdots$ are well defined. Applied to (1.6) and (1.7), De Morgan's laws take the form
$$\left(\bigcup_{\lambda \in \Lambda} A_\lambda\right)^c = \bigcap_{\lambda \in \Lambda} (A_\lambda)^c$$
$$\left(\bigcap_{\lambda \in \Lambda} A_\lambda\right)^c = \bigcup_{\lambda \in \Lambda} (A_\lambda)^c.$$

Phrases such as *if $\cdots$ then $\cdots$*, and *$\cdots$ if and only if $\cdots$*, are used quite frequently to connect *simple statements*. These are statements, such as $1 \in \mathbb{N}$ or $2 = 3$, which can be described as either *true* or *false* (but not both). In order to clarify what we mean by such phrases, and to take advantage of some of the conventional logical symbols for the sake of brevity and typographical convenience, we conclude this section with a brief word on logical connectives.

If P is a statement, $\sim P$ will denote the *negation* of P. Thus $\sim P$ is false when P is true, and vice-versa. Suppose Q is another statement. The compound statement

$$if\ P\ then\ Q$$

and

$$P\ implies\ Q$$

will be considered equivalent statements, meaning that *if P is true then Q is true*. This is written symbolically as

$$P \Rightarrow Q. \tag{1.8}$$

We also express this conditional statement by saying that *P is a sufficient condition for Q*, or that *Q is a necessary condition for P*. To prove the implication (1.8) we have to exclude the possibility that P is true and Q is false. This may be done by one of three ways:

1. Assume that P is true and prove that Q is true (*direct proof*).

2. Assume that Q is false and prove that P is false (*contrapositive proof*).

3. Assume that P is true and Q is false, then show that this leads to a contradiction (*proof by contradiction*).

In a theorem, corollary, lemma, etc., P usually stands for the hypothesis, or assumptions, and Q the conclusion, and we follow one of these procedures to go from P to Q.

When *P implies Q* and *Q implies P*, that is, when $P \Rightarrow Q$ and $Q \Rightarrow P$, we abbreviate this to

$$P \Leftrightarrow Q,$$

and we say that *P is equivalent to Q*, or the more common phrase: *P if and only if Q*. This also means *P is a necessary and sufficient condition for Q*. Thus $x = 1 \Rightarrow x^2 = 1$, $x = -1 \Rightarrow x^2 = 1$ and $x^2 = 1 \Rightarrow x \in \{1, -1\}$. Hence

$$x^2 = 1 \Leftrightarrow x \in \{1, -1\}.$$

EXERCISES 1.1

1. Prove that the operations $\cup$ and $\cap$ are *commutative* and *associative*, in the sense that $A \cup B = B \cup A$, $A \cap B = B \cap A$ for any pair of sets A and B, and that

$$A \cup (B \cup C) = (A \cup B) \cup C$$
$$A \cap (B \cap C) = (A \cap B) \cap C$$

    for any sets A, B and C.

2. Prove the *distributive laws*

$$A \cap (B \cup C) = (A \cap B) \cup (A \cap C),$$
$$A \cup (B \cap C) = (A \cup B) \cap (A \cup C).$$

3. Prove that $(A \cup B) \setminus B \subseteq A$. Under what conditions do we have $(A \cup B) \setminus B = A$?

4. Prove that
    (a) $A \setminus B = A \setminus (A \cap B)$

(b) $A \setminus (B \setminus C) = (A \setminus B) \cup (A \cap C)$
(c) $(A \setminus B) \setminus C = A \setminus (B \cup C)$
(d) $A \setminus (B \cup C) = (A \setminus B) \cap (A \setminus C)$
(e) $(A \cup B) \setminus C = (A \setminus C) \cup (B \setminus C)$.

5. The *symmetric difference* of the sets A and B is defined as $A \Delta B = (A \setminus B) \cup (B \setminus A)$. Draw a schematic diagram of $A \Delta B$ and prove

   (a) $A \Delta B = B \Delta A$ (commutative property of Δ)
   (b) $A \Delta \varnothing = A$
   (c) $A \Delta A = \varnothing$
   (d) $A \cap (B \Delta C) = (A \cap B) \Delta (A \cap C)$ (distributive property)
   (e) $A \Delta B = (A \cup B) \setminus (A \cap B)$.

6. Prove that if $A \cup B = A$ and $A \cap B = A$, then $A = B$.

7. Let A be a set. The set of all subsets of A is known as the *power set* of A, and is denoted by $\mathcal{P}(A)$. Thus $B \in \mathcal{P}(A)$ if, and only if, $B \subseteq A$. The sets A and $\varnothing$ clearly belong to $\mathcal{P}(A)$. Determine the power set of $\{1, 2, 3, 4, 5\}$.

8. If A is a finite set we shall use the symbol $|A|$ to denote the *number of elements* in A. Verify that $|\mathcal{P}(A)| = 32$ for the set given in exercise 1.1.7. More generally, prove that $|\mathcal{P}(A)| = 2^{|A|}$.

9. Let $\mathcal{A}$ be the collection of all sets. If $\mathcal{A}$ is a set then clearly $\mathcal{A} \in \mathcal{A}$, which is unusual for ordinary sets. We shall call a set X *ordinary* if $X \notin X$, and *extraordinary* if $X \in X$. Let $\mathcal{B}$ be the collection of all ordinary sets. Assuming that $\mathcal{B}$ is a set, is $\mathcal{B}$ ordinary or extraordinary? Show that if $\mathcal{B}$ is ordinary then it is extraordinary, and if $\mathcal{B}$ is extraordinary then it is ordinary.

   This is the famous paradox due to the English philosopher Bertrand Russel, which pointed to the need for some restrictions on what can be regarded as a "set". In our work such sets as $\mathcal{A}$ and $\mathcal{B}$ are excluded by the assumption that we work within a fixed universal set, in which such "large" sets have no place.

1.2 Functions

Let a and b be elements in any set. By the definition of equality between sets, we have $\{a, a, b\} = \{a, b\} = \{b, a\}$. This means that the repetition of an element in a set does not change the set, and that the order of the elements in the set $\{a, b\}$ is irrelevant. If we wish to give significance to their order we shall write (a, b), the *ordered pair* a, b, which is defined as the set $\{\{a\}, \{a, b\}\}$. Thus $(a, b) = (c, d)$ implies $\{\{a\}, \{a, b\}\} = \{\{c\}, \{c, d\}\}$, from which we can only conclude that $a = c$ and $b = d$. Hence $(a, b) = (b, a)$ only if $a = b$. a is called the *first coordinate* and b the *second coordinate* of (a, b).

Definition 1.1 Let A and B be sets. The *cartesian product* of A and B, denoted by $A \times B$, is the set of all ordered pairs (a, b) such that $a \in A$ and $b \in B$. Symbolically, we have $A \times B = \{(a, b) : a \in A, \ b \in B\}$.

From this definition it follows that $A \times B = \emptyset$ if either $A = \emptyset$ or $B = \emptyset$. For example

$$\{1, 2\} \times \{3, 4, 5\} = \{(1, 3), (1, 4), (1, 5), (2, 3), (2, 4), (2, 5)\}.$$

Note that

$$\{3, 4, 5\} \times \{1, 2\} = \{(3, 1), (3, 2), (4, 1), (4, 2), (5, 1), (5, 2)\}$$
$$\neq \{1, 2\} \times \{3, 4, 5\},$$

so, in general, $A \times B \neq B \times A$.

The set $\{1, 2\} \times \{3, 4, 5\}$ may be represented graphically by points in the cartesian plane, as shown in Figure 1.3.

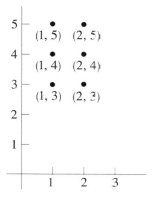

Figure 1.3

Definition 1.2 For any sets A and B, a subset of $A \times B$ is called a *binary relation from A to B*. A subset of $A \times A$ is called a *binary relation on A*.

We generally refer to a binary relation as simply a *relation* when there is no need to emphasize its binary nature. If R is a relation from A to B then $R \subseteq A \times B$. Here are some examples of relations from $\{1,2\}$ to $\{3,4,5\}$:

$$R_1 = \{(1,3),(1,4),(2,5)\} \tag{1.9}$$
$$R_2 = \{(1,3),(2,4)\} \tag{1.10}$$
$$R_3 = \{(1,5),(2,5)\} \tag{1.11}$$
$$R_4 = \{1,2\} \times \{3,4,5\} \tag{1.12}$$
$$R_5 = \{(2,3)\} \tag{1.13}$$

When $(a,b) \in R$ we say that a is *related to*, or *associated with*, b by R. At the risk of abusing our notation, we sometimes express this symbolically by writing aRb. One of the important binary relations on $\mathbb{N}$ is the relation "greater than", which is expressed by

$$R = \{(2,1),(3,1),(3,2),(4,1),(4,2),(4,3),(5,1),\cdots\}. \tag{1.14}$$

This is also expressed by writing $2 > 1, 3 > 1, 3 > 2, 4 > 1, 4 > 2, 4 > 3, 5 > 1, \cdots$, where $>$ is the conventional mathematical symbol for "greater than", which will be defined more precisely in the next chapter. The relation (1.14) is represented graphically in Figure 1.4.

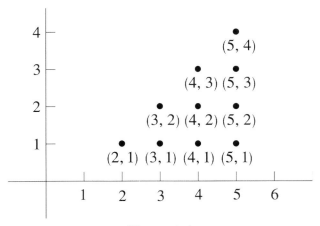

Figure 1.4

Definition 1.3 If R is a relation from A to B, the *inverse relation* R^{-1} is the relation defined by $\{(b,a) : (a,b) \in R\}$. Thus, referring to equations (1.9) to (1.13), we have

$$R_1^{-1} = \{(3,1),(4,1),(5,2)\}$$
$$R_2^{-1} = \{(3,1),(4,2)\}$$
$$R_3^{-1} = \{(5,1),(5,2)\}$$
$$R_4^{-1} = \{3,4,5\} \times \{1,2\}$$
$$R_5^{-1} = \{(3,2)\}.$$

Definition 1.4 A relation R from A to B is called a *function*, or a *mapping*, from A to B if
  (i) for every $x \in A$ there is a $y \in B$ such that $(x,y) \in R$
  (ii) whenever $(x,y) \in R$ and $(x,z) \in R$, then $y = z$.

In other words, a function from A to B is a relation in which *every* element of A is associated with *one, and only one*, element in B.

Among the relations in (1.9) to (1.13), we see that only R_2 and R_3 are functions from $\{1,2\}$ to $\{3,4,5\}$. In R_1 and R_4 the element $1 \in \{1,2\}$ is associated with more than one element in $\{3,4,5\}$, and in R_5 the element $1 \in \{1,2\}$ is not associated with any element in $\{3,4,5\}$.

By convention, we write

$$f : A \to B$$

to indicate that f is a function from A to B. In functional notation the statement that $(x,y) \in f$ is expressed as

$$f : x \mapsto y$$

or, more commonly, as

$$y = f(x).$$

We then say that y is the *image* of x, and that x is a *pre-image* of y, under f. Note that, by Definition 1.4, there is only one image for x, but there may be more than one pre-image of y. $f(x)$ is also referred to as the *value* of the function f at x. The set A is the *domain of definition*, or simply the *domain*, of the function f, frequently denoted by D_f, while B is its *co-domain*. Thus a function f is completely determined by its domain $D_f = A$, its co-domain B, and the rule $y = f(x)$ which assigns to every element $x \in A$ an image y in B. This is expressed concisely in the form

$$f : A \to B, \; y = f(x).$$

If C is a subset of A, the *image* of C under f is the set

$$f(C) = \{f(x) : x \in C\},$$

which is a subset of B. The set

$$f(A) = \{f(x) : x \in A\},$$

composed of the images under f of all the elements of the domain of f, is called the *range* of f and is denoted by R_f. Clearly $R_f \subseteq B$.

If $y \in B$, Definition 1.4 does not guarantee the existence of a pre-image $x \in A$ such that $f(x) = y$. In equation (1.10), for example, 5 has no pre-image under R_2. Furthermore, if a pre-image exists, it may not be unique, as may be seen in equation (1.11), where both 1 and 2 are pre-images of 5 under R_3. For any $E \subseteq B$, we define

$$f^{-1}(E) = \{x \in A : f(x) \in E\}.$$

Thus $f^{-1}(\{y\})$, which is usually simplified to $f^{-1}(y)$, is the set

$$\{x \in A : f(x) \in \{y\}\} = \{x \in A : f(x) = y\}.$$

Note that if $y \notin R_f$ then $f^{-1}(y) = \varnothing$, and we clearly have $f^{-1}(B) = f^{-1}(R_f) = D_f$. The reader is invited to verify the following statements for any function $f : A \to B$:

$$f(\varnothing) = \varnothing \tag{1.15}$$

$$\text{If } A_1 \subseteq A_2 \subseteq A \text{ then } f(A_1) \subseteq f(A_2) \tag{1.16}$$

$$f\left(\bigcup_{\lambda \in \Lambda} A_\lambda\right) = \bigcup_{\lambda \in \Lambda} f(A_\lambda) \tag{1.17}$$

$$f\left(\bigcap_{\lambda \in \Lambda} A_\lambda\right) \subseteq \bigcap_{\lambda \in \Lambda} f(A_\lambda) \tag{1.18}$$

$$f^{-1}(\varnothing) = \varnothing \tag{1.19}$$

$$\text{If } B_1 \subseteq B_2 \subseteq B \text{ then } f^{-1}(B_1) \subseteq f^{-1}(B_2) \tag{1.20}$$

$$f^{-1}\left(\bigcup_{\lambda \in \Lambda} E_\lambda\right) = \bigcup_{\lambda \in \Lambda} f^{-1}(E_\lambda) \qquad (1.21)$$

$$f^{-1}\left(\bigcap_{\lambda \in \Lambda} E_\lambda\right) = \bigcap_{\lambda \in \Lambda} f^{-1}(E_\lambda) \qquad (1.22)$$

$$f^{-1}(E^c) = [f^{-1}(E)]^c \qquad (1.23)$$

When A and B are subsets of the set of real numbers $\mathbb{R}$, the function $f : A \to B$ has a graphical representation in the cartesian plane, such as the one in Figure 1.5, where $D_f = A$ is shown on the horizontal real axis, the x-axis, and $R_f \subseteq B$ on the vertical real axis, the y-axis. The set of ordered pairs $\{(x, y) : x \in D_f, y = f(x)\}$, represented by the set of points in the cartesian plane with coordinates x and $f(x)$, is the *graph*, or the *curve*, of the function f.

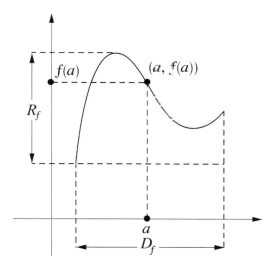

Figure 1.5

The function f, represented in Figure 1.5, is a typical example of a *real function f of a real variable x*, and is the subject of our study in this book. Examples of such functions are given by

$$f : \mathbb{R} \to \mathbb{R}, \ f(x) = 2x + 1 \qquad (1.24)$$

$$g : \mathbb{R} \to \mathbb{R}, \ g(x) = x^2, \qquad (1.25)$$

which are represented graphically in Figure 1.6.

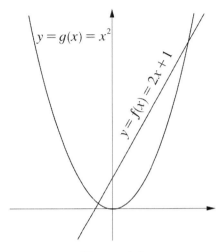

Figure 1.6

When a real function f is defined by the rule $x \mapsto f(x)$, the "largest" set A in $\mathbb{R}$ for which $x \in A$ implies $f(x) \in \mathbb{R}$ is called the *natural domain* of f. Consequently, the natural domain of the function defined by $f(x) = 1/x$ is the set $\mathbb{R}\backslash\{0\}$, and the function defined by $g(x) = \sqrt{x-1}$ has the natural domain $\{x \in \mathbb{R}, x \geq 1\} = [1, \infty)$.

According to Definition 1.4, a function f from A to B is a binary relation from A to B, and, as such, has an inverse relation f^{-1}. The question that comes to mind is: when is the relation f^{-1} a function? To answer that question we first present the following definition.

Definition 1.5 The function f is *injective*, or "one-to-one", if, for any x_1 and x_2 in D_f such that $f(x_1) = f(x_2)$, we have $x_1 = x_2$.

Thus an injective function takes distinct elements in the domain to distinct elements in the range.

The function f defined in (1.24) is injective because, if $2x_1 + 1 = 2x_2 + 1$, then we must have $x_1 = x_2$. But the function g in (1.25) is not injective because, for example, $g(-1) = g(1) = 1$ though $-1 \neq 1$.

Theorem 1.1
For any function f, the inverse f^{-1} is a function if and only if f is injective, in which case $D_{f^{-1}} = R_f$ and $R_{f^{-1}} = D_f$.

Proof
(a) First we shall assume that f is injective and prove that f^{-1} is a function from R_f to D_f.
(i) Let $y \in R_f$, then there is an $x \in D_f$ such that $y = f(x)$, that is $(x, y) \in f$. But this implies $(y, x) \in f^{-1}$.
(ii) Suppose $(y, x_1) \in f^{-1}$ and $(y, x_2) \in f^{-1}$, then $(x_1, y) \in f$ and $(x_2, y) \in f$, that is, $y = f(x_1)$ and $y = f(x_2)$. Since f is injective we must have $x_1 = x_2$.

(b) Suppose now that f^{-1} is a function. To show that f is injective, let $f(x_1) = f(x_2) = y$. Then $(x_1, y), (x_2, y) \in f$, or $(y, x_1), (y, x_2) \in f^{-1}$. But since f^{-1} is a function we must have $x_1 = x_2$. Hence f is injective. □

Corollary 1.1 *If the function $f : A \to B$ is injective, then $f^{-1} : R_f \to A$ is also injective.*

Proof. Let $y_1, y_2 \in R_f$ and suppose $f^{-1}(y_1) = f^{-1}(y_2) = x$. Then $f(x) = y_1$ and $f(x) = y_2$. Since f is a function, it follows that $y_1 = y_2$. □

Going back to the function $f : \mathbb{R} \to \mathbb{R}$ defined in (1.24) by the equation $f(x) = 2x+1$, we note that any point y in the co-domain $\mathbb{R}$ is the image of some point x in the domain $\mathbb{R}$, in fact it is a simple matter to calculate x from the equation $y = 2x+1$ as $x = (y-1)/2$. Hence the co-domain of f coincides with its range.

Definition 1.6 A function $f : A \to B$ is *surjective*, or "onto", if $R_f = B$.

Thus the function f defined in (1.24) is surjective since $R_f = \mathbb{R}$, but the function g defined in (1.25) is not, since $R_g = \{x^2 : x \in \mathbb{R}\} = [0, \infty) \neq \mathbb{R}$. Note that, had we defined g as a function from $\mathbb{R}$ to $[0, \infty)$, the function would have been surjective (but not injective). In fact, taking the co-domain of a function as its range automatically makes it surjective, but the range of a function is not always immediately obvious from its definition, or it may not be relevant to the discussion at hand.

Definition 1.7 A function which is both injective and surjective is called *bijective*. Such a function is also referred to as a *bijection*, or a *one-to-one correspondence*.

It is important at this point to emphasize once again that a function is not completely defined by the rule $y = f(x)$, but also by its domain and co-domain. Thus the function g defined by (1.25) is neither injective nor surjective. By changing its domain from $\mathbb{R}$ to $[0,\infty)$ it becomes injective (from $[0,\infty)$ to $\mathbb{R}$), and by shrinking its co-domain to $[0,\infty)$, it becomes surjective. The resulting function

$$h : [0, \infty) \to [0, \infty), \ h(x) = x^2 \qquad (1.26)$$

is now a bijection. Its inverse

$$h^{-1} : [0, \infty) \to [0, \infty), \ h^{-1}(x) = \sqrt{x},$$

is also a bijection, and it is a simple matter to conclude from Theorem 1.1 and its corollary that this is true of any function. In other words, *the inverse of a bijection is a bijection.*

The procedure by which we shrank the domain of the function g, defined in (1.25), from $\mathbb{R}$ to $[0,\infty)$ is an example of what is called a *restriction*. This is one of the many ways we use to modify and compose existing functions to obtain new ones. Here we give some of the more common methods:

1. Restriction and extension of the domain

If f is a function from A to B and $C \subset A$, the function $g : C \to B$, $g(x) = f(x)$, is called the *restriction* of f to C, and is denoted by $f|_C$. We also call f an *extension* of g to A. The function defined on $\mathbb{R}\setminus\{0\}$ by $x \mapsto 1/x$, for example, may be extended to $\mathbb{R}$ by the assignment $0 \mapsto 0$. Many significant developments in mathematics are the result of discovering extensions of known functions which preserve certain basic properties. The process is known as *generalization*, which is the driving force behind many developments in mathematical thought.

2. Composition

If we have two functions

$$f : A \to B, \ g : B \to C$$

then we can define the *composition* of f and g, denoted by $g \circ f$, as the function

$$g \circ f : A \to C, \ (g \circ f)(x) = g(f(x)).$$

Note that $D_{g \circ f} = D_f = A$. Going back to the functions defined in (1.24) and (1.25) we have

$$(g \circ f)(x) = g(f(x)) = (2x + 1)^2$$
$$(f \circ g)(x) = f(g(x)) = 2x^2 + 1,$$

so we conclude that $f \circ g \neq g \circ f$ in general. Note that the composition $g \circ f$ is well defined whenever $R_f \subseteq D_g$.

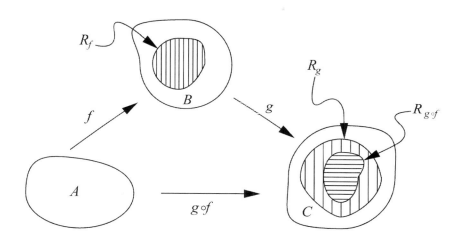

Figure 1.7

The following properties of composite functions easily follow from the definitions above:

(i) If $f : A \to B$, $g : B \to C$, and $h : C \to D$ then

$$(h \circ g) \circ f = h \circ (g \circ f).$$

This just says that composition is an associative binary operation on functions where it is defined.

(ii) If f and g are both injective, then so is $g \circ f$.
If f and g are both surjective, then so is $g \circ f$.
If f and g are both bijective, then so is $g \circ f$.

(iii) If f is injective then $(f^{-1} \circ f)(x) = x$ for all $x \in D_f$, and $(f \circ f^{-1})(x) = x$ for all $x \in R_f$.

3. Algebraic Combinations

In dealing with real functions of a real variable, we can use the algebraic properties of real numbers to define the *sum*, *product*, and *quotient* of two functions f and g in the usual manner:

(i) $f + g$ is the function defined on $D_f \cap D_g$ by $(f + g)(x) = f(x) + g(x)$.

(ii) $f \cdot g$ is the function defined on $D_f \cap D_g$ by $(f \cdot g)(x) = f(x) \cdot g(x)$.

(iii) f/g is the function defined on $D_f \cap D_g \setminus \{x \in D_g : g(x) = 0\}$ by $(f/g)(x) = f(x)/g(x)$.

Note that multiplication of a function by a constant is a special case of (ii), and that the difference of two functions, $f - g$, is obtained from (i) and (ii) as $f + (-1)g$.

Finally, a word about the functions that we shall use in our examples. We expect that the reader is familiar with the exponential, logarithmic, and trigonometric functions. Nevertheless, the exponential and logarithmic functions are introduced in Section 2.4, and the trigonometric functions are briefly sketched in Chapter 4, after which we use them quite freely in our examples in the first few chapters of the book. These transcendental functions will be defined later on through power series, but to avoid them until then is to deprive our examples of much needed variety and vitality.

EXERCISES 1.2

1. Prove that, for any three sets,

    (a) $A \times (B \cup C) = (A \times B) \cup (A \times C)$
    (b) $A \times (B \cap C) = (A \times B) \cap (A \times C)$.

2. A relation R in A is called *reflexive* if xRx for all $x \in A$, *symmetric* if xRy implies yRx for all $x, y \in A$, and *transitive* if xRy and yRz imply xRz. Determine the properties of each of the following relations:

    (a) m "divides" n, defined on the set $\mathbb{N}$
    (b) x "equals" y on $\mathbb{R}$
    (c) A "is a proper subset of" B on $\mathcal{P}(\mathbb{R})$.

    An *equivalence relation* is a relation which is reflexive, symmetric, and transitive. Which of the relations defined above is an equivalence relation?

3. Prove equations (1.15) through (1.18).

4. Prove equations (1.19) through (1.23).

5. Give an example of two functions f and g such that $f \neq g$ but $f \circ g = g \circ f$.

6. Given
$$f : \mathbb{R}\setminus\{1\} \to \mathbb{R} \quad f(x) = \frac{x+1}{x-1},$$
prove that $f^{-1} = f$.

7. For any two functions for which the composition $g \circ f$ is defined, prove that $D_{g \circ f} = f^{-1}(D_g)$.

8. Prove properties (i) to (iii) of the composition of functions.

9. Suppose $f : A \to B$. Prove the following:

   (a) $f(f^{-1}(E)) \subseteq E$ for all $E \subseteq B$. Show that equality holds for all $E \subseteq B$ if, and only if, f is surjective.

   (b) $f^{-1}(f(C)) \supseteq C$ for all $C \subseteq A$. Show that equality holds for all $C \subseteq A$ if, and only if, f is injective.

   (c) f is injective if, and only if, $f(C_1 \cap C_2) = f(C_1) \cap f(C_2)$ for all $C_1, C_2 \subseteq A$.

   (d) f is surjective if, and only if, $[f(C)]^c \subseteq f(C^c)$ for all $C \subseteq A$.

10. If A is a finite set and $f : A \to B$ is a bijective function, prove that B is also finite and $|B| = |A|$.

11. Let $f : A \to B$, $g : C \to D$. Determine the domain and range of both $f \circ g$ and $g \circ f$.

12. 11. Let A be any (non-empty) set. A function from $A \times A$ to A is called a *binary operation* on A. Show that $\cup, \cap$, and $\triangle$ are binary operations on $\mathcal{P}(A)$.

2

Real Numbers

This chapter looks at the *real number system* from an axiomatic point of view, and thereby lays the foundation for our future work. It may seem, at first sight, that we are being too pedantic in spending too much time discussing or proving intuitively obvious ideas. However, we should keep in mind that intuition, though often useful, can sometimes be misleading. A prime example of this is the intuitive idea that any measurable length can be expressed as a common fraction. This led to the assumption that $\sqrt{2}$ may be represented by a rational number, which was later proved to be false. This will be discussed in section 2.3.

Any attempt to build the real numbers by a constructive approach, first by defining the natural numbers, then the integers, then the rational numbers, and finally the real numbers, will not be an easy task. See, for example, [RUD] for an exposition of how the real numbers may be constructed from the rationals using *Dedekind cuts*. We shall, instead, adopt an approach as old as Euclid, whereby we assume the existence of a set $\mathbb{R}$, called the *set of real numbers*, which satisfies certain properties, called *axioms* or *postulates,* that will allow us to derive the other pertinent properties we seek in this course. This approach, it seems to us, is less ambiguous and more economical, and should not obstruct or diminish our intuitive capability if the axioms are chosen carefully enough.

There are twelve axioms to be imposed on $\mathbb{R}$, the first eleven of which are algebraic. These are the *field axioms* (A1 to A9) and the *order axioms* (A10 and A11). But the twelfth, the *completeness axiom,* is different, and it is this last axiom which provides $\mathbb{R}$ with a *topological closure property* that allows us to conduct analysis. The axioms will be presented in three stages in order to emphasize their separate implications.

2.1 Field Axioms

We shall assume that there are two *binary operations* on $\mathbb{R}$, that is, functions defined from $\mathbb{R} \times \mathbb{R}$ to $\mathbb{R}$

$$+ : (a,b) \mapsto a+b, \quad \cdot : (a,b) \mapsto a \cdot b,$$

the first called *addition*, and the second *multiplication*, such that the following properties are satisfied:

A1. $a + b = b + a$ *for all* $a, b \in \mathbb{R}$.

   This is called the *commutative property* of addition.

A2. $a + (b + c) = (a + b) + c$ *for all* $a, b, c \in \mathbb{R}$.

   (*associative property* of addition).

A3. *There is an element* 0 *in* $\mathbb{R}$ *such that* $a + 0 = 0 + a = a$ *for all* $a \in \mathbb{R}$.

   (the existence of a *neutral element* for addition).

A4. *For every element* a *in* $\mathbb{R}$ *there is an element in* $\mathbb{R}$, *denoted by* $-a$, *which satisfies* $a + (-a) = (-a) + a = 0$.

   (the existence of an *additive inverse*).

A5. $a \cdot b = b \cdot a$ *for all* $a, b \in \mathbb{R}$.

   (*commutative property* of multiplication).

A6. $a \cdot (b \cdot c) = (a \cdot b) \cdot c$ *for all* $a, b, c, \in \mathbb{R}$.

   (*associative property* of multiplication).

A7. *There is an element* $1 \neq 0$ *in* $\mathbb{R}$ *such that* $a \cdot 1 = 1 \cdot a = a$ *for all* $a \in \mathbb{R}$.

   (the existence of a *neutral element* for multiplication).

A8. *For every element* $a \neq 0$ *in* $\mathbb{R}$ *there is an element in* $\mathbb{R}$, *denoted by* a^{-1}, *which satisfies* $a \cdot a^{-1} = a^{-1} \cdot a = 1$.

   (the existence of a *multiplicative inverse*).

A9. $a \cdot (b + c) = a \cdot b + a \cdot c$ *for all* $a, b, c \in \mathbb{R}$.

   (*distributive property* of multiplication over addition).

In algebraic terminology, axioms A1 to A4 state that the set $\mathbb{R}$ under the operation $+$, or the *system* $(\mathbb{R}, +)$, is a *commutative group*. Similarly A5 to A8 state that the system $(\mathbb{R}\setminus\{0\}, \cdot)$ is a *commutative group*. The set $\mathbb{R}$, under the operations $+$ and $\cdot$, and subject to the nine axioms A1 to A9, is called the *field of real numbers*, and is denoted by $(\mathbb{R}, +, \cdot)$. Usually we write $\mathbb{R}$ when we mean the field $(\mathbb{R}, +, \cdot)$.

From the field axioms listed above we can derive other algebraic properties of the real numbers, as may be seen in the following examples.

Example 2.1 If $a, b, c \in \mathbb{R}$ and $a + b = a + c$, then $b = c$.

Proof. Suppose $a + b = a + c$. By A4, a has an additive inverse $-a$, and we have

$$-a + (a + b) = -a + (a + c)$$
$$(-a + a) + b = (-a + a) + c$$

by the associative property of addition. Now we use A4 once more to obtain

$$0 + b = 0 + c,$$

which, by A3, implies $b = c$. $\square$

Similarly, we can prove that if $a \cdot b = a \cdot c$ and $a \neq 0$, then $b = c$.

Example 2.2 (i) $a \cdot 0 = 0$ for all $a \in \mathbb{R}$.
(ii) If $a \cdot b = 0$ then either $a = 0$ or $b = 0$.

Proof. (i) $1 + 0 = 1$ by A3. Therefore

$$a \cdot (1 + 0) = a \cdot 1$$
$$= a \cdot 1 + 0. \qquad (A3)$$

But

$$a \cdot (1 + 0) = a \cdot 1 + a \cdot 0. \qquad (A9)$$

From Example 2.1, we conclude that $a \cdot 0 = 0$.

(ii) Suppose $a \cdot b = 0$ and $a \neq 0$. By A8, a^{-1} exists and gives

$$a^{-1} \cdot (a \cdot b) = a^{-1} \cdot 0 = 0. \qquad \text{by (i)}$$

But

$$a^{-1} \cdot (a \cdot b) = (a^{-1} \cdot a) \cdot b \qquad (A6)$$
$$= 1 \cdot b \qquad (A8)$$
$$= b. \qquad (A7)$$

Consequently, $b = 0$. □

We can also use A4 to define *subtraction* on $\mathbb{R}$ by
$$a - b = a + (-b) \quad \text{for all } a, b \in \mathbb{R},$$
and A8 to define *division* by
$$a \div b = a \cdot (b^{-1}) \quad \text{for all } a \in \mathbb{R}, \ b \in \mathbb{R}\setminus\{0\}.$$

It is a simple matter to verify that these two operations are not commutative. The quotient $a \div b$ is usually written as a ratio $\dfrac{a}{b}$ or a/b.

EXERCISES 2.1

Prove the following statements:

1. Given $a, b \in \mathbb{R}$, the equation $x + a = b$ has the unique solution $x = b - a$.

2. Given $a, b \in \mathbb{R}$, where $a \neq 0$, the equation $a \cdot x = b$ has the unique solution $x = a^{-1} \cdot b = \dfrac{b}{a}$.

3. $-(-a) = a$ for any $a \in \mathbb{R}$.

4. If $a \neq 0$ then $a^{-1} \neq 0$ and $(a^{-1})^{-1} = a$.

5. $(-1) \cdot a = -a$ for all $a \in \mathbb{R}$.

6. $(-a) \cdot b = -(a \cdot b)$ for all $a, b \in \mathbb{R}$.

7. $(-a) \cdot (-b) = a \cdot b$ for all $a, b \in \mathbb{R}$.

8. $(-a)^{-1} = -(a^{-1})$ for all $a \in \mathbb{R}\setminus\{0\}$.

9. If $a \neq 0$ and $b \neq 0$ then $a \cdot b \neq 0$ and $(a \cdot b)^{-1} = a^{-1} \cdot b^{-1}$.

10. Prove that the neutral elements for addition and multiplication are unique.

2.2 Order Axioms

We assume that there is a subset P of $\mathbb{R}$ with the following properties:

A10. *For any $a \in \mathbb{R}$ one, and only one, of the following alternatives holds: $a \in P$, or $a = 0$, or $-a \in P$.*

A11. *If $a, b \in P$ then $a + b \in P$ and $a \cdot b \in P$.*

P is called the set of *positive real numbers*. For any pair of real numbers a and b we now say a *is greater than* b (or b *is less than* a) and we write

$$a > b \tag{2.1}$$

(or $b < a$) if, and only if, $a - b \in P$. This defines the binary relation *greater than*, denoted by $>$, on $\mathbb{R}$. Thus

$$P = \{a \in \mathbb{R} : a > 0\},$$

and if we define the set $P^- = \{a \in \mathbb{R} : -a \in P\}$, then we have

$$P^- = \{a \in \mathbb{R} : a < 0\},$$

which is the set of *negative real numbers*. Now A10 implies $P \cap \{0\} = \{0\} \cap P^- = P^- \cap P = \varnothing$, and

$$\mathbb{R} = P \cup \{0\} \cup P^-.$$

In other words, any real number is either positive, negative, or 0. We also use the abbreviated notation $a \geq b$ (equivalently, $b \leq a$) when $a > b$ or $a = b$, that is, when $a - b \in P \cup \{0\}$. It now follows, from A10, that if $a \geq b$ and $a \leq b$ then $a = b$.

Axioms A10 and A11, added to A1-A9, introduce *order* into the field $(\mathbb{R}, +, \cdot)$, which now becomes an *ordered field*.

Let A be a subset of $\mathbb{R}$. $m \in A$ is called the *least element* of A, or the *minimum* of A, if $m \leq a$ for all $a \in A$. Clearly, if A has a minimum it cannot have more than one; for if m_1 and m_2 are both minima of A then, by definition, $m_1 \leq m_2$ and $m_2 \leq m_1$, hence $m_1 = m_2$. Similarly, $M \in A$ is the *greatest element*, or the *maximum*, of A if $M \geq a$ for all $a \in A$. If M exists, it is also unique. We shall often use the notation $m = \min A$ and $M = \max A$. The set

$$\{x \in \mathbb{R} : x > 1, x \leq 2\} = \{x \in \mathbb{R} : 1 < x \leq 2\},$$

for example, has no minimum. Its maximum is 2.

Example 2.3 Prove that the relation $<$ on $\mathbb{R}$ is *transitive*, in the sense that, if $a < b$ and $b < c$ then $a < c$.

Proof. Suppose $a < b$ and $b < c$. Then, by definition, $b - a \in P$ and $c - b \in P$. Using A11, we obtain
$$(c - b) + (b - a) \in P.$$
But $(c - b) + (b - a) = c - a$ by A2, A3, and A4. Hence $c - a \in P$, which means $a < c$. □

The *product* notation $a \cdot b$ is usually abbreviated to ab.

Example 2.4 For any $a, b, c \in \mathbb{R}$, with $a < b$, prove the following:

(i) $a + c < b + c$

(ii) If $c > 0$ then $ac < bc$

(iii) If $c < 0$ then $ac > bc$.

Proof. (i) Using axioms A1, A2, A3, and A4 we have
$$(b + c) - (a + c) = b - a.$$
Since $a < b$ it follows that $(b+c) - (a+c) \in P$, and hence $a+c < b+c$.

(ii) Here both c and $b - a$ are positive, i.e. in P, and by A11 so is their product $(b - a)c$. But
$$(b - a)c = bc - ac. \qquad (A9)$$
Hence $bc - ac \in P$, which just means that $ac < bc$.

(iii) In this case $-c$ and $b - a$ are both positive, so $-c(b - a) \in P$. But, using the results of Exercises 2.1.7 and 2.1.8,
$$-c(b - a) = ac - bc.$$
Thus $ac - bc \in P$, or $ac > bc$. □

Example 2.5 $x^2 \geq 0$ for any real number x.

Proof. If x is positive then, by A11, so is x^2. If $x = 0$ then, by Example 2.2(i), $x^2 = 0$. If x is negative, then $-x$ is positive and, by Exercise 2.1.8, $x^2 = (-x)^2$, which is again positive by A11. □

Remark 2.1 This last example implies, in particular, that $1 = 1^2 \geq 0$, and since $1 \neq 0$, we must have $1 > 0$.

Definition 2.1 For any real number x, the *absolute value*, or *modulus*, of x is defined as
$$|x| = \begin{cases} x, & x \geq 0 \\ -x, & x < 0. \end{cases}$$

Using part (iii) of Example 2.4, it is a simple matter to show that this definition implies $|x| \geq 0$, $|-x| = |x|$ and $|x| \geq x$ for any $x \in \mathbb{R}$. The next theorem states the more significant properties of the absolute value. Note that $\sqrt{0} = 0$, and $\sqrt{a}$ is the *positive* square root of a if $a > 0$. If $a < 0$ then, by Example 2.5, $\sqrt{a}$ cannot be a real number.

Theorem 2.1

For any real numbers x and y,
(i) $|x| = \sqrt{x^2}$
(ii) $|x| < |y| \Leftrightarrow x^2 < y^2$
(iii) $|xy| = |x||y|$
(iv) $|x| < a \Leftrightarrow -a < x < a$, where $a > 0$
(v) $|x + y| \leq |x| + |y|$
(vi) $||x| - |y|| \leq |x - y|$.

Proof. We shall only prove (v), which is known as the *triangle inequality*, leaving the other properties as exercises.
$$\begin{aligned}(x+y)^2 &= x^2 + y^2 + 2xy \\ &\leq |x^2| + |y^2| + 2|xy| \\ &= |x|^2 + |y|^2 + 2|x||y| \quad \text{by (iii)} \\ &= (|x| + |y|)^2\end{aligned}$$

From (ii) we now conclude that $|x + y| \leq |x| + |y|$. □

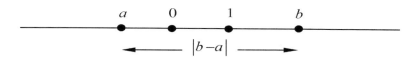

Figure 2.1

It is convenient to represent the real numbers as *points* on a straight line, which is usually taken horizontal, and which extends indefinitely in both directions. An arbitrary fixed point on this line is chosen as the point corresponding to 0, and is called the *origin*. The points to the *right* of the origin represent the *positive* numbers, while those to the *left* represent the *negative* numbers. By selecting the point which corresponds to the number 1 to the right of 0, the unit length on the line is determined. Now if $b > 0$ then the point corresponding to b is the point located b units of length to the right of the origin. Similarly, if $a < 0$, the point corresponding to a will lie to the left of the origin at a distance of $|a|$ units. The *distance* between a and b is then $|b - a|$. From an intuitive geometrical point of view we would expect this *correspondence* between $\mathbb{R}$ and the points on the real line to be a *bijective function* p from $\mathbb{R}$ to the points on the line defined by $x \mapsto p(x)$, such that $x_1 < x_2$ if, and only if, the point $p(x_2)$ is located $x_2 - x_1$ units to the right of $p(x_1)$. The resulting line on which the geometric points are assigned numerical values, or *coordinates*, in this fashion is called the *real line*. This correspondence between $\mathbb{R}$ and the points on the real line will often be taken as an *identification*, and hence we shall refer to the "point x" when we mean the "real number x" sometimes, and the "point whose coordinate is x" at other times.

Using the relations $<$ and $\leq$ we shall now define a class of subsets of $\mathbb{R}$ which we have already encountered and which will be used repeatedly, the *real intervals*. For any pair of real numbers a and b such that $a < b$, we define

(i) the *open interval*

$$(a,b) = \{x \in \mathbb{R} : a < x < b\},$$

(ii) the *closed interval*

$$[a,b] = \{x \in \mathbb{R} : a \leq x \leq b\},$$

(iii) the *half-open*, or *half-closed*, *intervals*

$$[a,b) = \{x \in \mathbb{R} : a \leq x < b\}$$
$$(a,b] = \{x \in \mathbb{R} : a < x \leq b\}$$

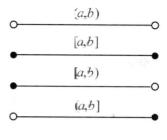

Figure 2.2

(iv) the *unbounded intervals*

$$(a, \infty) = \{x \in \mathbb{R} : x > a\}$$
$$[a, \infty) = \{x \in \mathbb{R} : x \geq a\}$$
$$(-\infty, a) = \{x \in \mathbb{R} : x < a\}$$
$$(-\infty, a] = \{x \in \mathbb{R} : x \leq a\}$$
$$(-\infty, \infty) = \mathbb{R}.$$

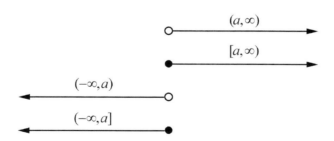

Figure 2.3

We can also use the relations $<$ and $\leq$ to define an important class of functions, the *monotonic functions*. A function $f : D \to \mathbb{R}$ is said to be *increasing* if

$$x, y \in D, \ y > x \Rightarrow f(y) \geq f(x),$$

and *strictly increasing* if

$$x, y \in D, \ y > x \Rightarrow f(y) > f(x).$$

On the other hand, f is *decreasing* if

$$x, y \in D, \ y > x \Rightarrow f(y) \leq f(x),$$

and *strictly decreasing* if

$$x, y \in D, \ y > x \Rightarrow f(y) < f(x).$$

A function is called *monotonic* if it is either increasing or decreasing, and *strictly monotonic* if it is either strictly increasing or strictly decreasing.

Exercises 2.2

1. Prove the following statements:

    (a) If $a > 0$ then $\dfrac{1}{a} > 0$.

    (b) If $a > b > 0$, then $\dfrac{1}{a} < \dfrac{1}{b}$.

    (c) If $ab > 0$, then either a and b are both positive, or they are both negative.

    (d) If $a \leq b$ and $c < d$, then $a + c < b + d$.

    (e) If $0 < a < b$ and $0 < c < d$, then $ac < bd$.

    (f) If $a < b$ and $c < d$, then $ad + bc < ac + bd$.

    (g) If $0 < a < 1$ then $a^2 < a$, and if $a > 1$ then $a^2 > a$.

2. (a) If $a \leq b + \varepsilon$ for all positive values of ε, prove that $a \leq b$.

    (b) If $|x| \leq \varepsilon$ for all $\varepsilon > 0$, prove that $x = 0$.

3. Prove parts (i) to (iv) and (vi) of Theorem 2.1.

4. Under what conditions does equality hold in Theorem 2.1(v)?

5. If $a, b \in \mathbb{R}$ and $b \neq 0$, prove that $\left|\dfrac{a}{b}\right| = \dfrac{|a|}{|b|}$.

6. For each of the following inequalities determine the solution set in $\mathbb{R}$, and show its representation on the real line:

(a) $-1 < 3x - 7 \leq 4$
(b) $|8x - 1| < 2$
(c) $5 \leq |4 - 6x|$
(d) $\dfrac{1}{x} < 1$
(e) $x^2 - 7x + 10 > 0$
(f) $|x + 4| < |2x - 1|$
(g) $|x| + |x + 1| < 3$
(h) $\dfrac{1}{x - 4} < \dfrac{5}{x + 5}$.

7. If
$$|x - x_0| < \min\left\{1, \frac{\varepsilon}{2(|y_0| + 1)}\right\}, \quad |y - y_0| < \frac{\varepsilon}{2(|x_0| + 1)},$$
prove that $|xy - x_0 y_0| < \varepsilon$.

8. For any pair of real numbers a and b, prove that
$$\max\{a, b\} = \frac{1}{2}(a + b + |b - a|),$$
$$\min\{a, b\} = \frac{1}{2}(a + b - |b - a|).$$

9. Prove that, for any $x_1, x_2, y_1, y_2 \in \mathbb{R}$,
$$|x_1 y_1| + |x_2 y_2| \leq \sqrt{x_1^2 + x_2^2}\sqrt{y_1^2 + y_2^2},$$
and then generalize to obtain the *Schwarz inequality*
$$\sum_{i=1}^{n} |x_i y_i| \leq \left(\sum_{i=1}^{n} x_i^2\right)^{1/2} \left(\sum_{i=1}^{n} y_i^2\right)^{1/2}.$$
Hint: Start with $2ab \leq a^2 + b^2$.

10. Prove that a strictly monotonic function is injective.

2.3 Natural Numbers, Integers, Rational Numbers

As a consequence of our axiomatic approach to the real numbers, we now need to use the same axioms listed above to extract some important subsets of $\mathbb{R}$. This we shall do before stating the last axiom, A12. The reason for that is two-fold: First, to emphasize that A12 plays no part in the definition of these subsets, and secondly, to show the role it plays in distinguishing $\mathbb{R}$ from them. Noting that the field properties postulate the existence of 1, we start with the following definition

Definition 2.2 A subset A of $\mathbb{R}$ is called *inductive* if
(i) $1 \in A$
(ii) $a \in A \Rightarrow a + 1 \in A$.

Denoting the collection of inductive sets by $\mathcal{I}$, we see that $\mathcal{I}$ is not empty, for it includes $\mathbb{R}$. Furthermore, it is a simple matter to prove that the intersection of any collection of inductive sets is an inductive set. Now we define the set of *natural numbers* as the intersection of all the inductive subsets of $\mathbb{R}$, that is,

$$\mathbb{N} = \bigcap \{A \subseteq \mathbb{R} : A \in \mathcal{I}\}.$$

In other words, $\mathbb{N}$ is the smallest inductive set in $\mathbb{R}$. This is in line with the intuitive idea that

$$\{1, 2, 3, \cdots\} = \{1, 1+1, 1+1+1, \cdots\}.$$

As a consequence, we obtain the following result, commonly referred to as the *principle of mathematical induction*.

Theorem 2.2
If $A \subseteq \mathbb{N}$ satisfies
(i) $1 \in A$
(ii) $n \in A \Rightarrow n+1 \in A$,
then $A = \mathbb{N}$.

This just says that any inductive subset of $\mathbb{N}$ is $\mathbb{N}$ itself.

Proof. A is inductive, hence, by definition of $\mathbb{N}$, $\mathbb{N} \subseteq A$. But since $A \subseteq \mathbb{N}$, we must have $A = \mathbb{N}$. $\square$

Example 2.6 If $m, n \in \mathbb{N}$ then $m + n \in \mathbb{N}$ and $mn \in \mathbb{N}$; that is, $\mathbb{N}$ is *closed* under addition and multiplication.

Proof. Let $m \in \mathbb{N}$ and $A = \{n \in \mathbb{N} : m + n \in \mathbb{N}\}$. Clearly, $A \subseteq \mathbb{N}$. We shall prove that A is inductive, and hence equals $\mathbb{N}$. Note that

(i) $m+1 \in \mathbb{N}$ by definition of $\mathbb{N}$, hence $1 \in A$.
(ii) If $n \in A$ then $m+n \in \mathbb{N}$ and, by definition of $\mathbb{N}$, $(m+n)+1 \in \mathbb{N}$. Using A2, we can write
$$(m+n)+1 = m+(n+1),$$
hence $n+1 \in A$. By the principle of mathematical induction, $A = \mathbb{N}$.

By defining $A = \{n \in \mathbb{N} : mn \in \mathbb{N}\} \subseteq \mathbb{N}$ we can similarly show that A is inductive, and therefore coincides with $\mathbb{N}$. □

A natural number n is *even* if it is a multiple of 2, otherwise it is *odd*. Any even number may be expressed as $2k$ for some $k \in \mathbb{N}$, and an odd number as $2k-1$ for some $k \in \mathbb{N}$.

Example 2.7 For any $n \in \mathbb{N}$ prove the following:
(a) $n \geq 1$.
(b) If $n \neq 1$, then $n-1 \in \mathbb{N}$.
(c) If $m \in \mathbb{N}$, $m > n$, then $m-n \in \mathbb{N}$.
(d) If $m \in \mathbb{N}$, $m > n$, then $m \geq n+1$.

Proof
(a) Let $A = \{n \in \mathbb{N} : n \geq 1\}$. It then follows that
(i) $1 \in A$.
(ii) If $n \in A$ then $n \geq 1$. Since $1 > 0$ (see Remark 2.1) we have $n+1 \geq 1+0 = 1$, hence $n+1 \in A$.
This proves that A is inductive, so $A = \mathbb{N}$.

(b) Let $A = \{1\} \cup \{m \in \mathbb{N} : m-1 \in \mathbb{N}\}$. Here, again, we note that $1 \in A$; and if $n \in A \subseteq \mathbb{N}$ then $n+1 \in \mathbb{N}$, and since $n+1-1 = n \in \mathbb{N}$, we have $n+1 \in A$, so $A = \mathbb{N}$. Thus, if $n \in \mathbb{N}$ and $n \neq 1$, then $n \in \{m \in \mathbb{N} : m-1 \in \mathbb{N}\}$ and therefore $n-1 \in \mathbb{N}$.

(c) Let $A = \{n \in \mathbb{N} : if\ m \in \mathbb{N},\ m > n\ then\ m-n \in \mathbb{N}\}$. To prove that $A = \mathbb{N}$ we shall prove that A is inductive. From part (b) we know that $1 \in A$. Suppose then that $n \in A$. By definition of A, $k-n \in \mathbb{N}$ for any natural number $k > n$. If $m \in \mathbb{N}$ and $m > n+1 > 1$, then $m-1 \in \mathbb{N}$ and $m-1 > n$. Consequently $m-(n+1) = (m-1)-n \in \mathbb{N}$, and hence $n+1 \in A$. Thus $A = \mathbb{N}$.

(d) Suppose $m, n \in \mathbb{N}$ and $m > n$. From (c) we know that $m-n \in \mathbb{N}$, and from (a) we conclude that $m-n \geq 1$, that is, $m \geq n+1$. □

We can go on in this manner and prove all the other familiar properties of the natural numbers, but we shall not do that. Instead, we shall prove one of the more important of these properties, known as the *well ordering property*, and assume the validity of the rest.

Theorem 2.3
Any non-empty subset of $\mathbb{N}$ has a least element.

Proof. Suppose $S \subseteq \mathbb{N}$ and $S \neq \varnothing$. We shall assume that S has no minimum, and show that this leads to a contradiction with our hypothesis. Define

$$A = \{n \in \mathbb{N} : \{1, 2, \cdots, n\} \cap S = \varnothing\}.$$

(i) $1 \notin S$, otherwise 1 would be the smallest element in S by Example 2.7(a), hence $1 \in A$.

(ii) Let $n \in A$. If $k \in S$ then $k \notin \{1, 2, \cdots, n\}$, hence $k > n$. By Example 2.7(d), $k \geq n+1$. Now if $n+1 \in S$ then $n+1 = \min S$, in contradiction with our assumption. Therefore $n+1 \notin S$, so $n+1 \in A$. Thus A is inductive and must therefore coincide with $\mathbb{N}$. But this implies $S = \varnothing$, which contradicts our initial assumption and proves the theorem. □

We have already seen how the principle of mathematical induction can be used to deduce the properties of the natural numbers. It can also be used effectively to prove certain statements involving the natural numbers. To that end, we present two equivalent versions of the principle.

First statement: Suppose that, for every $n \in \mathbb{N}$, we have a statement $P(n)$. If
  (i) $P(1)$ is true
  (ii) $P(n)$ is true $\Rightarrow P(n+1)$ is true,
then $P(n)$ is true for all $n \in \mathbb{N}$.

Second statement: For every $n \in \mathbb{N}$, let $P(n)$ be a statement. If
  (i) $P(1)$ is true
  (ii) $P(1), P(2), \cdots, P(n)$ are all true $\Rightarrow P(n+1)$ is true,
then $P(n)$ is true for all $n \in \mathbb{N}$.

Remark 2.2 If $P(1)$ in (i) is replaced by $P(k)$ for some natural number $k > 1$, and the implication in (ii) holds for all $n \geq k$, then clearly $P(n)$ will be true for all $n \in \{k, k+1, k+2, \cdots\}$. This follows from the observation that, if we define $Q(1)$ as $P(k)$, and $Q(m)$ as $P(k+m-1)$, then, by applying induction on $m \in \mathbb{N}$ to $Q(m)$, we establish the truth of $P(k+m-1)$ for all $m \in \mathbb{N}$. By the same token we can also start with $P(0)$ instead of $P(1)$ in step (i), and thereby prove $P(n)$ for any $n \in \mathbb{N} \cup \{0\} = \mathbb{N}_0$.

Example 2.8 Prove that, for any $n \in \mathbb{N}$,
$$\sum_{k=1}^{n} k^2 = \frac{1}{6}n(n+1)(2n+1).$$

Proof. Assume that $P(n)$ denotes the statement
$$\sum_{k=1}^{n} k^2 = \frac{1}{6}n(n+1)(2n+1).$$

Then, by setting $n = 1$, we see that $P(1)$ is true. Now if $P(n)$ is true, then
$$\begin{aligned}
\sum_{k=1}^{n+1} k^2 &= \sum_{k=1}^{n} k^2 + (n+1)^2 \\
&= \frac{1}{6}n(n+1)(2n+1) + (n+1)^2 \\
&= \frac{1}{6}(n+1)[n(2n+1) + 6(n+1)] \\
&= \frac{1}{6}(n+1)[2n^2 + 7n + 6] \\
&= \frac{1}{6}(n+1)(n+2)(2n+3),
\end{aligned}$$
which shows that $P(n+1)$ is true. Hence, by induction, $P(n)$ is true for all $n \in \mathbb{N}$. $\square$

A natural number greater than 1 which is divisible only by itself and 1 is called a *prime number*. The first few primes are $2, 3, 5, 7, 11, \cdots$. If a is a real number and n is a natural number, then $a^n = a \cdot a \cdots a$ (repeated n-times) is the *n-th power* of a, which is clearly a real number. The next result is known as the *prime factorization theorem*.

Theorem 2.4
Every natural number greater than 1 may be expressed as a product of prime numbers.

Proof. Suppose $P(n)$ denotes the statement: n may be expressed as a product of prime numbers. We shall use the second statement of the principle of induction:
(i) $P(2)$ is clearly true.
(ii) Suppose $P(2), P(3), P(4), \cdots, P(n)$ are all true. Either $n+1$ is prime and hence $P(n+1)$ is true, or there are two natural numbers k and m, both greater than 1, such that
$$n+1 = km.$$

Since k and m are both necessarily less than $n+1$, we have $k \leq n$ and $m \leq n$. From assumption (ii), both k and m can be expressed as products of primes, and therefore $n+1$ is also a product of primes, that is, $P(n+1)$ is true. Thus, by induction, $P(n)$ is true for all $n \in \{2, 3, 4, \cdots\}$. □

Remark 2.3 If the prime factors of n are $p_1, p_2, \cdots, p_k$ then there are natural numbers $m_1, m_2, \cdots, m_k$ such that

$$n = p_1^{m_1} p_2^{m_2} \cdots p_k^{m_k},$$

and this representation of n by the product of its prime factors is *unique* if the prime factors are arranged in increasing order $p_1 < p_2 < \cdots < p_k$.

Referring to the field properties once more, we use the existence of 0 and the additive inverse (A3 and A4) to define the integers.

Definition 2.3 An *integer* is a real number x such that either $x \in \mathbb{N}$, or $x = 0$, or $-x \in \mathbb{N}$. The set of integers, denoted by $\mathbb{Z}$, is therefore the subset of $\mathbb{R}$ defined by

$$\begin{aligned}\mathbb{Z} &= \mathbb{N} \cup \{0\} \cup \{-n : n \in \mathbb{N}\} \\ &= \{\cdots, -3, -2, -1, 0, 1, 2, 3, \cdots\}.\end{aligned}$$

It is worth noting that $\mathbb{Z}$ satisfies axioms A1 to A11 except A8, so it is closed under addition, multiplication, subtraction, but not division. If a and b are integers, we can always solve $x + a = b$, but not necessarily $ax = b$, in $\mathbb{Z}$. The set of integers, under addition and multiplication, is an example of an *algebraic ring*, which is defined as a set with two operations that satisfy axioms A1 to A7 and axiom A9. Now we use A8 to define the rational numbers.

Definition 2.4 A *rational number* is a real number which can be represented as $a \cdot b^{-1}$, where $a, b \in \mathbb{Z}$ and $b \neq 0$. The set of rational numbers, denoted by $\mathbb{Q}$, may therefore be expressed as

$$\mathbb{Q} = \left\{\frac{a}{b} : a \in \mathbb{Z}, b \in \mathbb{Z}\setminus\{0\}\right\}.$$

Using the ordinary rules of arithmetic, it is straightforward to verify that $\mathbb{Q}$ satisfies all the axioms A1 to A11; and an equation of the form $ax + b = c$, where $a, b, c \in \mathbb{Q}$ and $a \neq 0$, has a unique solution $a^{-1}(c - b)$ in $\mathbb{Q}$. This would seem to imply that the rational numbers are rich enough to support a wide range of mathematical activity, to the extent that the ancient Greek mathematicians thought that any physical length could be represented by a rational number. But in their attempt to express the length of the hypotenuse of the triangle shown in Figure 2.4 as a rational number they failed, for the equation $x^2 = 2$, which expresses the Pythagorean relation between the sides of the triangle, has no solution in $\mathbb{Q}$, as the following theorem demonstrates.

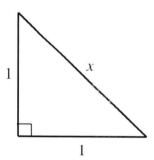

Figure 2.4

Theorem 2.5
There is no rational number x such that $x^2 = 2$.

Proof. Suppose there is an $x \in \mathbb{Q}$ such that $x^2 = 2$. Let

$$x = \frac{a}{b},$$

be the simplest representation of x, where $a, b \in \mathbb{N}$ and a and b have no common factors (except 1). Substituting into the equation $x^2 = 2$, we obtain

$$a^2 = 2b^2,$$

which implies that 2 is a prime factor of a^2. Therefore 2 is also a prime factor of a. By setting $a = 2m$ for some $m \in \mathbb{N}$, we obtain

$$a^2 = 4m^2 = 2b^2.$$

Therefore $b^2 = 2m^2$, and once again we conclude that 2 is a prime factor of b as well, which contradicts the assumption that a and b have no common factor. □

This surprising result pointed out quite early in the history of mathematics that the rational numbers, in spite of the fact that they satisfied all the field and order axioms, were inadequate for covering the real line; for, according to Theorem 2.5, the length of the hypotenuse of the triangle in Figure 2.4 cannot be expressed as a rational number. In other words, if the points on the real line are to represent the real numbers, then there are some points whose coordinates are not in $\mathbb{Q}$. So how do we arrive at these non-rational coordinates? In the next section we shall see how one more axiom, A12, fills these "gaps" and preserves the one-to-one correspondence between the points of the real line and the number system we seek.

EXERCISES 2.3

1. Prove the following statements:
   (a) The set of positive real numbers P is inductive.
   (b) The set $\{x \in P : x \neq 5\}$ is not inductive.
   (c) If the set A_λ is inductive for every $\lambda \in \Lambda$, then $\bigcap_{\lambda \in \Lambda} A_\lambda$ is inductive.

2. Prove the second part of Example 2.6.

3. Prove the equivalence of the principle of mathematical induction (Theorem 2.2) with the first and second statements of the principle.

4. Use induction to prove:
   (a) For all $n \in \mathbb{N}$,
   $$\sum_{k=1}^{n} k(k+1) = \frac{1}{3} n(n+1)(n+2).$$
   (b) For all $n \in \mathbb{N}$,
   $$\sum_{k=1}^{n} \frac{4}{k(k+1)(k+2)} = 1 - \frac{2}{(n+1)(n+2)}.$$
   (c) For all $n \in \mathbb{N}$,
   $$\sum_{k=1}^{n} k^3 = \frac{1}{4}[n(n+1)]^2$$

(d) If $u_{k+1} = 2u_k + 1$ for all $k \in \mathbb{N}$, show that
$$u_n = 2^{n-1}(u_1 + 1) - 1.$$
Given $u_1 = 1$, evaluate $\sum_{k=1}^{n} u_k$.

(e) $3^{2n} + 7$ is divisible by 8 for all $n \in \mathbb{N}$.

(f) $(3n+1)7^n - 1$ is divisible by 9 for all $n \in \mathbb{N}$.

5. If $x > -1$ prove that
$$(1+x)^n \geq 1 + nx \quad \text{for all } n \in \mathbb{N}.$$

6. Prove that $n < 2^n$ for all $n \in \mathbb{N}$.

7. Prove that $2^{n-1} \leq n!$ for all $n \in \mathbb{N}$. For what values of n is the inequality $2^n \leq n!$ valid?

8. What axioms does $\mathbb{N}$ satisfy?

9. Verify that $\mathbb{Z}$ satisfies all the axioms from A1 to A11, except for A8.

10. Verify that $\mathbb{Q}$ satisfies all the axioms from A1 to A11.

11. Prove that the numbers $\sqrt{3}, \sqrt{5}, \sqrt{6}, \sqrt[3]{2}, \sqrt[3]{3}$ are all not rational.

12. Prove that $\sqrt{p}$ is not a rational number for any prime number p.

13. Prove that
$$0 \leq a \leq 1 \Rightarrow 0 \leq a^n \leq a \quad \text{for all } n \in \mathbb{N},$$
and that
$$a \geq 1 \Rightarrow a^n \geq a \quad \text{for all } n \in \mathbb{N}.$$
Show that the relation $\leq$ can be replaced by $<$ in the first implication, and that $\geq$ can be replaced by $>$ in the second.

14. Use induction to prove the *binomial theorem*

$$(x+y)^n = x^n + nx^{n-1}y + \frac{n(n-1)}{2}x^{n-2}y^2 + \cdots$$
$$+ \frac{n(n-1)}{2}x^2 y^{n-2} + nxy^{n-1} + y^n$$
$$= \sum_{k=0}^{n} \binom{n}{k} x^{n-k} y^k$$

for any $n \in \mathbb{N}$ and $x, y \in \mathbb{R}$. Here $\binom{n}{k}$ is the *binomial coefficient*, defined by

$$\binom{n}{k} = \frac{n!}{(n-k)!k!},$$

where $k! = k(k-1)(k-2)\cdots(2)(1)$ for any natural number k, and $0! = 1$.

Hint: First prove that

$$\binom{n}{k} + \binom{n}{k+1} = \binom{n+1}{k+1}.$$

15. Prove that $(n+1)^n \leq n^{n+1}$ for all $n \geq 2$.

16. If a and b are positive real numbers and $m, n \in \mathbb{N}$, use induction to show that

   (i) $a^m a^n = a^{m+n}$,
   (ii) $(a^m)^n = a^{mn}$,
   (iii) $(ab)^n = a^n b^n$,
   (iv) $a < b \Leftrightarrow a^n < b^n$.

17. Using the convention $a^0 = 1$, $a^{-n} = 1/a^n$, where $a > 0$, check that the properties (i) to (iii) in Exercise 2.3.16 remain valid for all $m, n \in \mathbb{Z}$.

2.4 Completeness Axiom

We start by giving some preparatory definitions.

Definition 2.5 Let A be a subset of $\mathbb{R}$.

(i) If there is a real number b such that
$$x \leq b \text{ for all } x \in A,$$
then b is called an *upper bound* of A, and the set A is said to be *bounded above*.
(ii) If there is a real number a such that
$$x \geq a \text{ for all } x \in A,$$
then a is called a *lower bound* of A, and A is said to be *bounded below*.
(iii) A is *bounded* if it is bounded above and below, and *unbounded* if it is not bounded.

As an example, the interval $A = [0, 1)$ is bounded above by 1. In fact, any number $b \geq 1$ is an upper bound of A. A is also bounded below by 0, and by any number $a < 0$. Consequently the interval $[0, 1)$ is a bounded subset of $\mathbb{R}$. But the interval $[0, \infty)$, which is bounded below, has no upper bound and is therefore unbounded.

It is clear from Definition 2.5 that, if b is an upper bound of A, any number greater than b will also be an upper bound of A. It would therefore make sense to seek the "smallest" among such upper bounds, for it would be the most useful. Similarly, if A is bounded below, the "largest" number which qualifies as a lower bound for A will yield all the other lower bounds.

Definition 2.6 Let A be a subset of $\mathbb{R}$. An element $b \in \mathbb{R}$ is called a *least upper bound*, or *supremum*, of A if
(i) b is an upper bound of A, that is, $b \geq x$ *for all* $x \in A$, and
(ii) there is no upper bound of A which is less than b, that is,
$$u \geq x \text{ for all } x \in A \Rightarrow u \geq b.$$

Similarly, we call $a \in \mathbb{R}$ a *greatest lower bound*, or *infimum*, if
(i) a is a lower bound of A, and
(ii) there is no lower bound of A which is greater than a, that is,
$$u \leq x \text{ for all } x \in A \Rightarrow u \leq a.$$

It should first be noted that, if there is a least upper bound for A, it must be unique; for if b and b' are both least upper bounds of A, then by applying condition (ii) with b as a least upper bound and b' as an upper bound, we obtain $b' \geq b$, and by reversing the roles of b and

b' we obtain $b \geq b'$. Consequently $b = b'$. Similarly, the greatest lower bound of a subset of $\mathbb{R}$, if it exists, is unique.

The symbol $\sup A$ will denote the least upper bound of A, and $\inf A$ will denote its greatest lower bound. When it exists, $\sup A \in \mathbb{R}$, and if $\sup A \in A$ then we must have $\sup A = \max A$. Similarly, $\inf A = \min A$ if $\inf A \in A$.

Example 2.9 If A is any of the intervals (a,b), $[a,b)$, $(a,b]$, or $[a,b]$ then
$$\sup A = b, \ \inf A = a.$$

Proof. Take $A = (a,b)$. The proof is similar for the other cases. Note first that, by definition of (a,b), $b > a$ and b is an upper bound of A. If u is an upper bound of A such that $u < b$, then
$$a < u < \frac{u+b}{2} < b.$$
This implies
$$\frac{u+b}{2} \in A \ \text{ and } \ u < \frac{u+b}{2},$$
which contradicts the assumption that u is an upper bound of A. Hence $u \geq b$ and $b = \sup A$.

To prove that $\inf A = a$ we can either follow a similar line of argument, i.e. proof by contradiction, or use the next lemma. $\square$

Lemma 2.1 *If we define* $-A = \{-x : x \in A\}$, *then the set A is bounded below if, and only if, $-A$ is bounded above. Furthermore,*
$$\inf A = -\sup(-A)$$
when either side exists.

Proof. Using the result of Example 2.4, we have
$$x \geq c \Leftrightarrow -x \leq -c,$$
and this implies A is bounded below (by c) if, and only if, $-A$ is bounded above (by $-c$).

Now if $a = \inf A$ then $-a$ is an upper bound of $-A$. If u is another upper bound of $-A$ then $-u$ is a lower bound of A. By the definition of $\inf A$, $-u \leq a$, hence $u \geq -a$. We have therefore proved that
$$-a = \sup(-A),$$

or, $\inf A = a = -\sup(-A)$. We arrive at the same equality if we assume the existence of $\sup(-A)$ and follow a similar argument. □

Now we are ready to state the twelfth axiom, also known as the *completeness axiom*.

A12. *If A is a non-empty subset of $\mathbb{R}$ which is bounded above, then it has a least upper bound in $\mathbb{R}$.*

Before discussing the implications of this axiom, we should note that, in view of lemma 2.1, it is equivalent to

A12'. *If A is a non-empty subset of $\mathbb{R}$ which is bounded below, then it has a greatest lower bound in $\mathbb{R}$.*

The next theorem demonstrates how the set of real numbers $\mathbb{R}$, equipped with axiom A12, is now large enough to provide solutions to such equations as $x^2 = 2$, an equation which had no solution in $\mathbb{Q}$ as we saw in Theorem 2.4.

Theorem 2.6
If $n \in \mathbb{N}$ and $a > 0$, then there is a number $x \in \mathbb{R}$ such that $x^n = a$.

Proof. We shall consider the two cases $a \geq 1$ and $a < 1$ separately.
(a) Assume $a \geq 1$, and let
$$A = \{t \in \mathbb{R} : t \geq 0, \ t^n \leq a\}.$$
We first note that A is not empty, for $1 \in A$. Secondly, A is bounded above by a, for
$$a \geq 1 \Rightarrow a^n \geq a$$
(see Exercise 2.3.12), and
$$t > a \Rightarrow t^n > a^n \geq a \Rightarrow t \notin A.$$
By axiom A12, A has a least upper bound, say $x = \sup A$. We shall now prove that $x^n = a$ by excluding the two possibilities, $x^n < a$ and $x^n > a$.

(i) Suppose $x^n < a$. We shall exhibit a positive number u such that $x + u \in A$, thereby contradicting the fact that x is an upper bound of A. By the binomial theorem,
$$(x+u)^n = \sum_{k=0}^{n-1} \binom{n}{k} x^k u^{n-k} + x^n,$$

where
$$\binom{n}{k} = \frac{n!}{(n-k)!k!}.$$
Let
$$c = \sum_{k=0}^{n-1}\binom{n}{k}x^k \quad \text{and} \quad u = \min\{1, \frac{a-x^n}{c}\}.$$
Then $0 < u \leq 1$, and hence
$$(x+u)^n \leq x^n + cu \leq x^n + c\frac{a-x^n}{c} = a.$$
By definition of A, this means $x + u \in A$.

(ii) Now suppose $x^n > a$. We shall contradict the fact that x is a least upper bound of A by producing an upper bound of A smaller than x. Define
$$\varepsilon = x^n - a > 0 \quad \text{and} \quad y = x - \frac{\varepsilon}{nx^{n-1}}.$$
It then follows that $0 < y < x$, and
$$\begin{aligned} x^n - y^n &= (x-y)(x^{n-1} + yx^{n-2} + \cdots + y^{n-2}x + y^{n-1}) \\ &< \frac{\varepsilon}{nx^{n-1}}(x^{n-1} + x^{n-1} + \cdots + x^{n-1}) \\ &= \frac{\varepsilon}{nx^{n-1}} \cdot nx^{n-1} = \varepsilon. \end{aligned}$$
Hence
$$y^n > x^n - \varepsilon = a,$$
which makes y an upper bound of A; for if $z \in A$ is such that $z > y$ then $z^n > y^n > a$, which contradicts the definition of A. We have therefore obtained an upper bound of A which is less than $\sup A$, a clear contradiction. Thus we must conclude that $x^n = a$.

(b) To complete the proof we have to deal with the case $0 < a < 1$. Take $b = 1/a$. Since $b > 1$ we know, from (a), that there is a real number y such that $y^n = b$. Now $y \neq 0$ and $1/y$ satisfies $x^n = a$. □

This theorem proves the existence of the *positive n-th root* of a, denoted by $\sqrt[n]{a}$ or $a^{1/n}$, for any $a > 0$ and any $n \in \mathbb{N}$. Taking $n = a = 2$, we see that $\sqrt{2} \in \mathbb{R}$. Since we already know that $\sqrt{2} \notin \mathbb{Q} \subseteq \mathbb{R}$, we must conclude that $\mathbb{Q} \subset \mathbb{R}$. Numbers such as $\sqrt{2}$ are called *irrational*, and Exercises 2.3.11 and 2.3.12 exhibit other examples of irrational numbers. But we shall soon discover that the set of irrational numbers $\mathbb{R}\backslash\mathbb{Q}$ is much larger than the collection of roots of natural numbers. In a certain sense, which will soon be clarified, we shall see that there are actually "more" irrational numbers than rational ones in $\mathbb{R}$.

The next result is known as the *theorem of Archimedes*.

Theorem 2.7
The set of natural numbers is not bounded above.

Proof. $\mathbb{N}$ is not empty, since $1 \in \mathbb{N}$, so if $\mathbb{N}$ is bounded above it must, by A12, have a least upper bound. Assuming $\alpha = \sup \mathbb{N}$, we have $\alpha \geq n$ for all $n \in \mathbb{N}$. Now let $n \in \mathbb{N}$ be arbitrary. By the inductive property of $\mathbb{N}$, we also have $n + 1 \in \mathbb{N}$. Hence $\alpha \geq n + 1$, which implies $\alpha - 1 \geq n$. Since n was arbitrary, we must conclude that $\alpha - 1$ is an upper bound on $\mathbb{N}$. But this contradicts the defining property of α. $\square$

Corollary 2.7.1 *For every $x > 0$ there is a natural number n such that $x > 1/n$.*

Proof. By Theorem 2.7, $1/x$ is not an upper bound of $\mathbb{N}$. So there is a number $n \in \mathbb{N}$ such that $n > 1/x$, which implies $x > 1/n$. $\square$

Thus, if $x \in \mathbb{R}$ satisfies $x \leq 1/n$ for every $n \in \mathbb{N}$ then $x \leq 0$; and if $0 \leq x \leq 1/n$ for every $n \in \mathbb{N}$ then we can only conclude that $x = 0$. This is an improvement on the result of Exercise 2.2.2, where it is required to prove that

$$x < y + \varepsilon \text{ for all } \varepsilon > 0 \Rightarrow x \leq y$$
$$|x| < \varepsilon \text{ for all } \varepsilon > 0 \Rightarrow x = 0.$$

Corollary 2.7.2 *For every $x \geq 0$ there is an $n \in \mathbb{N}$ such that $n - 1 \leq x < n$.*

Proof. By Theorem 2.7 we know that the set $\{m \in \mathbb{N} : x < m\}$ is not empty. By the well-ordering property (Theorem 2.3) it has a smallest element, call it n. We then have $x < n$, but $x \not< n - 1$, hence $n - 1 \leq x < n$. $\square$

The next theorem establishes a fundamental property of $\mathbb{Q}$ as a subset of $\mathbb{R}$. It is referred to as the *density of $\mathbb{Q}$ in $\mathbb{R}$*, and expresses the "pervasiveness" of the rational numbers in $\mathbb{R}$, in the sense that between any two real numbers there is a rational number, in fact an infinite number of rational numbers.

Theorem 2.8
If x and y are real numbers and $x < y$, then there is a rational number r such that $x < r < y$.

Proof
(i) We first consider the case where $x \geq 0$. By Corollary 2.7.1, since $y - x > 0$ there is a natural number n such that

$$y - x > \frac{1}{n},$$

or,
$$nx + 1 < ny. \qquad (2.2)$$

Since $nx \geq 0$ we can find an $m \in \mathbb{N}$, by Corollary 2.7.2, such that

$$m - 1 \leq nx < m. \qquad (2.3)$$

Now (2.2) and (2.3) imply

$$nx < m \leq nx + 1 < ny,$$

from which we obtain
$$x < \frac{m}{n} < y.$$

Take $r = m/n$.

(ii) Let $x < 0$. Using theorem 2.7, we can find a $k \in \mathbb{N}$ such that $k > -x$. Consequently,

$$0 < k + x < k + y.$$

In view of (i), there is a rational number q which satisfies

$$k + x < q < k + y.$$

Since $r = q - k$ is also rational and satisfies $x < r < y$, the proof is complete. □

By repeated application of Theorem 2.8, we see that between any two distinct real numbers there is an infinite number of rational numbers. The irrational numbers are also dense in $\mathbb{R}$, as the next corollary shows.

Corollary 2.8 *If x and y are real numbers and $x < y$, then there is an irrational number t such that $x < t < y$.*

Proof. If $x < y$ then $\sqrt{2}x < \sqrt{2}y$, and we can use Theorem 2.8 to conclude the existence of a rational number $r \neq 0$ which satisfies

$$\sqrt{2}x < r < \sqrt{2}y,$$

and this implies
$$x < \frac{r}{\sqrt{2}} < y.$$
Setting $t = r/\sqrt{2}$, we see that t cannot be rational, otherwise $\sqrt{2} = r/t$ would be rational, in violation of Theorem 2.5. $\square$

We end this section with a brief demonstration of how the completeness postulate can be used to define the *exponential function* a^x for any real number x. Based on Theorem 2.6 and the rules for manipulating integral powers of a positive number (see Exercises 2.3.16 and 2.3.17), we can define a^r for any $a > 0$ and any $r \in \mathbb{Q}$ (see Exercises 2.4.10). It can also be shown that

(i) $a^r a^s = a^{r+s}$ and $(a^r)^s = a^{rs}$ for all $r, s \in \mathbb{Q}$

(ii) If $a > 1$, the function $r \mapsto a^r$ is strictly increasing on $\mathbb{Q}$.

Let $a > 1$. For any $x \in \mathbb{R}$ we define the set
$$A_a(x) = \{a^r : r \in \mathbb{Q}, r \leq x\}.$$
$A_a(x)$ is clearly non-empty and bounded above (by a^N, for example, where N is any integer greater than x), so $\sup A_a(x)$ exists in $\mathbb{R}$. Now we define the function $\exp_a : \mathbb{R} \to \mathbb{R}$ by
$$\exp_a(x) = \sup A_a(x),$$
and note that $\exp_a(x) > 0$ for all $x \in \mathbb{R}$, since $a^r > 0$ for all $r \in \mathbb{Q}$.

Since a^r is a strictly increasing function on $\mathbb{Q}$ (Exercise 2.4.11), we clearly have $\sup\{a^r : r \in \mathbb{Q}, r \leq q\} = a^q$ for any $q \in \mathbb{Q}$, hence
$$a^q = \exp_a(q) \quad \text{for all } q \in \mathbb{Q}.$$
This allows us to extend the exponential function a^x from $\mathbb{Q}$ to $\mathbb{R}$ by setting
$$a^x = \exp_a(x) \quad \text{for all } x \in \mathbb{R},$$
and it is not difficult to show that a^x is also a strictly increasing function of $x \in \mathbb{R}$ (Exercise 2.4.12).

If $0 < a < 1$, we define
$$a^x = \frac{1}{(1/a)^x},$$
and it follows, in this case, that the function a^x is strictly decreasing. Since, for all $x \in \mathbb{R}$, $1^x = 1$, we have $a^x > 1$ when $a > 1$ and $a^x < 1$

when $0 < a < 1$. If $a > 0$ and $b > 0$, then $(ab)^x = a^x b^x$, and if $0 < a < b$, then $a^x < b^x$ for all $x > 0$ (Exercise 2.4.13). The following relations can also be shown to hold for any $a > 0$ and all $x, y \in \mathbb{R}$ (Exercise 2.4.14):

(i) $a^x a^y = a^{x+y}$.
(ii) $a^x / a^y = a^{x-y}$.
(iii) $(a^x)^y = a^{xy}$.

Based on these properties, we can show that the function $\exp_a(x) = a^x$, for $a > 0$ and $a \neq 1$, maps $\mathbb{R}$ onto $(0, \infty)$ (Exercise 2.4.15). Being strictly monotonic, and hence injective, $\exp_a$ has an inverse which maps $(0, \infty)$ onto $\mathbb{R}$, called the *logarithmic function* to the base a, and denoted by $\log_a$. Thus

$$\log_a : (0, \infty) \to \mathbb{R}$$

$$y = \log_a x \Leftrightarrow x = a^y.$$

Using the properties of the exponential function listed above, it is straightforward to deduce the corresponding properties of the logarithmic function:

(i) $\log_a(xy) = \log_a x + \log_a y$
(ii) $\log_a(x/y) = \log_a x - \log_a y$
(iii) $\log_a(x^y) = y \log_a x$.

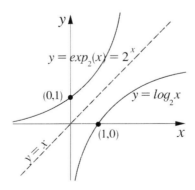

Figure 2.5

EXERCISES 2.4

1. Determine sup A and inf A where they exist:
   (a) $A = \{x \in \mathbb{R} : x^2 - 9 > 0\}$
   (b) $A = \left\{n \in \mathbb{N} : 1 - \dfrac{(-1)^n}{n}\right\}$
   (c) $A = \mathbb{Q}$
   (d) $A = \mathbb{N}_0 = \mathbb{N} \cup \{0\}$.

2. If b is an upper bound of A, prove that $b = \sup A$ if, and only if, for every $\varepsilon > 0$ there is an element $a \in A$ such that $a > b - \varepsilon$.
   State and prove the corresponding result for inf A.

3. If the sets A and B are bounded above and $A \subseteq B$, prove that $\sup A \leq \sup B$. What can you say about inf A and inf B?

4. If both A and B are bounded above, prove that $A \cup B$ is bounded above and
   $$\sup(A \cup B) = \max\{\sup A, \sup B\}.$$
   State and prove a corresponding result for $\inf(A \cup B)$.

5. Let A and B be subsets of $\mathbb{R}$, and define
   $$A + B = \{a + b : a \in A, b \in B\}.$$
   Prove that
   $$\sup(A + B) = \sup A + \sup B$$
   provided A and B are bounded above. State and prove the corresponding result for $\inf(A + B)$.

6. Let $A \subseteq \mathbb{R}$. For any $k \in \mathbb{R}$ define
   $$kA = \{ka : a \in A\}.$$
   If A is bounded above and $k > 0$, prove that
   $$\sup(kA) = k \cdot \sup A.$$
   What can be said if $k \leq 0$?

7. If $x \in \mathbb{Q}$ and $y \notin \mathbb{Q}$, prove that $x + y \notin \mathbb{Q}$. When is $xy \in \mathbb{Q}$?

8. If $x > 0$ prove that for, every $y \in \mathbb{R}$, there is an $n \in \mathbb{N}$ such that $nx > y$.

9. Suppose X and Y are non-empty sets which satisfy
   (i) $X \cup Y = \mathbb{R}$
   (ii) $x \in X, \ y \in Y \Rightarrow x < y$.
   Prove the existence of a number $c \in \mathbb{R}$ such that
   $$z < c \Rightarrow z \in X,$$
   $$z > c \Rightarrow z \in Y.$$

10. If $m, n \in \mathbb{N}$, define $a^{m/n} = (a^{1/n})^m = (\sqrt[n]{a})^m$ for any $a \geq 0$, and $a^{-m/n} = (a^{1/n})^{-m} = 1/(\sqrt[n]{a})^m$ for any $a > 0$. Show that $a^{m/n} = (a^m)^{1/n}$ for all $m, n \in \mathbb{Z}$ and $a > 0$. Thus a^r is a well defined positive number for any $r \in \mathbb{Q}$ and $a > 0$.

11. If $a > 0$ and $r, s \in \mathbb{Q}$, show that
    (a) $a^r a^s = a^{r+s}$ and $(a^r)^s = a^{rs}$.
    (b) $r < s \Leftrightarrow a^r < a^s$ when $a > 1$, and $r < s \Leftrightarrow a^r > a^s$ when $0 < a < 1$.

12. Prove that $\exp_a x = a^x$ is a strictly increasing function on $\mathbb{R}$ when $a > 1$.

13. Prove that the exponential function satisfies the relations
    (a) If $a > 0$ and $b > 0$, then $(ab)^x = a^x b^x$ for all $x \in \mathbb{R}$.
    (b) If $0 < a < b$, then $a^x < b^x$ for all $x \in \mathbb{R}$.

14. For any $a > 0$, prove that a^x satisfies the following relations for all $x, y \in \mathbb{R}$:
    (a) $a^x a^y = a^{x+y}$
    (b) $a^x / a^y = a^{x-y}$
    (c) $(a^x)^y = a^{xy}$.

15. Prove that $\exp_a$ maps $\mathbb{R}$ onto $(0, \infty)$.
    Hint: For any $y \in (0, \infty)$, let $B(y) = \{t \in \mathbb{R} : a^t < y\}$ and $x = \sup B(y)$. Show that $a^x = y$.

2.5 Decimal Representation of Real Numbers

In this section we exhibit a constructive method for obtaining $\mathbb{R}$ from $\mathbb{Q}$. Anyone familiar with "long division" is aware of the usual decimal representation of a positive common fraction p/q as

$$\frac{p}{q} = x_0.x_1x_2x_3\cdots = x_0 + \frac{x_1}{10} + \frac{x_2}{10^2} + \frac{x_3}{10^3} + \cdots,$$

where $x_0 \in \mathbb{N}_0$, $x_i \in \{0, 1, 2, 3, \cdots, 9\}$, and $i \in \mathbb{N}$. We shall now show what this representation means, how it can be justified, and how it extends to any positive real number x. Since negative numbers are obtained from positive numbers by multiplication by -1, we thereby obtain the decimal representation of any real number.

Let x be a real number such that $x \geq 0$. We shall prove that there is an integer $x_0 \geq 0$ and numbers $x_1, x_2, x_3, \cdots, x_n$ in the set $\{0, 1, 2, \cdots, 9\}$ such that

$$0 \leq x - \sum_{i=0}^{n} \frac{x_i}{10^i} < \frac{1}{10^n} \quad \text{for all } n \in \mathbb{N}_0. \tag{2.4}$$

When the inequality (2.4) holds for *all* non-negative integers n, we shall write

$$x = x_0.x_1x_2x_3\cdots. \tag{2.5}$$

The right-hand side (2.5) is then called the *decimal expansion* of x.

Now we shall show how we arrive at (2.4) by using the principle of induction. According to Corollary 2.7.2, given any $x \geq 0$ we can always find a (unique) $n \in \mathbb{N}_0$ which satisfies

$$n \leq x < n + 1.$$

n is called the *integral part* (or the *integer part*) of x, and will be denoted by $[x]$. By setting $x_0 = [x]$, we obtain

$$0 \leq x - x_0 < 1,$$

which is the relation (2.4) when $n = 0$. Suppose the numbers $x_1, x_2, \cdots, x_n$ are chosen from the set $\{0, 1, 2, \cdots, 9\}$ so that (2.4) holds. Multiplication by 10^{n+1} yields

$$0 \leq 10^{n+1}\left(x - \sum_{i=0}^{n} \frac{x_i}{10^i}\right) < 10.$$

By defining x_{n+1} as the *integer*

$$x_{n+1} = \left[10^{n+1}\left(x - \sum_{i=0}^{n} \frac{x_i}{10^i}\right)\right],$$

which is clearly in $\{0, 1, 2, \cdots, 9\}$, we obtain

$$0 \leq 10^{n+1}\left(x - \sum_{i=0}^{n} \frac{x_i}{10^i}\right) - x_{n+1} < 1,$$

or

$$0 \leq x - \sum_{i=0}^{n+1} \frac{x_i}{10^i} < \frac{1}{10^{n+1}}.$$

Thus (2.4) is true for $n+1$, and hence for all $n \in \mathbb{N}_0$.

We can use this procedure to obtain the expansions of some common fractions, such as

$$\frac{1}{2} = 0.5000\cdots$$
$$\frac{11}{8} = 1.375000\cdots$$
$$\frac{122}{990} = 0.12323\cdots,$$

which can be verified by using long division. The following points are noteworthy as they shed more light on this procedure.

1. For a real number $x \geq 0$, the integer part of x, $x_0 = [x]$, also has a representation based on 10, given by

$$x_0 = \xi_m \cdot 10^m + \xi_{m-1} \cdot 10^{m-1} + \cdots + \xi_2 \cdot 10^2 + \xi_1 10 + \xi_0,$$

where $m \in \mathbb{N}_0$ and $\xi_i \in \{0, 1, 2, \cdots, 9\}$, $i = 0, 1, 2, \cdots, m$.

2. Two numbers having the same expansions are identical. Suppose

$$x = x_0.x_1x_2x_3\cdots$$
$$y = x_0.x_1x_2x_3\cdots,$$

then, for every $n \in \mathbb{N}$,

$$|x - y| \leq \left|x - \sum_{i=0}^{n} \frac{x_i}{10^i}\right| + \left|\sum_{i=0}^{n} \frac{x_i}{10^i} - y\right|$$
$$< \frac{1}{10^n} + \frac{1}{10^n} = \frac{2}{10^n}.$$

But it is a simple exercise to show, by induction, that $2/10^n < 1/n$, and then use Corollary 2.7.1 to conclude that $|x - y| = 0$, or $x = y$.

3. Every $x \geq 0$ has a unique decimal expansion. To see that, suppose

$$x = x_0.x_1x_2x_3\cdots$$
$$x = y_0.y_1y_2y_3\cdots.$$

If $x_i \neq y_i$ for some $i \in \mathbb{N}_0$, let $n = \min\{i \in \mathbb{N}_0 : x_i \neq y_i\}$; that is, let $x_i = y_i$ for all $i = 0, 1, 2, \cdots, n-1$ and $x_n \neq y_n$. By definition of the decimal expansion, we have

$$0 \leq x - \sum_{i=0}^{n} \frac{x_i}{10^i} < \frac{1}{10^n}$$
$$0 \leq x - \sum_{i=0}^{n} \frac{y_i}{10^i} < \frac{1}{10^n},$$

$$\Rightarrow \left| \sum_{i=0}^{n} \frac{x_i}{10^i} - \sum_{i=0}^{n} \frac{y_i}{10^i} \right| < \frac{1}{10^n}$$
$$\Rightarrow \left| \frac{x_n}{10^n} - \frac{y_n}{10^n} \right| < \frac{1}{10^n}$$
$$\Rightarrow |x_n - y_n| < 1,$$

which is impossible since x_n and y_n were assumed to be distinct integers.

Note that, relaxing the condition (2.4) to

$$0 \leq x - \sum_{i=0}^{n} \frac{x_i}{10^i} \leq \frac{1}{10^n},$$

results in some numbers having two expansions, such as

$$\frac{1}{2} = 0.5000\cdots, \quad \frac{1}{2} = 0.4999\cdots.$$

4. If $x_0 \in \mathbb{N}_0$ and $x_i \in \{0, 1, 2, \cdots, 9\}$, there is a real number $x \geq 0$ such that $x = x_0.x_1x_2x_3\cdots$. The number x is in fact none other than

$$\sup \left\{ \sum_{i=0}^{n} \frac{x_i}{10^i} : n \in \mathbb{N} \right\}.$$

(See Exercise 2.5.4.)

5. The decimal expansion of a rational number is characterized by the fact that, after a certain point, the digits in the expansion start repeating themselves. This is due to the fact that, after dividing one natural number by another, the remainder is always less than the divisor. Consequently, to arrive at the decimal expansion of p/q by long division, the remainder at each step of the division will be an integer in the set $\{0, 1, 2, \cdots, q-1\}$, so at a certain point a remainder will reappear, at which point the expansion will be repeated.

6. So far we have only discussed decimal expansions, that is, expansions based on 10. But we can follow a similar procedure to obtain expansions based on any integer $b \in \mathbb{N}\setminus\{1\}$. The digits appearing in such an expansion would belong to the set $\{0, 1, 2, \cdots, b-1\}$. The most important among such expansions is the *binary expansion*, where $b = 2$. Its importance stems from the fact that it is particularly suited to electronic calculators. Because only the digits 0 and 1 appear in the expansion, one can be associated with an open circuit and the other with a closed circuit. *Ternary expansions* are based on 3, and so on. Thus the number 5 in decimal notation has the binary expansion $1 \cdot 2^2 + 0 \cdot 2^1 + 1 \cdot 2^0 = 101$ and the ternary expansion $1 \cdot 3^1 + 2 \cdot 3^0 = 12$. The rational number $3/7$ has the binary expansion $0.011011\cdots$ and the ternary expansion $0.1021210212\cdots$.

7. We can also obtain the expansion of any real number x to any base geometrically. Here we present the case of the ternary expansion which we shall have occasion to use when we discuss the Cantor set in Chapter 3, but the procedure can be adapted to any base b.

    For each $x \in [0, \infty)$, we set $x_0 = [x]$ and express the integer x_0 as a sum $\sum_{i=0}^{n} t_i \cdot 3^i$ with $t_i \in \{0, 1, 2\}$, which is represented by the ternary expansion $t_n t_{n-1} \cdots t_1 t_0$. Since $x - x_0 \in [0, 1)$ we need only indicate how to obtain the ternary expansion of $x = 0.x_1 x_2 x_3 \cdots$ in $[0, 1)$:

    (i) Divide the interval $[0,1)$ into 3 equal, disjoint sub-intervals $[0, 1/3)$, $[1/3, 2/3)$ and $[2/3, 1)$, each of length $1/3$. Label these intervals 0,1 and 2, respectively, as shown in Figure 2.6.

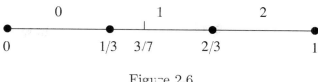

Figure 2.6

x belongs to one, and only one, sub-interval. If $x \in [i/3, (i+1)/3)$, where $i = 0, 1$ or 2, we set $x_1 = i$. Consider, for example, $x = 3/7$. This lies in the sub-interval $[1/3, 2/3)$ and therefore $x_1 = 1$.

(ii) Now divide the sub-interval containing x into three equal, disjoint sub-intervals of length $1/3^2$ and label them $0, 1, 2$ from left to right. Set x_2 equal to the label of the sub-interval where x lies. Taking $x = 3/7$ again, we see that

$$\left[\frac{1}{3}, \frac{2}{3}\right) = \left[\frac{1}{3}, \frac{1}{3} + \frac{1}{3^2}\right) \cup \left[\frac{1}{3} + \frac{1}{3^2}, \frac{1}{3} + \frac{2}{3^2}\right) \cup \left[\frac{1}{3} + \frac{2}{3^2}, \frac{2}{3}\right)$$
$$= [1/3, 4/9) \cup [4/9, 5/9) \cup [5/9, 2/3).$$

Since $3/7 \in [1/3, 4/9)$, we set $x_2 = 0$.

(iii) Once again we divide the sub-interval containing x into 3 disjoint sub-intervals of length $1/3^3$ and set x_3 equal to the label of the sub-interval where x lies. Since $3/7 \in [11/27, 12/27)$, we have $x_3 = 2$. Continuing in this fashion we obtain the ternary expansion of $3/7$ as $0.10212\cdots$ (see Figure 2.7).

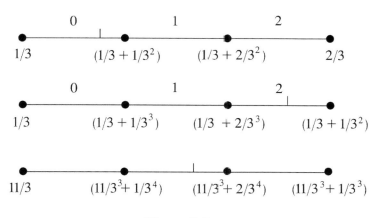

Figure 2.7

Note that the division into semi-closed sub-intervals results in a unique expansion of x. Had we used closed sub-intervals instead, we would not necessarily obtain a unique expansion. A number of the form $m/3^n$, for example, would lie in two adjacent sub-intervals at the n-th stage. Choosing the left label for x_n gives $x_{n+k} = 2$ for all $k \geq 1$, and choosing the right label makes $x_{n+k} = 0$ for all $k \geq 1$. Thus we end up with two expansions, one ending in a string of 2's and the other in a string of 0's.

EXERCISES 2.5

1. Determine the decimal, binomial and ternary expansions of each of the following numbers:
   (a) 1/4 (b) 5/8 (c) 11/15.

2. Find the rational number having
   (a) the decimal expansion $0.37212121\cdots$
   (b) the ternary expansion $0.020121212\cdots$
   (c) the binary expansion $1.01010101\cdots$

3. Use the decimal expansion to prove the density of $\mathbb{Q}$ in $\mathbb{R}$.

4. If $x_0 \in \mathbb{N}_0$ and $x_i \in \{0, 1, 2, \cdots, 9\}$, $i \in \mathbb{N}$, prove that the set
$$\left\{ \sum_{i=0}^{n} \frac{x_i}{10^i} : n \in \mathbb{N} \right\}$$
is bounded above, and if x is its least upper bound then x has the decimal expansion
$$x = x_0.x_1x_2x_3\cdots.$$

2.6 Countable Sets

We can compare two finite sets A and B in terms of "size" by comparing $|A|$ and $|B|$, the number of elements in A and B respectively (see Exercise 1.1.8). Thus $\{1, 2, 3, 4\}$ is "larger" than $\{a, b, c\}$, and $\{a, b, c\}$

is "equal in size" to $\{-1, 0, 1\}$. Things are less clear when we deal with infinite sets, such as $\mathbb{N}$, $\mathbb{Z}$, $\mathbb{Q} \cap [0, 1]$, or $\mathbb{R} \backslash \mathbb{Q}$ for example. We have no clear method for deciding which of these infinite sets is more "numerous". One of the early achievements of set theory was that it provided a basis for extending this comparison between finite sets to infinite sets.

Definition 2.7 A set A is *equivalent* to a set B, expressed symbolically by $A \sim B$, if there is a bijection from A to B.

Theorem 2.9
For any three sets A, B, C
(i) $A \sim A$
(ii) If $A \sim B$ then $B \sim A$
(iii) If $A \sim B$ and $B \sim C$ then $A \sim C$.

Proof
(i) Since the *identity function*
$$I_A : A \to A, \ I_A(x) = x$$
is bijective, we have $A \sim A$.
(ii) Since $A \sim B$, there is a bijection $f : A \to B$. Now $f^{-1} : B \to A$ is also a bijection, hence $B \sim A$.
(iii) Since $A \sim B$ and $B \sim C$, there are two bijections
$$f : A \to B, \ g : B \to C.$$
As the composition $g \circ f : A \to C$ is also bijective (see Section 1.2), we must have $A \sim C$. $\square$

Properties (i), (ii), and (iii) of the relation $\sim$ are called, respectively, *reflexivity*, *symmetry*, and *transitivity* (see Exercise 1.2.2). A binary relation, such as $\sim$, possessing these three properties, is referred to as an *equivalence relation*.

Definition 2.8 A set A is *finite* if it is empty or if there is a natural number n such that $A \sim \{1, 2, \cdots, n\}$. A is *infinite* if it is not finite.

If $A = \varnothing$ we define $|A| = 0$, and if $A \sim \{1, 2, \cdots, n\}$, then A has exactly n elements (see Exercise 1.2.9), that is, $|A| = n$. For any two finite sets A and B, it therefore follows from the transitivity of $\sim$ that $A \sim B$ if and only if $|A| = |B|$; and we can also show (Exercise 2.6.1) that if $A \subset B$ then $|A| < |B|$, and hence A and B are not equivalent.

Thus the relation $\sim$ partitions finite sets into *equivalence classes*. The sets in any one class have the same number of elements.

Now let us turn to infinite sets.

Example 2.10 Let $\mathbb{N}_1 = \{1, 3, 5, \cdots\}$ be the set of odd integers, and $\mathbb{N}_2 = \{2, 4, 6, \cdots\}$ be the set of even integers. The bijection

$$f : \mathbb{N}_1 \to \mathbb{N}_2, \ f(n) = n + 1$$

implies $\mathbb{N}_1 \sim \mathbb{N}_2.$ The bijection

$$g : \mathbb{N} \to \mathbb{N}_2, \ g(n) = 2n$$

also implies $\mathbb{N} \sim \mathbb{N}_2$. Thus we see that the sets $\mathbb{N}_1, \mathbb{N}_2$, and $\mathbb{N}$ are all equivalent, though both $\mathbb{N}_1$ and $\mathbb{N}_2$ are *proper* subsets of $\mathbb{N}$, a situation which we have just seen cannot arise with finite sets.

Example 2.11 $\mathbb{Z} \sim \mathbb{N}$.

Proof. Let $f : \mathbb{Z} \to \mathbb{N}$ be defined by

$$f(n) = \begin{cases} 2n & if \ n > 0 \\ -2n + 1 & if \ n \leq 0. \end{cases}$$

Note that f maps the positive integers to the even numbers, and the non-positive integers to the odd numbers, that is

$$n > 0 \Leftrightarrow f(n) \in \mathbb{N}_2$$
$$n \leq 0 \Leftrightarrow f(n) \in \mathbb{N}_1.$$

To show that f is injective, suppose $f(n_1) = f(n_2) \in \mathbb{N}_2$. This implies $2n_1 = 2n_2$, and so $n_1 = n_2$. If, on the other hand, $f(n_1) = f(n_2) \in \mathbb{N}_1$, then $-2n_1 + 1 = -2n_2 + 1$, and again we obtain $n_1 = n_2$.

To show that f is surjective, suppose $m \in \mathbb{N} = \mathbb{N}_1 \cup \mathbb{N}_2$. If $m \in \mathbb{N}_1$, then there is a non-negative integer k such that $m = 2k + 1$, in which case $f(-k) = m$. If $m \in \mathbb{N}_2$ then $m = 2k$ for some positive integer k, and $f(k) = m$. Thus f is bijective. $\square$

Definition 2.9 A set A is *denumerable* if $A \sim \mathbb{N}$, and *countable* if it is either denumerable or finite.

From Examples 2.10 and 2.11 we know that the sets $\mathbb{N}_1, \mathbb{N}_2$, and $\mathbb{Z}$ are all denumerable. Since a denumerable set A has a one-to-one correspondence with the natural numbers, $f : \mathbb{N} \to A$, we can label the elements of A according to this correspondence, and hence write

$$A = \{a_1, a_2, a_3, \cdots\} = \{a_i : i \in \mathbb{N}\},$$

where $a_i = f(i)$.

Lemma 2.2 *Every subset of $\mathbb{N}$ is countable.*

Proof. Suppose $S \subseteq \mathbb{N}$. If S is finite then it is countable by Definition 2.9. If S is infinite then it is not empty and, by the well ordering property of $\mathbb{N}$ (Theorem 2.3), S has a minimal element, call it n_1. The set $S \setminus \{n_1\}$ is not empty, otherwise S would be finite, so it also has a minimal element n_2. Continuing in this manner, we form the set

$$T = \{n_1, n_2, n_3, \cdots\},$$

where

$$n_k = \min S \setminus \{n_1, n_2, n_3, \cdots, n_{k-1}\}.$$

Note that $n_1 < n_2 < n_3 < \cdots$, so the function

$$f : \mathbb{N} \to T, \ f(k) = n_k$$

is clearly bijective, and $T \sim \mathbb{N}$. To complete the proof we shall show that $T = S$; but since $T \subseteq S$ it suffices to show that $S \subseteq T$.

Since $n_1 < n_2 < n_3 < \cdots$, we have $k \leq n_k$ for all $k \in \mathbb{N}$. Now suppose $m \in S \subseteq \mathbb{N}$. We know from the well ordering property that the non-empty set $\{k \in \mathbb{N} : m \leq n_k\}$ has a minimal element, say j, and, by definition of n_j, $m = n_j \in T$. Thus $S \subseteq T$. □

Theorem 2.10
Every subset of a countable set is countable.

Proof. Suppose A is a countable set and $B \subseteq A$. If B is finite then, by definition, it is countable. If B is infinite then so is A, and therefore $A \sim \mathbb{N}$. Hence there is a bijection $f : \mathbb{N} \to A$. But then

$$f^{-1}(B) \subseteq f^{-1}(A) = \mathbb{N},$$

and, by Lemma 2.2, $f^{-1}(B)$ is countable. Now the restriction of f to $f^{-1}(B)$ is a bijective function from $f^{-1}(B)$ to B, which establishes the equivalence $f^{-1}(B) \sim B$. Hence B is countable. □

Theorem 2.11
If the sets A and B are both countable, then $A \times B$ is countable.

Proof. Let us first assume that both A and B are infinite. We then have two bijective functions

$$f : A \to \mathbb{N}$$
$$g : B \to \mathbb{N}.$$

Define the function

$$h : A \times B \to \mathbb{N}, \ h(x, y) = 2^{f(x)} \cdot 3^{g(y)},$$

and note that

$$h(x_1, y_1) = h(x_2, y_2) \ \Rightarrow \ 2^{f(x_1)} \cdot 3^{g(y_1)} = 2^{f(x_2)} \cdot 3^{g(y_2)}.$$

By Theorem 2.4 and Remark 2.3, $f(x_1) = f(x_2)$ and $g(y_1) = g(y_2)$, and, since f and g are bijective, this implies $x_1 = x_2$ and $y_1 = y_2$. Thus h is injective and therefore

$$A \times B \sim R_h \subseteq \mathbb{N},$$

where R_h is the range of h. By Theorem 2.10, the set $A \times B$ is countable.

If both A and B are finite then clearly $A \times B$ is finite and hence countable. To cover the case when either A or B is finite we can add more elements to the finite set to make it denumerable. For example, if

$$A = \{a_1, a_2, \cdots, a_n\},$$

then we define

$$C = \{a_1, a_2, \cdots, a_n, n+1, n+2, \cdots\} \sim \mathbb{N}.$$

Now $C \times B$ is countable and contains $A \times B$, so again $A \times B$ is countable by Theorem 2.10. □

Theorem 2.12
If A_n is a countable set for every $n \in \mathbb{N}$ then $\bigcup_{n=1}^{\infty} A_n$ is countable.

Proof. We can assume, without loss of generality, that each A_n is denumerable, otherwise we add more elements where necessary as we did in the proof of Theorem 2.11. We therefore have

$$A_n = \{a_{n1}, a_{n2}, a_{n3}, \cdots\}.$$

For any $x \in \bigcup_{n=1}^{\infty} A_n$, let
$$n_x = \min\{n \in \mathbb{N} : x \in A_n\},$$
and choose $m_x \in \mathbb{N}$ such that
$$x = a_{n_x m_x}.$$
Now the function
$$f : \bigcup_{n=1}^{\infty} A_n \to \mathbb{N} \times \mathbb{N}, \quad f(x) = (n_x, m_x)$$
is injective, so, by Theorems 2.10 and 2.11, $\bigcup_{n=1}^{\infty} A_n$ is countable. $\square$

For the purposes of this study, the next two theorems are among the more significant results of set theory.

Theorem 2.13
The set of rational numbers is denumerable.

Proof. Let $f : \mathbb{Q} \to \mathbb{Z} \times \mathbb{Z}$ be the function defined by the rule
$$f\left(\frac{m}{n}\right) = (m, n),$$
where the rational number m/n is in simplest form, that is, $m \in \mathbb{Z}$ and $n \in \mathbb{N}$ have no common factors. f is clearly injective, hence
$$\mathbb{Q} \sim R_f \subseteq \mathbb{Z} \times \mathbb{Z}.$$
Since $\mathbb{Z} \times \mathbb{Z}$, by Theorems 2.10 and 2.11, is countable, $\mathbb{Q}$ is also countable. Being infinite, $\mathbb{Q}$ is therefore denumerable. $\square$

Figure 2.8 shows a possible listing of the rational numbers as a sequence.

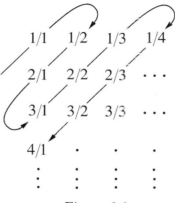

Figure 2.8

But not every infinite set is denumerable, as we shall now see.

Theorem 2.14
The set of real numbers is not denumerable.

Proof. By Theorem 2.10, It suffices to prove that the interval $[0, 1) \subset \mathbb{R}$ is not countable. Since $[0,1)$ is not finite we have to show that it is not denumerable. Let us assume that $[0,1)$ is denumerable, and hence can be represented as a sequence of numbers x_i, with $i \in \mathbb{N}$,

$$[0, 1) = \{x_1, x_2, x_3, \cdots\}.$$

Each $x_i \in [0, 1)$ has a decimal representation (which does not end in a sequence of 9's) given by

$$x_i = 0.x_{i1}x_{i2}x_{i3}\cdots.$$

For every $i \in \mathbb{N}$, choose $a_i \in \{0, 1, 2, \cdots, 8\}$ such that $a_i \neq x_{ii}$. Now the number

$$a = 0.a_1 a_2 a_3 \cdots$$

clearly belongs to $[0,1)$, but $a \neq x_i$ for any $i \in \mathbb{N}$, thereby contradicting the assumption that $[0, 1)$ is denumerable. $\square$

Since the union of two countable sets is a countable set (Theorem 2.12), the irrational numbers $\mathbb{Q}^c$ are uncountable, $\mathbb{R}$ being the union of $\mathbb{Q}$ and $\mathbb{Q}^c$. This result, that $\mathbb{Q}^c$ is somehow more "numerous" than $\mathbb{Q}$, may seem surprising in view of the density of $\mathbb{Q}$ in $\mathbb{R}$, as expressed by Theorem 2.8.

For a finite set A we used the symbol $|A|$ to denote the number of elements in A, or any set equivalent to A. If A is infinite, we could of course write $|A| = \infty$ and leave it at that. But we have just seen how the two infinite sets $\mathbb{Q}$ and $\mathbb{R}$ have different "degrees" of infinity. Loosely speaking, we could say $\mathbb{Q}$ is "denumerably infinite" while $\mathbb{R}$ is "continuously infinite". It is therefore tempting to extend the definition of $|A|$ to infinite sets, as is done in [CAN], and assign a *value* to $|A|$ depending on the *equivalence class* to which A belongs. These values are called *cardinal numbers*. Thus the cardinal number of a finite set is a non-negative integer equal to the number of elements in the set. The cardinal number of a denumerable set is conventionally denoted by $\aleph_0$, and the cardinal number for $\mathbb{R}$ is c (for *continuum*). Based on the results we have, we can therefore write

$$|\mathbb{N}| = |\mathbb{Z}| = |\mathbb{Q}| = \aleph_0.$$

It is not difficult to show that any open interval in $\mathbb{R}$ is equivalent to $\mathbb{R}$ (see Exercises 2.6.3 and 2.6.4), so we also have

$$|\mathbb{R}| = |(a,b)| = |\mathbb{Q}^c| = c.$$

Cardinal numbers can be ordered according to the following definition: For any two sets A and B, the relation $|A|<|B|$ means that A is equivalent to a subset of B but B is not equivalent to any subset of A. Thus $\aleph_0 < c$ because $\mathbb{N} \subset \mathbb{R}$ and $\mathbb{R}$ is not countable. The question as to whether there is a cardinal number between $\aleph_0$ and c is not yet known, but has been shown to be independent of the axioms of set theory (see [COH]). In other words it cannot be proved nor disproved using those axioms. But the assumption that there is no cardinal number between $\aleph_0$ and c is known as the *continuum hypothesis*.

Another question that can be raised is: are there any cardinals greater than c? The answer is positive, as we shall now explain. For a finite set A we saw in Exercise 1.1.8 that the cardinal number of the set of subsets of A is $|\mathcal{P}(A)| = 2^{|A|}$, which is greater than $|A|$. If A is infinite, we can also prove (see Exercise 2.6.9) that $|A| < |\mathcal{P}(A)|$. Thus $\mathcal{P}(\mathbb{R})$, the power set of $\mathbb{R}$, has a cardinal number greater than c. By *defining* $2^{|A|}$ to be the cardinal number of $\mathcal{P}(A)$ even when A is infinite, and using the notation

$$2^{\aleph_0} = \aleph_1, \ 2^{\aleph_1} = \aleph_2, \ 2^{\aleph_2} = \aleph_3, \ \cdots,$$

we arrive at the sequence of cardinals

$$0 < 1 < 2 < \cdots < \aleph_0 < \aleph_1 < \aleph_2 < \aleph_3 < \cdots.$$

We know that $\aleph_0$, the cardinality of denumerable sets, is the first infinite cardinal number. It can be proved that $c = \aleph_1$ by showing that $\mathbb{R}$ is equivalent to $\mathcal{P}(\mathbb{N})$ (see [B&M]). The assumption that there are no cardinal numbers between $\aleph_n$ and $\aleph_{n+1}$ is known as the *generalized continuum hypothesis*, and that is also known to be independent of the axioms of set theory.

Exercises 2.6

1. The purpose of this exercise is to prove that a set A is equivalent to one of its proper subsets if, and only if, A is an infinite set.

    (a) If A is finite and $B \subseteq A$, prove that $B \sim A$ if and only if $B = A$.

    (b) If A is denumerable prove that it has an equivalent proper subset.

    (c) If A is infinite prove that it has a denumerable subset.

    (d) If A is infinite prove that it has an equivalent proper subset.

2. Prove that $[0, 1) \sim (0, 1)$.

3. Prove that $(a, b) \sim (0, 1)$ for any $a < b$.

4. Prove that $\mathbb{R} \sim (0, 1)$.

5. Prove that $[0, 1] \times [0, 1] \sim [0, 1]$.

    Hint: Use decimal expansions.

6. Prove that the zeroes of all polynomials with integer coefficients is a denumerable set. This is called the set of *algebraic numbers*, and it clearly includes $\mathbb{Q}$ as a proper subset (why?). What can you say about the zeroes of all polynomials with rational coefficients?

    Hint: First prove that the set of all polynomials of degree n with integer coefficients is denumerable.

7. If there is a surjective function $f : A \to B$, prove that $|A| \geq |B|$.

8. If there is an injective function $f : A \to B$, prove that $|A| \leq |B|$.

9. Let A be any set. Prove that $|A| < |\mathcal{P}(A)|$ by the following procedure:

    (i) Define an injective function from A to $\mathcal{P}(A)$.

    (ii) Prove that there is no surjective function from A to $\mathcal{P}(A)$ by contradiction: Assume the existence of such a function, call it f, and consider the set $B = \{x \in A : x \notin f(x)\}$. Show that there is no element $a \in A$ such that $f(a) = B$ by showing that if $a \in A \backslash B$ we get a contradiction, and if $a \in B$ we get another contradiction.

10. Discuss the cardinality of the set of functions from A to B where the sets A and B are: (i) countable, (ii) uncountable.

3

Sequences

This chapter is devoted to sequences of real numbers and their convergence properties. This will serve as a bridge to the concept of the limit of a function, which is the fundamental concept in analysis.

3.1 Sequences and Convergence

Definition 3.1 A *sequence* is a function whose domain is $\mathbb{N}$. A *real sequence* is a sequence whose range is in $\mathbb{R}$.

A real sequence is therefore a function

$$x : \mathbb{N} \to \mathbb{R}, \ x(n) = x_n, \tag{3.1}$$

which is commonly denoted by the sequence of real numbers $(x_1, x_2, x_3, \cdots, x_n, \cdots)$, or $(x_n : n \in \mathbb{N})$. Note that the *sequence* $(x_n : n \in \mathbb{N})$ is not the same as the *set* $\{x_n : n \in \mathbb{N}\}$, for the former is the *ordered* list of values of the function x, some of which may be repeated, while the latter is merely its *range*. Thus the constant sequence $x_n = 1$ is denoted by $(1, 1, 1, \cdots)$, and its range is the set $\{1\}$.

We can also talk about a *sequence of closed intervals*

$$[0, 1], [0, 1/2], [0, 1/3], \cdots, [0, 1/n], \cdots,$$

or a *sequence of powers of* x

$$1, x, x^2, \cdots, x^n, \cdots.$$

But, in this chapter, when we talk about a *sequence* we generally mean a *sequence of real numbers*, that is, a *real sequence*. We shall often abbreviate $(x_n : n \in \mathbb{N})$ to (x_n).

The sequence (3.1) is usually defined by giving the rule for obtaining its n-th term x_n. Here are some examples:

(a) $(2n)$ is the sequence of even natural numbers $(2, 4, 6, \cdots)$. Its range is $\mathbb{N}_2$.

(b) $((-1)^n)$ is the sequence $(-1, 1, -1, 1, \cdots)$ whose range is $\{1, -1\}$.

(c) $(1, 1/2, 1/3, \cdots)$ is the sequence whose n-th term is $1/n$.

But a sequence can also be defined by induction. For example, if we define
$$x_1 = 0, \ x_n = \frac{1}{2}(1 + x_{n-1}) \ \text{ for all } n \geq 2,$$
then the resulting sequence is $(0, 1/2, 3/4, 7/8, 15/16, \cdots)$.

In our study of sequences we are really interested in the behaviour of x_n for large values of n. Consider the two sequences
$$x_n = \frac{n-1}{n}, \ y_n = (-1)^n.$$

In the first sequence $(0, 1/2, 3/4, 4/5, \cdots)$ the n-th term x_n gets closer to 1 as n increases (see Figure 3.1). In the second sequence $(-1, 1, -1, 1, \cdots)$, even though the n-th term y_n for even n is 1, no matter how large we take n there will be terms that are not close to 1, namely the terms for which n is odd.

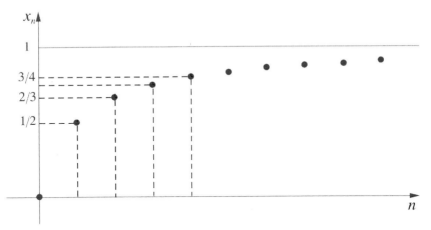

Figure 3.1

Suppose we agree that two numbers x and y are "close to each other" if $|x - y|$ is less than 0.0001, then $x_n = \dfrac{n-1}{n}$ is "close to 1" provided $n > 1000$. Instead of 0.0001 we can choose *any* other small positive number ε as a measure of *proximity*, and $|x_n - 1|$ can be made smaller than ε by taking n large enough. In this case, where $|x_n - 1| = 1/n$, "large enough" means $1/n$ should be less than ε, or, equivalently, n

should be greater than $1/\varepsilon$. More generally we have the following definition.

Definition 3.2 A sequence (x_n) is said to be *convergent* if there exists a real number x with the following property: For every $\varepsilon > 0$ there is an integer N such that $n \geq N$ implies $|x_n - x| < \varepsilon$. If (x_n) is not convergent, it is called *divergent*.

When (x_n) is convergent, the number x specified in this definition is called a *limit* of the sequence (x_n), and x_n is said to *converge to* x. This is expressed symbolically by writing

$$\lim_{n \to \infty} x_n = x,$$

which is often abbreviated to

$$\lim x_n = x,$$

or

$$x_n \to x.$$

Definition 3.2 is the initial, and crucial, step into analysis, and it is worth contemplating its meaning and its implications at this point. To that end we make the following definition, followed by some observations:

Definition 3.3 Let $x \in \mathbb{R}$. A set $V \subseteq \mathbb{R}$ is called a *neighborhood* of x if there is a number $\varepsilon > 0$ such that

$$(x - \varepsilon, x + \varepsilon) \subseteq V.$$

Remarks 3.1

1. According to Definition 3.2, the statement $\lim x_n = x$ means that, no matter what measure ε of proximity we use, we can always find an $N \in \mathbb{N}$ such that *all* terms of the sequence following x_{N-1} are within a distance less than ε of x, that is, $-\varepsilon < x_n - x < \varepsilon$ for all $n \geq N$.

2. Using the language of neighborhoods, we can state Definition 3.2 in the following equivalent version: The sequence (x_n) converges to x if every neighborhood of x contains all but a finite number of terms of the sequence.

To see the equivalence, suppose $x_n \to x$ according to the first version. If V is any neighborhood of x then, by the definition of a neighborhood, there is a positive number ε such that $(x - \varepsilon, x + \varepsilon) \subseteq V$. Since $x_n \to x$

there is an integer N such that $|x_n - x| < \varepsilon$ for all $n \geq N$, that is, $x_n \in (x - \varepsilon, x + \varepsilon)$ for all $n \geq N$. Thus we see that all the terms of the sequence, with the possible exception of $x_1, x_2, \cdots, x_{N-1}$, lie in V.

Suppose, on the other hand, that every neighborhood of some point x contains all the terms of (x_n) except for a finite number of them. To prove that $x_n \to x$ according to Definition 3.2, suppose $\varepsilon > 0$ and let $V = (x - \varepsilon, x + \varepsilon)$. Now V is a neighborhood of x, so all the terms of (x_n), except for a finite number, must lie in V. Hence there is an integer N such that $x_n \in V$, i.e., $|x_n - x| < \varepsilon$, for all $n \geq N$.

3. One would expect the integer N to depend on ε, so that a smaller choice for ε may require a larger value of N in order to satisfy the implication

$$n \geq N \Rightarrow |x_n - x| < \varepsilon. \tag{3.2}$$

We can therefore write $N = N(\varepsilon)$ when we wish to emphasize this dependence, but it should be kept in mind that ε does not determine a *unique* N, for if (3.2) is true for a certain integer N, then it is certainly also true for any larger integer. This point will be exploited repeatedly in many proofs which rely on Definition 3.2.

4. Suppose the sequence (x_n) satisfies the following statement: For every $\varepsilon > 0$ there is an integer N such that $|x_n - x| < c\varepsilon$ for all $n \geq N$, where c is a positive constant which does not depend on ε (or n). From this we may safely conclude that $\lim x_n = x$; for if we are given any positive ε and we set $\varepsilon' = \varepsilon/c$ then ε' is also positive, and using ε' in the above statement, we conclude the existence of an integer N' such that $|x_n - x| < c\varepsilon' = \varepsilon$ for all $n \geq N'$.

Thus, to prove the convergence $x_n \to x$, it suffices in Definition 3.2 to show that $|x_n - x| < c\varepsilon$, provided c is a positive constant independent of ε.

5. Clearly, whether a sequence (x_n) converges or diverges depends on the behaviour of its terms for large values of n, and is not affected by what happens in the beginning of the sequence. For any fixed integer $m \geq 0$ we shall call the sequence

$$(x_{m+1}, x_{m+2}, x_{m+3}, \cdots)$$

the *m-tail* of (x_n). In other words, the m-tail of (x_n) is the sequence $(y_n : y_n = x_{m+n}, n \in \mathbb{N})$.

If an m-tail of (x_n) converges to x, then (x_n) itself converges to x. To see that, suppose we are given $\varepsilon > 0$, then there is an N such that

$$n \geq N \Rightarrow |y_n - x| < \varepsilon.$$

Consequently, if $n \geq m + N$, i.e., $n - m \geq N$, then $|y_{n-m} - x| = |x_n - x| < \varepsilon$. This means $x_n \to x$. Conversely we can also prove that, if a sequence converges, then all its m-tails converge, and to the same limit.

6. From Definition 3.2 we clearly have

$$x_n \to x \Leftrightarrow |x_n - x| \to 0.$$

Example 3.1
$$\lim \frac{n-1}{n} = 1.$$

This equality simply asserts our ability to ensure, for any choice of proximity measure, that we can proceed far enough in the sequence so that the "late terms" are close to 1. Given $\varepsilon > 0$ as a measure of the proximity of $\frac{n-1}{n}$ to 1, our burden is to find N such that all the terms $x_N, x_{N+1}, x_{N+2}, \cdots$ are within a distance ε of 1. In other words we have to show that

$$\left| \frac{n-1}{n} - 1 \right| = \frac{1}{n} < \varepsilon \qquad (3.3)$$

for "large enough" values of n. From Corollary 2.7.1 to the theorem of Archimedes, we know that there is a natural number N such that $1/N < \varepsilon$. One possible choice is $N = [1/\varepsilon] + 1$. Now for every $n \geq N$ we obtain

$$\frac{1}{n} \leq \frac{1}{N} < \varepsilon,$$

which proves (3.3).

Example 3.2
$$\lim \frac{1}{2^n} = 0.$$

Suppose we are given $\varepsilon > 0$. We need to make $\left| \frac{1}{2^n} - 0 \right| = \frac{1}{2^n} < \varepsilon$. Since, by induction, $2^n > n$ for all $n \in \mathbb{N}$ (Exercise 2.3.6), it suffices to make $\frac{1}{n} < \varepsilon$. From Corollary 2.7.1 there is an $N \in \mathbb{N}$ such that $\frac{1}{N} < \varepsilon$, and hence

$$n \geq N \Rightarrow \frac{1}{n} < \varepsilon \Rightarrow \frac{1}{2^n} < \varepsilon.$$

70   Elements of Real Analysis

Example 3.3 The sequence $((-1)^n)$ is divergent.

Suppose, for the sake of argument, that the sequence is convergent and that $\lim(-1)^n = x$. This implies, by Remark 3.1.2, that any neighborhood V of x contains all but a finite number of terms of the sequence $((-1)^n)$. Hence V must contain the two numbers 1 and -1, since they come up infinitely often. Take $V = (x - 1/2, x + 1/2)$. Since $-1, 1 \in V$, the distance between -1 and 1 (which is 2) is less than the length of V (which is 1). This implies $2 < 1$, which is impossible. Thus the assumption that the sequence is convergent is false.

Figure 3.2

Example 3.4 The sequence (n) is divergent.

Once again, suppose that $\lim n = x$. Take $\varepsilon = 1$. There is now an N such that

$$n \geq N \Rightarrow |n - x| < 1 \Rightarrow x - 1 < n < x + 1,$$

and this implies that $n < x + 1$ for all $n \geq N$. In particular, $N < x + 1$, so we must have $n < x + 1$ for all $n \leq N$. Consequently,

$$n < x + 1 \text{ for all } n \in \mathbb{N},$$

that is, $\mathbb{N}$ is bounded by $x + 1$, thereby contradicting the theorem of Archimedes.

We end this section by settling a question which may have occurred to the reader: Can a sequence have more than one limit?

Theorem 3.1
If the sequence (x_n) is convergent, its limit is unique.

Proof. Suppose that $\lim x_n = x$ and $\lim x_n = y$. Take any $\varepsilon > 0$. Since $x_n \to x$, there is an $N_1 \in \mathbb{N}$ such that

$$|x_n - x| < \varepsilon/2 \text{ for all } n \geq N_1.$$

Similarly, since $x_n \to y$, there is an $N_2 \in \mathbb{N}$ such that
$$|x_n - y| < \varepsilon/2 \text{ for all } n \geq N_2.$$
If we now take $n \geq \max\{N_1, N_2\}$, then both inequalities $|x_n - x| < \varepsilon/2$ and $|x_n - y| < \varepsilon/2$ hold, and hence
$$|x - y| \leq |x - x_n| + |x_n - y|$$
$$< \frac{\varepsilon}{2} + \frac{\varepsilon}{2} = \varepsilon.$$
Since $\varepsilon > 0$ was arbitrary, we must conclude that $|x - y| = 0$, i.e., $x = y$. $\square$

There is another, more elegant, proof of Theorem 3.1 which runs as follows: Assume $x \neq y$, and pick a neighborhood V_1 of x and a neighborhood V_2 of y such that $V_1 \cap V_2 = \varnothing$. For example take $V_1 = (x - \varepsilon, x + \varepsilon)$ and $V_2 = (y - \varepsilon, y + \varepsilon)$, where $\varepsilon = |x - y|/2$. Since $x_n \to x$, all but a finite number of terms of the sequence lie in V_1; and since $x_n \to y$, all but a finite number of its terms also lie in V_2. But this contradicts the fact that $V_1 \cap V_2 = \varnothing$. $\square$

Figure 3.3

We end this section with a result which we shall have occasion to use later on.

Example 3.5 Suppose that the two sequences (x_n) and (y_n) converge to a common limit c, and that the *shuffled* sequence (z_n) is defined by
$$(z_1, z_2, z_3, z_4, \cdots) = (x_1, y_1, x_2, y_2, \cdots).$$
We shall show that the sequence (z_n) also converges to c.

Let ε be any positive number. Since $x_n \to c$ and $y_n \to c$, there are two positive integers N_1 and N_2 such that
$$n \geq N_1 \Rightarrow |x_n - c| < \varepsilon,$$
$$n \geq N_2 \Rightarrow |y_n - c| < \varepsilon.$$

Define $N = \max\{N_1, N_2\}$, and note that $x_k = z_{2k-1}$, $y_k = z_{2k}$ for all $k \in \mathbb{N}$. Then

$$k \geq N \Rightarrow |x_k - c| = |z_{2k-1} - c| < \varepsilon,$$
$$|y_k - c| = |z_{2k} - c| < \varepsilon.$$

Hence
$$n \geq 2N - 1 \Rightarrow |z_n - c| < \varepsilon,$$
which just means $\lim z_n = c$.

Exercises 3.1

1. Determine the n-th term, x_n, for each of the following sequences:

   (a) $(5, 8, 11, 14, \cdots)$

   (b) $\left(\dfrac{1}{2}, \dfrac{2}{3}, \dfrac{3}{4}, \dfrac{4}{5}, \cdots\right)$

   (c) $\left(\dfrac{1}{2}, 1, \dfrac{5}{4}, \dfrac{7}{5}, \cdots\right)$

   (d) $\left(1, -\dfrac{1}{4}, 9, -\dfrac{1}{16}, \cdots\right)$.

2. Write the first five terms for each of the following sequences:

   (a) $x_1 = 2$, $x_{n+1} = \dfrac{1}{2}\left(x_n + \dfrac{2}{x_n}\right)$

   (b) $x_1 = 1$, $x_{n+1} = \sqrt{2x_n}$

   (c) $x_1 = 2$, $x_2 = -1$, $x_{n+2} = x_{n+1} - x_n$

   (d) $x_1 = x_2 = 1$, $x_{n+1} = x_{n-1} + x_n$ for all $n \geq 2$. This is known as the *Fibonacci sequence*.

3. For each of the following sequences, assume $\varepsilon = 0.005$ as a measure of proximity, and determine N so that x_n is "close" to x for all $n \geq N$:

   (a) $x = 0$, $x_n = \dfrac{1}{n}$

   (b) $x = 0$, $x_n = \dfrac{1}{2n - 1}$

(c) $x = 0$, $x_n = \dfrac{(-1)^n}{n}$

(d) $x = 3$, $x_n = \dfrac{3n-1}{n+2}$.

4. Use Definition 3.2 to prove

   (a) $\lim \dfrac{1}{n^2+2} = 0$

   (b) $\lim \left(\dfrac{1}{n} + \dfrac{n}{n+1}\right) = 1$

   (c) $\lim \dfrac{k}{n} = 0$ for any $k \in \mathbb{R}$.

5. Prove that a constant sequence is convergent.

6. Prove that $\lim x_n = 0$ if, and only if, $\lim |x_n| = 0$.

7. Which of the following sets is a neighborhood of 0?

   (a) $[-1, 1)$.
   (b) $[0, 1)$.
   (c) $\{0\} \cup \{1/n : n \in \mathbb{N}\}$.
   (d) $\bigcap_{n=1}^{\infty} [-1/n, 1/n]$.
   (e) $\bigcap_{n=1}^{\infty} (-1/n, 1/n)$.

8. Prove that the open interval (a, b) is a neighborhood of all its points.

9. Prove that $\mathbb{R}\setminus\{c\}$ is a neighborhood of all $x \neq c$.

3.2 Properties of Convergent Sequences

In this section we look at convergence in the light of the known properties of $\mathbb{R}$, its operations and its order relation. This will allow us to break up a sequence into simpler components and determine its convergence (or divergence), and its limit, based on the behaviour of these components.

Definition 3.4 A sequence (x_n) is *bounded* if there is a positive number K such that
$$|x_n| \leq K \quad \text{for all } n \in \mathbb{N}.$$

Thus a sequence is bounded if, and only if, its range is a bounded set.

Theorem 3.2
If a sequence is convergent, then it is bounded.

Proof. Suppose $x_n \to x$. If we choose $\varepsilon = 1$, we know that there is an integer N such that
$$|x_n - x| < 1 \quad \text{for all } n \geq N.$$
Since $|x_n| - |x| \leq |x_n - x|$, it follows that
$$|x_n| - |x| < 1 \quad \text{for all } n \geq N,$$
or
$$|x_n| \leq 1 + |x| \quad \text{for all } n \geq N.$$
Setting $K = \max\{|x_1|, |x_2|, \cdots, |x_{N-1}|, 1 + |x|\}$, we see that
$$|x_n| \leq K \quad \text{for all } n \in \mathbb{N}. \quad \square$$

Note that the converse of Theorem 3.2 is false, for the sequence $((-1)^n)$, though bounded, is divergent, as we just saw in Example 3.3.

Theorem 3.3
If $x_n \to x$ and $x \neq 0$, then there is a positive number M and an integer N such that
$$|x_n| > M \quad \text{for all } n \geq N.$$

Proof. Let $\varepsilon = |x|/2$, which is a positive number. Hence there is an integer N such that
$$n \geq N \Rightarrow |x_n - x| < \varepsilon$$
$$\Rightarrow ||x_n| - |x|| < \varepsilon.$$
Consequently,
$$|x| - \varepsilon < |x_n| < |x| + \varepsilon \quad \text{for all } n \geq N.$$
From the left inequality, we see that $|x_n| > |x|/2$, and we can take $M = |x|/2$ to complete the proof. $\square$

By slightly modifying the proof we can also show that, if $x_n \to x$ and $x > 0$, then there is an $M > 0$ and an $N \in \mathbb{N}$ such that $x_n > M$ for all $n \geq N$.

Now we investigate the effect of various algebraic operations on sequences, in particular on the convergence properties of the resulting sequences.

Theorem 3.4

Let (x_n) and (y_n) be two convergent sequences with limits x and y, respectively. Then
(i) $(x_n + y_n)$ converges to $x + y$,
(ii) $(x_n y_n)$ converges to to xy,
(iii) (x_n/y_n) converges to x/y, provided $y \neq 0$.

Proof

(i) Since x_n and y_n are convergent, given $\varepsilon > 0$, there are integers N_1 and N_2 such that
$$n \geq N_1 \Rightarrow |x_n - x| < \varepsilon$$
$$n \geq N_2 \Rightarrow |y_n - y| < \varepsilon.$$

Let $N = \max\{N_1, N_2\}$. If we take $n \geq N$ then both inequalities $n \geq N_1$ and $n \geq N_2$ are satisfied, and it follows, using the triangle inequality, that
$$|(x_n + y_n) - (x + y)| = |(x_n - x) + (y_n - y)|$$
$$\leq |x_n - x| + |y_n - y|$$
$$\leq 2\varepsilon$$

Hence (see Remark 3.1.4) $\lim(x_n + y_n) = x + y$.

(ii) Let $\varepsilon > 0$. For any $n \in \mathbb{N}$, we have
$$|x_n y_n - xy| = |x_n y_n - x_n y + x_n y - xy|$$
$$\leq |x_n y_n - x_n y| + |x_n y - xy|$$
$$= |x_n| |y_n - y| + |y| |x_n - x|. \qquad (3.4)$$

Since the sequence (x_n) is convergent, it is bounded (by Theorem 3.2), hence there is a positive constant K such that
$$|x_n| \leq K \quad \text{for all } n \in \mathbb{N}.$$

Taking the integers N_1, N_2, N as in part (i), we have, in view of (3.4),
$$n \geq N \Rightarrow |x_n y_n - xy| \leq K |y_n - y| + |y| |x_n - x|$$
$$< K\varepsilon + |y| \varepsilon = c\varepsilon,$$

where $c = K + |y|$. This proves that $x_n y_n \to xy$.

(iii) Since $y_n \to y \neq 0$, we know from Theorem 3.3 that there is a positive constant M and an integer N_0 such that

$$|y_n| > M \quad \text{for all } n \geq N_0.$$

Consequently x_n/y_n is defined for all $n \geq N_0$, and the limit of (x_n/y_n) is taken to mean the limit of the tail $(x_n/y_n : n \geq N_0)$. We have

$$\left|\frac{x_n}{y_n} - \frac{x}{y}\right| = \frac{|x_n y - xy_n|}{|y_n||y|}$$
$$\leq \frac{|x_n y - xy| + |xy - xy_n|}{|y_n||y|}$$
$$= \frac{|x_n - x|}{|y_n|} + \frac{|x||y - y_n|}{|y_n||y|}. \tag{3.5}$$

Suppose we are given a positive ε. As in (i), there are integers N_1 and N_2 such that $|x_n - x| < \varepsilon$ for all $n \geq N_1$ and $|y_n - y| < \varepsilon$ for all $n \geq N_2$. Taking $N = \max\{N_0, N_1, N_2\}$, and recalling (3.5), we conclude that

$$n \geq N \Rightarrow n \geq N_0, \; n \geq N_1, \; n \geq N_2$$
$$\Rightarrow \left|\frac{x_n}{y_n} - \frac{x}{y}\right| < \frac{\varepsilon}{M} + \frac{|x|}{|y|M}\varepsilon = c\varepsilon,$$

where $c = \dfrac{1}{M} + \dfrac{|x|}{|y|M}$. $\square$

Theorem 3.5
Suppose $x_n \to x$ and $y_n \to y$. If $x_n \leq y_n$ for all n, then $x \leq y$.

Proof. Let ε be any positive number. Since $x_n \to x$ and $y_n \to y$, there exist $N_1, N_2 \in \mathbb{N}$ such that

$$|x_n - x| < \frac{\varepsilon}{2} \quad \text{for all } n \geq N_1$$
$$|y_n - y| < \frac{\varepsilon}{2} \quad \text{for all } n \geq N_2.$$

Thus, for all $n \geq \max\{N_1, N_2\}$, we have

$$x - \frac{\varepsilon}{2} < x_n < x + \frac{\varepsilon}{2}$$
$$y - \frac{\varepsilon}{2} < y_n < y + \frac{\varepsilon}{2}.$$

Using $x_n \leq y_n$, we arrive at
$$x - \frac{\varepsilon}{2} < x_n \leq y_n < y + \frac{\varepsilon}{2},$$
which implies
$$x < y + \varepsilon.$$
But since $\varepsilon > 0$ is arbitrary, we must have $x \leq y$ (Exercise 2.2.2). □

Remarks 3.2

1. If $x_n < y_n$ for all $n \in \mathbb{N}$, it would not necessarily follow that $x < y$. Can you think of an example where $x_n < y_n$ but $x = y$?

2. Clearly, the theorem remains true if the condition $x_n \leq y_n$ for all $n \in \mathbb{N}$ is replaced by $x_n \leq y_n$ for all $n \geq N$, where N is some natural number.

Theorem 3.6
Suppose the sequences $(x_n), (y_n)$, and (z_n) satisfy
$$x_n \leq y_n \leq z_n \text{ for all } n \geq N_0.$$
If $\lim x_n = \lim z_n = \ell$, then (y_n) is convergent and its limit is also ℓ.

Proof. Given $\varepsilon > 0$, there are $N_1, N_2 \in \mathbb{N}$ such that
$$|x_n - \ell| < \varepsilon \text{ for all } n \geq N_1$$
$$|z_n - \ell| < \varepsilon \text{ for all } n \geq N_2.$$
Now we define $N = \max\{N_0, N_1, N_2\}$, and note that
$$\begin{aligned}
n \geq N &\Rightarrow n \geq N_0,\ n \geq N_1,\ n \geq N_2 \\
&\Rightarrow x_n \leq y_n \leq z_n,\ \ell - \varepsilon < x_n < \ell + \varepsilon,\ \ell - \varepsilon < z_n < \ell + \varepsilon \\
&\Rightarrow \ell - \varepsilon < x_n \leq y_n \leq z_n < \ell + \varepsilon \\
&\Rightarrow \ell - \varepsilon < y_n < \ell + \varepsilon \\
&\Rightarrow |y_n - \ell| < \varepsilon,
\end{aligned}$$
which means $y_n \to \ell$. □

Example 3.6 If $x_n \to x$ then $|x_n| \to |x|$.

Proof. We know that
$$0 \leq ||x_n| - |x|| \leq |x_n - x| \text{ for all } n \in \mathbb{N}.$$

Since $x_n \to x$ implies $|x_n - x| \to 0$, it follows from Theorem 3.6 that $|x_n| - |x| \to 0$, or $|x_n| \to |x|$. □

Example 3.7 If $0 < a < 1$ then $\lim a^n = 0$.

Proof. We can always write $a = \dfrac{1}{1+b}$, where $b > 0$. By the binomial theorem (see Exercise 2.3.14),

$$(1+b)^n = 1 + nb + \frac{n(n-1)}{2}b^2 + \cdots + b^n > nb,$$

hence,

$$0 < a^n = \frac{1}{(1+b)^n} < \frac{1}{nb}.$$

Since $1/n \to 0$, $1/nb \to 0$, and we apply Theorem 3.6 to complete the proof. □

Since $-|a| \leq a \leq |a|$, Theorem 3.6 implies $\lim a^n = 0$ for all $a \in (-1, 1)$.

Example 3.8 If $c > 0$, then $\lim c^{1/n} = 1$.

Proof. First assume that $c > 1$, which implies $c^{1/n} > 1$, so there is a positive d_n such that

$$c^{1/n} = 1 + d_n.$$

Using the binomial theorem once again, we obtain

$$c = (1 + d_n)^n > nd_n,$$

and therefore

$$0 < d_n < \frac{c}{n}.$$

From Theorem 3.6 we conclude that $d_n \to 0$, and hence

$$c^{1/n} = 1 + d_n \to 1.$$

Now assume that $0 < c < 1$ and let $b = 1/c$. In this case $b > 1$ so, using the above result, we conclude that $b^{1/n} \to 1$. Now

$$c^{1/n} = \frac{1}{b^{1/n}} \to \frac{1}{1} = 1$$

by Theorem 3.4.

Finally, if $c = 1$ then $\lim c^{1/n} = \lim 1 = 1$. □

Example 3.9 If $x_n \geq 0$ for all $n \in \mathbb{N}$ and $\lim x_n = x$, then $\lim \sqrt{x_n} = \sqrt{x}$.

Proof. Let $x = 0$. If ε is a positive number then, working with ε^2, we can find an $N \in \mathbb{N}$ such that

$$|x_n - 0| = x_n < \varepsilon^2 \quad \text{for all } n \geq N.$$

But this implies $\left|\sqrt{x_n} - 0\right| = \sqrt{x_n} < \varepsilon$ for all $n \geq N$. Hence $\sqrt{x_n} \to 0$. If $x \neq 0$ then $x > 0$ and hence $\sqrt{x} > 0$. In this case

$$0 \leq \left|\sqrt{x_n} - \sqrt{x}\right| = \frac{|x_n - x|}{\left|\sqrt{x_n} + \sqrt{x}\right|} \leq \frac{|x_n - x|}{\sqrt{x}},$$

and, since $\lim |x_n - x| = 0$, Theorem 3.6 implies $\lim \left|\sqrt{x_n} - \sqrt{x}\right| = 0$, or $\lim \sqrt{x_n} = \sqrt{x}$. □

Example 3.10
$$\lim n^{1/n} = 1.$$

Proof. For any $n > 1$ we know that $n^{1/n} > 1$, so there is a positive number h_n such that
$$n^{1/n} = 1 + h_n.$$

Consequently,

$$\begin{aligned} n &= (1 + h_n)^n \\ &= 1 + nh_n + \frac{n(n-1)}{2}h_n^2 + \cdots + h_n^n \\ &> \frac{n(n-1)}{2}h_n^2, \end{aligned}$$

from which we can conclude that

$$0 < h_n^2 < \frac{2}{n-1}$$

for every $n > 1$. By Theorem 3.6, $h_n^2 \to 0$, hence $h_n \to 0$. □

EXERCISES 3.2

1. Determine whether the sequence (x_n) is convergent or divergent, and find the limit where it exists.

(a) $x_n = \dfrac{2n^3 + 3}{n^2 + 4}$

(b) $x_n = \dfrac{(-1)^n n}{2n + 1}$

(c) $x_n = \sqrt{n+1} - \sqrt{n}$

(d) $x_n = \dfrac{\sin n}{n}$

(e) $x_n = \begin{cases} 1/n & \text{if } n \in \mathbb{N}_1 \\ 0 & \text{if } n \in \mathbb{N}_2, \end{cases}$

where $\mathbb{N}_1$ and $\mathbb{N}_2$ are the odd and even positive integers, respectively.

2. If the sequences (x_n) and $(x_n + y_n)$ are both convergent, prove that (y_n) is also convergent and determine its limit. Can you state a corresponding result for the sequence $(x_n \cdot y_n)$?

3. Give an example of two divergent sequences (x_n) and (y_n) such that $(x_n + y_n)$ is convergent.

4. Give an example of a divergent sequence (x_n) such that $(|x_n|)$ is convergent. When does the convergence of $(|x_n|)$ imply the convergence of (x_n), and what is the relation between $\lim |x_n|$ and $\lim x_n$ when both exist?

5. If $\lim \dfrac{x_n - 1}{x_n + 1} = 0$ prove that $\lim x_n = 1$.

6. If $0 < b < 1$ prove that $nb^n \to 0$. Hint: See Example 3.7.

7. If $0 < a < b$ prove that $\lim \sqrt[n]{a^n + b^n} = b$.

8. Let $x_n > 0$ for every $n \in \mathbb{N}$, and $\lim \dfrac{x_{n+1}}{x_n} = L$. If $L < 1$ prove that $\lim x_n = 0$ using the following steps:

   (a) Assume $L < r < 1$ and let $\varepsilon = r - L$. Show that there is an integer N such that
   $$\dfrac{x_{n+1}}{x_n} < r \quad \text{for all } n \geq N.$$

   (b) Prove that
   $$x_n < x_N \cdot r^{n-N} \quad \text{for all } n \geq N.$$

(c) Conclude that $x_n \to 0$.

9. Under the hypothesis of Exercise 8, if $L > 1$ prove that (x_n) is unbounded and therefore divergent.

10. Give an example of a convergent sequence (x_n) and a divergent sequence (y_n) such that
$$\lim \frac{x_{n+1}}{x_n} = \lim \frac{y_{n+1}}{y_n} = 1.$$
Use the results of Exercises 8 and 9 to make a general statement on the implication of the value of L to the convergence of (x_n).

11. Use the result of Exercise 8 to evaluate $\lim x_n$, where it exists, for each of the following sequences:

(a) $x_n = \dfrac{a^n}{3^n}$, where $a > 0$.

(b) $x_n = \dfrac{n!}{2^n}$.

(c) $x_n = \dfrac{a^n}{n^2}$, where $a > 0$.

(d) $x_n = \dfrac{a^n}{n^n}$.

12. Prove that (a^n) converges if, and only if, $-1 < a \le 1$.

13. Given a sequence (x_n), the sequence (σ_n), where
$$\sigma_n = \frac{1}{n}(x_1 + x_2 + \cdots + x_n) = \frac{1}{n}\sum_{k=1}^{n} x_k,$$
is the *sequence of arithmetic averages* of (x_n). Prove that if (x_n) converges then (σ_n) converges to the same limit, but not vice-versa.

14. Prove that, for every $x \in \mathbb{R}$, there is a sequence (x_n) in $\mathbb{Q}$ such that $x_n \to x$.

15. Prove that, for every $x \in \mathbb{R}$, there is a sequence (x_n) in $\mathbb{Q}^c$ such that $x_n \to x$.

16. If $a_n > 0$ for all n, and $\lim \dfrac{a_{n+1}}{a_n} = L$, prove that $\sqrt[n]{a_n} \to L$.

Hint: Follow the procedure used in Exercise 8.

3.3 Monotonic Sequences

At this point the attentive reader may wonder what role the completeness axiom plays in the study of convergence. This question will now be addressed, and will take up the major part of the remainder of this chapter.

Definition 3.5 The sequence (x_n) is said to be
(i) *monotonically increasing*, or simply *increasing*, if $x_{n+1} \geq x_n$ for all $n \in \mathbb{N}$,
(ii) *strictly increasing* if $x_{n+1} > x_n$ for all $n \in \mathbb{N}$,
(iii) *monotonically decreasing*, or *decreasing*, if $x_{n+1} \leq x_n$ for all $n \in \mathbb{N}$, and
(iv) *strictly decreasing* if $x_{n+1} < x_n$ for all $n \in \mathbb{N}$.
An increasing or a decreasing sequence is called a *monotonic sequence*.

Note that (x_n) is increasing if, and only if, $(-x_n)$ is decreasing, so the properties of monotonic sequences may be fully investigated by restricting ourselves to increasing (or decreasing) sequences only. Here are some examples:

(i) $(1/n)$ is a (strictly) decreasing sequence.

(ii) (2^n) is (strictly) increasing.

(iii) $((-1)^n)$ is not monotonic.

Theorem 3.7
A monotonic sequence is convergent if, and only if, it is bounded. More specifically,
(i) if (x_n) is increasing and bounded above, then

$$\lim x_n = \sup\{x_n : n \in \mathbb{N}\};$$

(ii) if (x_n) is decreasing and bounded below, then

$$\lim x_n = \inf\{x_n : n \in \mathbb{N}\}.$$

Proof. If (x_n) is convergent then, by Theorem 3.2, it must be bounded.
(i) Assume (x_n) is increasing and bounded. The set $A = \{x_n : n \in \mathbb{N}\}$ will then be non-empty and bounded, so, by the completeness axiom, it has a least upper bound. Let $\sup A = x$. By Definition 2.6, given $\varepsilon > 0$ there is an $x_N \in A$ such that

$$x_N > x - \varepsilon.$$

Since x_n is increasing,
$$x_n \geq x_N \text{ for all } n \geq N. \tag{3.6}$$
But x is an upper bound for A, hence
$$x \geq x_n \text{ for all } n \in \mathbb{N}. \tag{3.7}$$
From (3.6) and (3.7) we arrive at
$$|x_n - x| = x - x_n \leq x - x_N < \varepsilon \text{ for all } n \geq N,$$
which means $x_n \to x$.

(ii) If (x_n) is decreasing and bounded, then $(-x_n)$ is increasing and bounded, and, from (i), we have
$$\lim(-x_n) = \sup(-A).$$
But since $\sup(-A) = -\inf A$, it follows that $\lim x_n = \inf A$. $\square$

In order to extend the concept of the limit to unbounded sequences, it is convenient at this stage to introduce the *extended real number system*. This is defined as the set of real numbers to which two additional elements, denoted by $\pm \infty$, are adjoined,
$$\bar{\mathbb{R}} = \mathbb{R} \cup \{-\infty, +\infty\},$$
such that the following properties hold:

(i) If x is a real number, then $-\infty < x < +\infty$, hence we also write $\bar{\mathbb{R}} = [-\infty, +\infty]$, $(x, +\infty] = (x, +\infty) \cup \{+\infty\}$, etc. Furthermore, $x \pm \infty = \pm \infty$, and $\dfrac{x}{\pm \infty} = 0$.

(ii) If $x > 0$, then $x \cdot (\pm \infty) = \pm \infty$.

(iii) If $x < 0$, then $x \cdot (\pm \infty) = \mp \infty$.

An *extended real number* is therefore either a real number, $+\infty$, or $-\infty$. If a set of real numbers A is not bounded above we shall define $\sup A$ to be $+\infty$, and if A is not bounded below then $\inf A$ is defined to be $-\infty$. Thus, in the extended number system, every non-empty set has a supremum and an infimum. The symbol $+\infty$ is often abbreviated to ∞.

We define a *neighborhood* of $+\infty$ to be any subset of $\bar{\mathbb{R}}$ which contains an interval $(M, +\infty]$ for some real number M. We then say that

$\lim x_n = +\infty$ if every neighborhood of $+\infty$ contains all but a finite number of terms of the sequence (x_n). This is equivalent to saying that, given any $M \in \mathbb{R}$, there is an $N \in \mathbb{N}$ such that $x_n > M$ for all $n \geq N$. Consequently every monotone sequence has a limit in $\bar{\mathbb{R}}$. In particular, part (i) of Theorem 3.7 may be restated as

If (x_n) is increasing, then $\lim x_n = \sup\{x_n : n \in \mathbb{N}\}$.

Note how the fact that the "point" $+\infty$ lies at the "upper edge" of the extended real numbers, so to speak, means that the sequence can only approach $+\infty$ "from below", and this in turn dictates the definition of the neighborhood of $+\infty$. Similarly, a corresponding definition for a neighborhood of $-\infty$ and a meaning to the statement $\lim x_n = -\infty$ can also be given (see Exercise 3.3.5). But it should be kept in mind that both $+\infty$ and $-\infty$ are not real numbers, and that, when $\lim x_n = +\infty$ or $\lim x_n = -\infty$, the sequence (x_n) is divergent according to Definition 3.2. More precisely, we should say it is convergent in $\bar{\mathbb{R}}$ but divergent in $\mathbb{R}$.

Example 3.11 Given $x_1 = 1$ and $x_{n+1} = \sqrt{2x_n}$ for all $n \geq 1$, prove that the sequence (x_n) is convergent and calculate its limit.

Proof. First we shall prove that x_n is monotonically increasing by showing that

$$x_{n+1} \geq x_n \quad \text{for all } n \in \mathbb{N}. \tag{3.8}$$

(i) Since $\sqrt{2} > 1$ we have $x_2 > x_1$, so (3.8) is true for $n = 1$.
(ii) If (3.8) is true for n, that is, $x_{n+1} \geq x_n$, then

$$x_{n+2} = \sqrt{2x_{n+1}} \geq \sqrt{2x_n} = x_{n+1},$$

hence (3.8) is also true for $n+1$. By induction, (3.8) is therefore true.

Secondly, we shall prove that x_n is bounded above by 2. Once again we use induction:
(i) $x_1 = 1 < 2$.
(ii) If $x_n \leq 2$ then

$$x_{n+1} = \sqrt{2x_n} \leq \sqrt{2 \cdot 2} = 2.$$

Hence $x_n \leq 2$ for all $n \in \mathbb{N}$.

By Theorem 3.7, the sequence (x_n) is convergent. Suppose $\lim x_n = x$. Since the sequence (x_{n+1}) is a tail of (x_n) we also have $\lim x_{n+1} = x$. By taking the limits of both sides of the equation $x_{n+1} = \sqrt{2x_n}$, and using the result of Example 3.9, we arrive at

$$x = \sqrt{2x},$$

which implies
$$x^2 = 2x.$$
Thus $x = 0$ or $x = 2$. But since $x \geq x_1 = 1$, we must have $x = 2$. □

Example 3.12 The sequence $\left(1 + \dfrac{1}{n}\right)^n$ is convergent.

Proof. We shall establish the convergence of this sequence by proving that it is monotonic and bounded. Let $x_n = \left(1 + \dfrac{1}{n}\right)^n$. By the binomial theorem,

$$x_n = 1 + n\frac{1}{n} + \frac{n(n-1)}{2!}\frac{1}{n^2} + \frac{n(n-1)(n-2)}{3!}\frac{1}{n^3} + \cdots + \frac{1}{n^n}$$

$$= 1 + 1 + \frac{1}{2!}\left(1 - \frac{1}{n}\right) + \frac{1}{3!}\left(1 - \frac{1}{n}\right)\left(1 - \frac{2}{n}\right) + \cdots$$

$$+ \frac{1}{n!}\left(1 - \frac{1}{n}\right)\left(1 - \frac{2}{n}\right)\cdots\left(1 - \frac{n-1}{n}\right).$$

Hence,

$$x_{n+1} = 1 + 1 + \frac{1}{2!}\left(1 - \frac{1}{n+1}\right) + \frac{1}{3!}\left(1 - \frac{1}{n+1}\right) + \cdots$$

$$+ \frac{1}{n!}\left(1 - \frac{1}{n+1}\right)\left(1 - \frac{2}{n+1}\right)\cdots\left(1 - \frac{n-1}{n+1}\right)$$

$$+ \frac{1}{(n+1)!}\left(1 - \frac{1}{n+1}\right)\left(1 - \frac{2}{n+1}\right)\cdots\left(1 - \frac{n}{n+1}\right).$$

Comparing these expressions for x_n and x_{n+1}, we note that every term in x_n is less than or equal to the corresponding term in x_{n+1}, and that, in addition, x_{n+1} has an extra positive term. Consequently,

$$x_{n+1} \geq x_n \quad \text{for all } n \in \mathbb{N},$$

i.e., the sequence (x_n) is increasing.

Furthermore, for every $n \in \mathbb{N}$, we have

$$x_n \leq 1 + 1 + \frac{1}{2!} + \frac{1}{3!} + \cdots + \frac{1}{n!}.$$

Using the inequality $2^{n-1} \leq n!$ (see Exercise 2.3.7), we conclude that

$$x_n \leq 1 + 1 + \frac{1}{2} + \frac{1}{2^2} + \cdots + \frac{1}{2^{n-1}}$$
$$= 1 + \frac{1 - (1/2)^n}{1 - 1/2}$$
$$< 1 + \frac{1}{1 - 1/2} = 3.$$

Thus (x_n) is bounded above by 3 and so, by Theorem 3.7, it is convergent. Its limit, the universal constant denoted by e, clearly satisfies the inequality $2 < e < 3$. □

Example 3.13 Let $a > 0$. If $x_1 = 1$, and

$$x_{n+1} = \frac{1}{2}\left(x_n + \frac{a}{x_n}\right) \quad \text{for all } n \in \mathbb{N}, \tag{3.9}$$

then (x_n) converges to $\sqrt{a}$.

Proof. First note that $x_n > 0$ for all $n \in \mathbb{N}$, and that x_n satisfies the quadratic equation
$$t^2 - 2x_{n+1}t + a = 0,$$
so its discriminant $4x_{n+1}^2 - 4a$ cannot be negative. Therefore $x_{n+1}^2 \geq a$ for all $n \in \mathbb{N}$, that is, (x_n) is bounded below.

On the other hand, we also have

$$x_{n+1} - x_n = \frac{1}{2}x_n + \frac{a}{2x_n} - x_n$$
$$= \frac{a}{2x_n} - \frac{x_n}{2}$$
$$= \frac{a - x_n^2}{2x_n} \leq 0 \quad \text{for all } n \geq 2.$$

This implies that (x_n) is decreasing, and hence convergent.

Assuming that $\lim x_n = x$, we clearly have $\lim x_{n+1} = x$, and $x \neq 0$ because $x_{n+1} \geq \sqrt{a}$ for all n. Taking the limits of both sides of equation (3.9) now yields

$$x = \frac{1}{2}\left(x + \frac{a}{x}\right)$$
$$2x^2 = x^2 + a$$
$$x = \sqrt{a},$$

where the root $-\sqrt{a}$ is excluded, since the sequence is bounded below by 0. □

If a is taken as a positive integer in Example 3.13, then $x_n \in \mathbb{Q}$ for all $n \in \mathbb{N}$. This example provides a numerical method for approximating irrational roots of the form $\sqrt{2}$, $\sqrt{3}$, $\sqrt{5}$, $\cdots$ by rational numbers. It was used by Arab mathematicians in the Middle Ages to approximate such roots.

EXERCISES 3.3

1. Prove that each of the following sequences is monotonic and bounded, then determine its limit:

    (a) $x_1 = 1$, $x_{n+1} = \dfrac{1}{7}(4x_n + 5)$ for all $n \in \mathbb{N}$.

    (b) $x_1 = 1$, $x_{n+1} = \sqrt{3 + x_n}$ for all $n \in \mathbb{N}$.

    (c) $x_1 = 1$, $x_{n+1} = \dfrac{4x_n + 2}{x_n + 3}$ for all $n \in \mathbb{N}$.

2. Given $x_n = \dfrac{1}{n} + \dfrac{1}{n+1} + \cdots + \dfrac{1}{2n}$, prove that (x_n) is convergent.

3. Suppose that $0 < x_1 < y_1$ and that the sequences (x_n) and (y_n) are defined as follows:
$$x_{n+1} = \sqrt{x_n y_n}, \quad y_{n+1} = \frac{1}{2}(x_n + y_n) \quad \text{for all } n \in \mathbb{N}.$$
Prove that (x_n) and (y_n) converge to the same limit.

4. Given $x_n = \dfrac{1 \cdot 3 \cdot 5 \cdots (2n-1)}{2 \cdot 4 \cdot 6 \cdots 2n}$,

    (a) Prove that (x_n) is convergent.
    (b) Prove that $\big((2n+1)x_n^2\big)$ is convergent.
    (c) Evaluate $\lim x_n$.

5. Define a neighborhood for $-\infty$. What does the statement $\lim x_n = -\infty$ mean? Restate part (ii) of Theorem 3.7 to include limits in $\overline{\mathbb{R}}$.

6. If the set A is non-empty and bounded above, prove that A contains an increasing sequence (x_n) such that $\lim x_n = \sup A$.

3.4 The Cauchy Criterion

To test the convergence of a general, non-monotonic, sequence we have only Definition 3.2 to rely on, and there we must first have a candidate for the limit before we can apply the definition. Needless to say, it would be much more convenient if we could decide on the convergence, or otherwise, of a sequence without having to first guess its limit, and then test the sequence for convergence to that limit. This is achieved by applying the so-called *Cauchy criterion*, which will be developed in this section.

Definition 3.6 The sequence (x_n) is called a *Cauchy sequence* if, for every positive number ε, there is a positive integer N such that

$$m, n \geq N \ \Rightarrow \ |x_n - x_m| < \varepsilon.$$

This means that, as we go far enough into a Cauchy sequence, the terms become as close to each other as we choose. We would expect this to be the case if the sequence were convergent, since the terms of such a sequence would have to approach each other as they approach a common limit.

Theorem 3.8
If (x_n) is convergent, then it is a Cauchy sequence.

Proof. Suppose $\lim x_n = x$ and ε is any positive number. We know there is an $N \in \mathbb{N}$ such that

$$|x_n - x| < \varepsilon/2 \ \text{ for all } n \geq N.$$

Now if we take $m, n \geq N$ then

$$|x_n - x| < \varepsilon/2, \ |x_m - x| < \varepsilon/2.$$

Using the triangle inequality (Theorem 2.1), we obtain

$$|x_n - x_m| \leq |x_n - x| + |x_m - x| < \varepsilon,$$

which means (x_n) is a Cauchy sequence. □

The question is: Is the converse of Theorem 3.8 true? In other words, is every Cauchy sequence convergent? The affirmative answer to this question is a consequence of the completeness of $\mathbb{R}$ expressed by axiom A12.

Theorem 3.9 (Cauchy's Criterion)
A sequence of real numbers is convergent if, and only if, it is a Cauchy sequence.

This is one of the fundamental theorems of real sequences. In view of Theorem 3.8, we need only prove that every Cauchy sequence is convergent in order to complete the proof of Theorem 3.9. This turns out to involve a more lengthy procedure than the previous proof, but, in the process, we shall establish some useful notions and results of no less significance than the criterion itself.

Definition 3.7
(i) A point $x \in \mathbb{R}$ is called a *cluster point*, or an *accumulation point*, of a set $A \subseteq \mathbb{R}$ if every neighborhood V of x contains an element $a \in A$ different from x. The symbol $\hat{A}$ will be used to denote the set of cluster points of A.
(ii) A point in A which is not a cluster point of A is called an *isolated point* of A.

Remarks 3.3
1. According to this definition, a cluster point of A need not belong to A, but an isolated point of A must lie in A.
2. $x \in \hat{A}$ if, and only if, for every $\varepsilon > 0$ there is a point $a \in A$ such that $a \neq x$ and $|x - a| < \varepsilon$, i.e., such that

$$0 < |a - x| < \varepsilon.$$

This follows from the definition of a neighborhood.

3. $x \in \hat{A}$ if, and only if, every neighborhood of x contains an infinite number of points of A. For if a neighborhood $V = (x - \delta, x + \delta)$ contains only a finite number of such points, say $a_1, a_2, \cdots, a_n$, where $a_i \neq x$, then

$$\varepsilon = \min_{1 \leq i \leq n} |a_i - x|$$

is a positive number. Since x is a cluster point of A, there is a point $a \in A$ such that $a \neq x$ and $|x - a| < \varepsilon$. By the definition of ε, it is clear that a lies in V but $a \neq a_i$ for any i, which contradicts our assumption. The implication in the other direction is obvious.

4. $x \in A$ is an isolated point of A if, and only if, there is a neighborhood V of x which does not intersect A except in x, i.e., such that $V \cap A = \{x\}$.

Example 3.14 The set of cluster points of any open interval (a, b) is the closed interval $[a, b]$.

Proof. Here we provide some schematic hints, and leave the details to the reader.

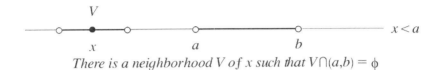

Figure 3.4

Example 3.15 $\hat{\mathbb{Z}} = \emptyset$.

Proof. Take any $x \in \mathbb{Z}$. Clearly $V = (x-1, x+1)$ is a neighborhood of x which contains no elements of $\mathbb{Z}$ except x, hence $x \notin \hat{\mathbb{Z}}$. Now suppose $x \notin \mathbb{Z}$, and let
$$\varepsilon = \min\{|n - x| : n \in \mathbb{Z}\}.$$
The interval $(x - \varepsilon, x + \varepsilon)$ contains no integers, hence $x \notin \hat{\mathbb{Z}}$. Thus $\mathbb{Z}$ is composed of isolated points. $\square$

Example 3.16 $\hat{\mathbb{Q}} = \widehat{\mathbb{Q}^c} = \mathbb{R}$.

Proof. This is a restatement of the density of the rational and the irrational numbers in $\mathbb{R}$ (Theorem 2.8 and its corollary), and we leave the details to the reader.

Some books refer to a cluster point as a *limit point*, and the next theorem provides some justification for this nomenclature.

Theorem 3.10
(i) If $x \in \hat{A}$ then there is a sequence (x_n) of distinct elements of A such that $x_n \to x$.
(ii) If the sequence (x_n) is convergent to x and the set $A = \{x_n : n \in \mathbb{N}\}$ is infinite, then $x \in \hat{A}$.

Proof
(i) Let $x \in \hat{A}$. Taking $\varepsilon = 1$ we can find $x_1 \in A$ such that
$$0 < |x_1 - x| < 1.$$
With $\varepsilon = 1/2$ there is an $x_2 \in A$ such that $x_2 \neq x_1$ and
$$0 < |x_2 - x| < \frac{1}{2}.$$
Suppose we can choose $x_1, x_2, \cdots, x_n$ in A such that
$$x_i \neq x_j \quad \text{for all } i \neq j, \tag{3.10}$$
$$|x_i - x| < \frac{1}{i} \tag{3.11}$$
Since the open interval $\left(x - \dfrac{1}{n+1}, x + \dfrac{1}{n-1}\right)$ is a neighborhood of x, it must contain an infinite number of points of A, hence we can choose $x_{n+1} \in A$ in this neighborhood such that
$$x_{n+1} \neq x_i \quad \text{for all } 1 \leq i \leq n.$$
Thus the set $\{x_1, x_2, \cdots, x_{n+1}\}$ satisfies the above relations (3.10) and (3.11). By mathematical induction, we therefore have a sequence (x_n) which satisfies these two relations. Now (3.10) implies the elements of the sequence are distinct, while (3.11) implies $x_n \to x$.
(ii) Suppose $x_n \to x$ and the set $A = \{x_n : n \in \mathbb{N}\}$ is infinite. Let V be a neighborhood of x. We shall prove that V contains an infinite number of elements of A. Since $x_n \to x$, there is an $N \in \mathbb{N}$ such that
$$x_n \in V \quad \text{for all } n \geq N.$$

that is,
$$A\setminus\{x_1, x_2, \cdots, x_{N-1}\} \subseteq V.$$
But the set $A\setminus\{x_1, x_2, \cdots, x_{N-1}\}$ is infinite. □

A finite set cannot have a cluster point, since any neighborhood of a cluster point contains an infinite number of elements of the set (Remark 3.3.3). On the other hand, an infinite set need not have a cluster point, as Example 3.15 demonstrates. So the question arises: When does a set possess a cluster point? The following result paves the way to answering this question. For any closed interval $I = [a, b]$, where $a \leq b$, we define the *length* of I to be $l(I) = b - a$.

Theorem 3.11 (Cantor)
Let (I_n) be a sequence of non-empty, closed and bounded intervals. If $I_{n+1} \subseteq I_n$ for every $n \in \mathbb{N}$, then the intersection $\bigcap_{n=1}^{\infty} I_n$ is not empty. Furthermore, if
$$\inf\{l(I_n) : n \in \mathbb{N}\} = 0,$$
then $\bigcap_{n=1}^{\infty} I_n$ is a single point.

Proof. Suppose $I_n = [a_n, b_n]$ and $I = \bigcap_{n=1}^{\infty} I_n$. Using the nested property of the intervals I_n, we have
$$\begin{aligned}m \geq p &\Rightarrow I_m \subseteq I_p \\ &\Rightarrow a_p \leq a_m < b_m \leq b_p.\end{aligned} \quad (3.12)$$

The set $A = \{a_n : n \in \mathbb{N}\} \subset \mathbb{R}$ is clearly not empty and bounded above by b_1, so by the completeness of $\mathbb{R}$ it has a least upper bound, which we shall call x. We shall prove that $x \in I$ by proving that $x \in I_k$, i.e., $a_k \leq x \leq b_k$, for all $k \in \mathbb{N}$.

By the definition of x, $a_k \leq x$ for all k. To prove that $x \leq b_k$, it suffices to prove that b_k is an upper bound of A, i.e., that $a_n \leq b_k$ for all $n \in \mathbb{N}$. If $n \leq k$ then, by (3.12), $a_n \leq a_k \leq b_k$; and if $n > k$ then, again by (3.12), $a_n \leq b_n \leq b_k$. b_k is therefore an upper bound of A, hence
$$a_k \leq x \leq b_k \quad \text{for all } k \in \mathbb{N}.$$

Now suppose that $\inf\{l(I_n) : n \in \mathbb{N}\} = 0$ and let $x, y \in I$. It then follows that $x, y \in I_n$ for every n, which implies
$$|x - y| \leq l(I_n),$$
or $|x - y|$ is a lower bound of the set $\{l(I_n) : n \in \mathbb{N}\}$. Hence

$$|x - y| \leq \inf\{l(I_n) : n \in \mathbb{N}\} = 0$$
$$\Rightarrow x = y. \quad \square$$

The next theorem gives sufficient conditions for the existence of a cluster point in a set.

Theorem 3.12 (Bolzano-Weierstrass)
Every infinite and bounded subset of $\mathbb{R}$ has at least one cluster point in $\mathbb{R}$.

Proof. Let A be an infinite and bounded set of real numbers. Being bounded, A is contained in some bounded closed interval $I_0 = [a_0, b_0]$. First we bisect I_0 into the two subintervals

$$I_0' = \left[a_0, \frac{a_0 + b_0}{2}\right], \quad I_0'' = \left[\frac{a_0 + b_0}{2}, b_0\right].$$

Since $A \subseteq I_0' \cup I_0''$ is infinite at least one of the sets $A \cap I_0'$ and $A \cap I_0''$ is infinite. Let $I_1 = [a_1, b_1] = I_0'$ if $A \cap I_0'$ is infinite, otherwise set $I_1 = I_0''$. We then have

(i) $I_1 \subseteq I_0$
(ii) $l(I_1) = \frac{1}{2}(b_0 - a_0)$
(iii) $A \cap I_1$ is infinite.

Now we bisect I_1 into the subintervals

$$I_1' = \left[a_1, \frac{a_1 + b_1}{2}\right], \quad I_1'' = \left[\frac{a_1 + b_1}{2}, b_1\right],$$

one of which necessarily intersects A in an infinite set. Define $I_2 = [a_2, b_2]$ to be I_1' if $A \cap I_1'$ is infinite, otherwise let $I_2 = I_1''$. Continuing in this fashion, we obtain, in the n-th step, the intervals $I_i = [a_i, b_i]$, $0 \leq i \leq n$, which satisfy

(i) $I_i \subseteq I_{i-1}$
(ii) $l(I_i) = \frac{1}{2^i}(b_0 - a_0)$
(iii) $A \cap I_i$ is infinite.

Let us again bisect I_n to obtain

$$I_n' = \left[a_n, \frac{a_n + b_n}{2}\right], \quad I_n'' = \left[\frac{a_n + b_n}{2}, b_n\right].$$

Since $I_n = I_n' \cup I_n''$ and $A \cap I_n$ is infinite, either $A \cap I_n'$ is infinite, in which case we set $I_{n+1} = [a_{n+1}, b_{n+1}] = I_n'$, or $A \cap I_n'$ is finite and $A \cap I_n''$ is infinite, in which case we choose $I_{n+1} = I_n''$. Now we see that

94 Elements of Real Analysis

(i) $I_{n+1} \subseteq I_n$
(ii) $l(I_{n+1}) = \frac{1}{2} l(I_n) = \frac{1}{2^{n+1}}(b_0 - a_0)$
(iii) $A \cap I_{n+1}$ is infinite.

By induction we therefore conclude the existence of a sequence $(I_n : n \in \mathbb{N})$ of closed, bounded intervals that satisfies conditions (i), (ii), and (iii) above. By Cantor's theorem there is a (single) point $x \in \bigcap_{n=1}^{\infty} I_n$. We complete our proof by showing that $x \in \hat{A}$.

Given any $\varepsilon > 0$ we can choose $n \in \mathbb{N}$ so that

$$\frac{b_0 - a_0}{2^n} < \varepsilon,$$

or, equivalently, $|I_n| < \varepsilon$. This, together with the fact that $x \in I_n$ for all n, implies

$$I_n \subset (x - \varepsilon, x + \varepsilon).$$

Since I_n contains an infinite number of elements of A, so does $(x - \varepsilon, x + \varepsilon)$, hence $x \in \hat{A}$. □

Lemma 3.1 *Every Cauchy sequence is bounded.*

Proof. Let (x_n) be any Cauchy sequence. Taking $\varepsilon = 1$, there is an integer N such that

$$|x_n - x_N| < 1 \text{ for all } n \geq N.$$

Since $|x_n| - |x_N| \leq |x_n - x_N|$, we have

$$|x_n| < |x_N| + 1 \text{ for all } n \geq N.$$

Thus $|x_n|$ is bounded by $\max\{|x_1|, \cdots, |x_{N-1}|, |x_N| + 1\}$. □

Now we are in a position to prove the Cauchy criterion.

Proof of Theorem 3.9. Let (x_n) be a Cauchy sequence and $A = \{x_n : n \in \mathbb{N}\}$. We have two cases:

(i) The set A is finite. In this case one of the terms of the sequence, say x, is repeated infinitely often. We shall prove that $x_n \to x$. Given $\varepsilon > 0$, there is an integer N such that

$$m, n \geq N \Rightarrow |x_n - x_m| < \varepsilon.$$

Since the term x is repeated infinitely often in the sequence, there is an $m > N$ such that $x_m = x$. Hence

$$|x_n - x| = |x_n - x_m| < \varepsilon \text{ for all } n \geq N,$$

which just means $x_n \to x$.

(ii) A is infinite. In view of Lemma 3.1, the set A is bounded and therefore, by the Bolzano-Weierstrass Theorem, it has a cluster point x. We shall prove that $x_n \to x$. Given $\varepsilon > 0$, there is an integer N such that
$$|x_n - x_m| < \varepsilon \text{ for all } n, m \geq N.$$
Since $x \in \hat{A}$, the interval $(x - \varepsilon, x + \varepsilon)$ contains an infinite number of terms of the sequence (x_n). Hence there is an $m \geq N$ such that $x_m \in (x - \varepsilon, x + \varepsilon)$, i.e., such that $|x_m - x| < \varepsilon$. Now, if $n \geq N$, then
$$|x_n - x| \leq |x_n - x_m| + |x_m - x| < \varepsilon + \varepsilon = 2\varepsilon,$$
which proves $x_n \to x$. □

Example 3.17 Let
$$x_1 = 1, \ x_2 = 2, \ x_n = \frac{1}{2}(x_{n-1} + x_{n-2}) \text{ for all } n > 2.$$
Prove that (x_n) is convergent.

Proof. Using induction on n, it is a simple exercise to show that
$$x_n - x_{n+1} = \frac{(-1)^n}{2^{n-1}} \text{ for all } n \in \mathbb{N}. \tag{3.13}$$
Hence, if $m > n$, then
$$|x_n - x_m| \leq |x_n - x_{n+1}| + |x_{n+1} - x_{n+2}| + \cdots + |x_{m-1} - x_m|$$
$$= \sum_{r=n}^{m-1} \frac{1}{2^{r-1}}$$
$$= \frac{1}{2^{n-1}} \sum_{r=0}^{m-n-1} \frac{1}{2^r}$$
$$= \frac{1}{2^{n-1}} \frac{1 - (1/2)^{m-n}}{1 - 1/2}$$
$$< \frac{1}{2^{n-1}} \frac{1}{1 - 1/2}$$
$$= \frac{1}{2^{n-2}}.$$

Now, given any $\varepsilon > 0$, we can choose N so that $\frac{1}{2^{N-2}} < \varepsilon$, and thereby conclude that
$$m > n \geq N \implies |x_n - x_m| < \frac{1}{2^{n-2}} \leq \frac{1}{2^{N-2}} < \varepsilon.$$

This means (x_n) is a Cauchy sequence, so, by the Cauchy criterion, it is convergent. □

In this example, note that the Cauchy criterion does not provide us with the limit of (x_n). To calculate $\lim x_n$ we can use induction again to obtain

$$x_{2n+1} = 1 + \sum_{k=1}^{n} \frac{1}{2^{2k-1}} \qquad (3.14)$$

$$= 1 + 2 \sum_{k=1}^{n} \frac{1}{4^k}$$

$$= 1 + 2 \cdot \frac{1}{4} \frac{1-(1/4)^n}{1-(1/4)}$$

$$= 1 + \frac{2}{3}\left[1 - \left(\frac{1}{4}\right)^n\right] \to \frac{5}{3}.$$

Now equation (3.13) implies

$$x_{2n} = x_{2n+1} + \frac{1}{2^{2n-1}} \to \frac{5}{3}.$$

Using Example 3.5, we therefore conclude that $\lim x_n = 5/3$. But there is a more general approach we can take to arrive at this result, based on subsequences. That will be the subject of the following section (see also Exercise 3.4.10).

EXERCISES 3.4

1. Determine $\hat{A}$ in each of the following:

    (a) $A = \mathbb{N}$
    (b) $A = [0,1] \cup (3,4) \cup (5,6]$
    (c) $A = \left\{3^n + \frac{1}{k} : k, n \in \mathbb{N}\right\}$
    (d) $A = \mathbb{Q}\setminus\{1/n : n \in \mathbb{N}\}$.

2. If $A \subseteq B$ what is the relation between $\hat{A}$ and $\hat{B}$?

3. Give an example of $A \subseteq \mathbb{R}$ in each case:

    (a) A has exactly four cluster points in $\mathbb{R}$

(b) $\hat{A}$ is denumerable
(c) A has a denumerable set of isolated points.

4. Given $A, B \subseteq \mathbb{R}$, prove that
   (a) $\widehat{A \cup B} = \widehat{A \cup B}$,
   (b) $\widehat{A \cap B} \subseteq \hat{A} \cap \hat{B}$

   Give an example of two sets A and B such that $\hat{A} \cap \hat{B} \neq \widehat{A \cap B}$.

5. Suppose $A \subseteq \mathbb{R}$ is not empty and bounded above. If $\sup A \notin A$, prove that $\sup A \in \hat{A}$.

6. Give an example of a sequence of open intervals (I_n) such that $I_{n+1} \subseteq I_n$ and $\bigcap_{n=1}^{\infty} I_n = \varnothing$.

7. If $x_n = \sum_{i=1}^{n} \frac{1}{i^2}$, prove that (x_n) is a Cauchy sequence (and hence convergent).

8. If (x_n) and (y_n) are Cauchy sequences, prove that both $(x_n + y_n)$ and $(x_n y_n)$ are also Cauchy sequences.

9. Prove the relations (3.13) and (3.14) in Example 3.17.

10. Summing over n in (3.13), prove that
$$x_{n+1} = 1 + \sum_{k=0}^{n-1} \frac{(-1)^k}{2^k} \to \frac{5}{3}.$$

3.5 Subsequences

Definition 3.8 Let $(x_n : n \in \mathbb{N})$ be a sequence of real numbers. If (n_k) is a strictly increasing sequence of natural numbers,
$$n_1 < n_2 < n_3, \cdots,$$
then the sequence
$$(x_{n_k} : k \in \mathbb{N}) = (x_{n_1}, x_{n_2}, x_{n_3}, \cdots)$$
is called a *subsequence* of (x_n).

Thus, given a sequence, we obtain a subsequence by omitting some of the terms of the given sequence and re-indexing without disturbing the order of the original sequence. Here are some examples:

(i) Every tail of (x_n), such as $(x_4, x_5, x_6, \cdots)$, is a subsequence of (x_n). By setting $n_k = k+3$ it is clear that the tail in question can be expressed as the subsequence (x_{n_k}), or (x_{k+3}).

(ii) Given a sequence (x_n), the terms with odd subscripts arranged in increasing order, $x_1, x_3, x_5, \cdots$, form a subsequence of (x_n). Here $n_k = 2k - 1$. The terms with even subscripts, $x_2, x_4, x_6, \cdots$, form another subsequence (x_{2k}).

(iii) $(1, 1/4, 1/9, \cdots) = (1/k^2)$ is a subsequence of $(1/n)$, but $(1/2, 1, 1/3, 1/4, \cdots)$ is not a subsequence of $(1/n)$, nor is $(0, 1, 1/2, 1/3, \cdots)$.

Remarks 3.4

1. Recalling the definition of the real sequence (x_n) as a function

$$x : \mathbb{N} \to \mathbb{R}, \quad x(n) = x_n,$$

we can say that a subsequence (x_{n_k}) of (x_n) is a *composition* $x \circ g$ of the function x with a strictly increasing function $g : \mathbb{N} \to \mathbb{N}$. Here, of course, we have $g(k) = n_k$.

2. A subsequence of a subsequence of (x_n) is a subsequence of (x_n), since the composition of two strictly increasing functions is a strictly increasing function.

3. Note that the condition

$$n_1 < n_2 < n_3 < \cdots$$

implies that $n_k \geq k$ for all $k \in \mathbb{N}$.

The next theorem should come as no surprise.

Theorem 3.13
If the sequence (x_n) converges to x, then every subsequence of (x_n) also converges to x.

Proof. Let (x_{n_k}) be a subsequence of (x_n), where $\lim_{n \to \infty} x_n = x$. We wish to prove that $\lim_{k \to \infty} x_{n_k} = x$. Suppose ε is any positive number. From the convergence $x_n \to x$ there is an $N \in \mathbb{N}$ such that

$$n \geq N \implies |x_n - x| < \varepsilon.$$

But since $n_k \geq k$ for every $k \in \mathbb{N}$,
$$k \geq N \implies n_k \geq N \implies |x_{n_k} - x| < \varepsilon,$$
which means $x_{n_k} \to x$ as $k \to \infty$. $\square$

This theorem has a very useful corollary.

Corollary 3.13 *If the sequence (x_n) is convergent and has a subsequence which converges to x, then $\lim x_n = x$.*

Proof. Suppose $x_n \to y$ and $x_{n_k} \to x$. From Theorem 3.13 we know that $x_{n_k} \to y$, and from the uniqueness of the limit we must have $x = y$. $\square$

Since any sequence (x_n) can be considered a subsequence of itself, the converse of Theorem 3.13 is also true; that is, if every subsequence of (x_n) converges to x then $x_n \to x$. In fact, if we further assume that (x_n) is bounded, we obtain a stronger result, expressed by Theorem 3.15 below. We start by proving the following version of the Bolzano-Weierstrass theorem.

Theorem 3.14
Every bounded sequence has a convergent subsequence.

Proof. Let (x_n) be a bounded sequence. If the set $A = \{x_n : n \in \mathbb{N}\}$ is finite, then there is at least one term in the sequence which is repeated infinitely often. The subsequence defined by that term is a constant sequence which is therefore convergent. If A is infinite, then, by the Bolzano-Weierstrass theorem, it has at least one cluster point $x \in \hat{A}$. By a slight modification of part (i) of the proof of Theorem 3.10, we can construct a subsequence (x_{n_k}) which converges to x. $\square$

Theorem 3.15
Let (x_n) be a bounded sequence. If every convergent subsequence of (x_n) has the same limit, then (x_n) converges to that same limit.

Proof. Note that we are not assuming that all the subsequences of (x_n) converge. By Theorem 3.14 we know that (x_n) has at least one convergent subsequence. Call its limit x.

Assuming $x_n \nrightarrow x$, there is a positive number ε such that, for any $N \in \mathbb{N}$, $|x_n - x| \geq \varepsilon$ for some $n \geq N$. Let $N = 1$. Then there is an integer $n_1 \geq 1$ such that $|x_{n_1} - x| \geq \varepsilon$. Now, with $N = n_1 + 1$, there is an $n_2 \geq N > n_1$ such that $|x_{n_2} - x| \geq \varepsilon$. Continuing in this fashion we obtain, by induction on k, a subsequence (x_{n_k}) which satisfies

$$|x_{n_k} - x| \geq \varepsilon \quad \text{for all } k \in \mathbb{N}. \tag{3.15}$$

The sequence (x_{n_k}) is clearly bounded, hence, by Theorem 3.14, it has a convergent subsequence, call it $(x_{n_{k_j}})$. But $(x_{n_{k_j}})$ is also a subsequence of (x_n) and, being convergent, its limit must be x by hypothesis. Thus

$$\lim_{j \to \infty} x_{n_{k_j}} = x. \tag{3.16}$$

However, (3.15) implies

$$\left| x_{n_{k_j}} - x \right| \geq \varepsilon \text{ for all } j \in \mathbb{N},$$

which is in contradiction with (3.16). The assumption $x_n \not\to x$ is therefore false and $x_n \to x$. $\square$

EXERCISES 3.5

1. Give an example of

   (a) a sequence with no convergent subsequence,
   (b) an unbounded sequence which has a convergent subsequence.

2. If every subsequence of (x_n) has a subsequence which converges to 0, prove that $\lim x_n = 0$.

3. Let $x_n \geq 0$ for all $n \in \mathbb{N}$. If $((-1)^n x_n)$ is convergent, prove that (x_n) is convergent, and determine its limit.

4. Suppose (x_n) is a bounded sequence of distinct terms, i.e., $x_n \neq x_m$ for all $n \neq m$. If the set $\{x_n : n \in \mathbb{N}\}$ has a single cluster point x, prove that $x_n \to x$.

5. In the last two sections of this chapter we could have followed a different procedure by first proving Theorem 3.14, then proving Cauchy's criterion and the Bolzano-Weierstrass theorem. This may be achieved by the following steps:

   (a) For any bounded sequence (x_n), we call $k \in \mathbb{N}$ a *summit point* if $x_k \geq x_n$ for all $n \geq k$. If the set of summit points of (x_n) is finite, show how you can choose an increasing subsequence of (x_n). If, on the other hand, the set of summit points is infinite, show that you can form a decreasing subsequence. From this you can deduce the existence of a convergent subsequence of (x_n), thereby proving Theorem 3.14.

(b) If (x_n) is a Cauchy sequence which has a convergent subsequence, prove that (x_n) converges to the same limit.

(c) Using (a) and (b), together with Lemma 3.1, prove Cauchy's criterion.

(d) Use Theorems 3.10 and 3.14 to prove the Bolzano-Weierstrass theorem.

3.6 Upper and Lower Limits

In this section we define two, more general, limits which exist for any bounded sequence. Let (x_n) be a bounded sequence. We define two sequences (y_n) and (z_n) as follows:

$$y_n = \sup\{x_k : k \geq n\} \tag{3.17}$$
$$z_n = \inf\{x_k : k \geq n\}. \tag{3.18}$$

Since
$$y_1 = \sup\{x_k : k \geq 1\} \geq y_2 = \sup\{x_k : k \geq 2\} \geq \cdots$$
is a decreasing sequence which is bounded below, by $\inf x_n$, it must be convergent. Its limit is called the *limit superior* (or the *upper limit*) of (x_n), and is denoted by $\limsup x_n$ (or $\overline{\lim} \, x_n$). Thus, by Theorem 3.7,

$$\limsup x_n = \lim y_n = \inf\{y_n : n \in \mathbb{N}\} = \inf_{n \in \mathbb{N}} \{\sup x_k : k \geq n\}.$$

Likewise (z_n) is an increasing sequence which is bounded above by $\sup x_n$. Its limit exists and is called the *limit inferior* (or *lower limit*) of x_n, and is denoted by $\liminf x_n$ (or $\underline{\lim} \, x_n$). Hence

$$\liminf x_n = \lim z_n = \sup\{z : n \in \mathbb{N}\} = \sup_{n \in \mathbb{N}} \{\inf x_k : k \geq n\}.$$

In the extended real number system, if (x_n) is not bounded above, then we write $\limsup x_n = +\infty$; and if (x_n) is not bounded below, then $\liminf x_n = -\infty$.

Observe that

$$\limsup x_n \geq \liminf x_n,$$
$$\limsup(-x_n) = -\liminf x_n.$$

Example 3.18 We know that the sequence

$$x_n = (-1)^n + \frac{1}{n}$$

is divergent because $(-1)^n$ is divergent while $1/n$ is convergent. The upper and lower limits are, respectively,

$$\limsup x_n = \lim_{n \to \infty} \left(1 + \frac{1}{n}\right) = 1,$$

$$\liminf x_n = \lim_{n \to \infty} (-1) = -1.$$

The following theorem provides some useful properties of the upper and lower limits, specifically those which we shall have occasion to use in the coming chapters. Let x^* and x_* denote $\limsup x_n$ and $\liminf x_n$ respectively.

Theorem 3.16
(i) Given any $\varepsilon > 0$, there is a positive integer N such that

$$x_n < x^* + \varepsilon \quad \text{for all } n \geq N,$$

that is, $x_n \in (-\infty, x^* + \varepsilon)$ for all but a finite number of $n \in \mathbb{N}$.
(ii) Given any $\varepsilon > 0$ and any $m \in \mathbb{N}$, there is an integer $n > m$ such that $x_n > x^* - \varepsilon$. In other words, $x_n \in (x^* - \varepsilon, \infty)$ for infinitely many $n \in \mathbb{N}$.
(iii) (x_n) converges to x if and only if $x^* = x_* = x$.
(iv) There is a subsequence of (x_n) which converges to x^*, and a subsequence of (x_n) which converges to x_*.
(v) If a subsequence of (x_n) converges to x, then $x_* \leq x \leq x^*$; that is, x^* is the greatest limit that can be attained by a convergent subsequence of (x_n), and x_* is the least such limit.

Proof
(i) Using the relation $x^* = \lim y_n$, where y_n is defined in equation (3.17), there is an integer N such that

$$n \geq N \Rightarrow x^* - \varepsilon < y_n < x^* + \varepsilon.$$

But $y_n \geq x_n$ for all $n \in \mathbb{N}$. Therefore

$$n \geq N \Rightarrow x_n < x^* + \varepsilon.$$

(ii) Suppose $m \in \mathbb{N}$. If $x_n \leq x^* - \varepsilon$ for all $n > m$, then, for all $k \geq n > m$, $x_k \leq x^* - \varepsilon$, so that

$$y_n \leq x^* - \varepsilon \quad \text{for all } n > m.$$

In the limit as $n \to \infty$, we obtain $x^* \leq x^* - \varepsilon$, a contradiction.

(iii) Applying property (i) to the sequence (x_n), we conclude that $(-\infty, x^* + \varepsilon)$ contains all but a finite number of terms of (x_n). The same property applied to $(-x_n)$ implies that $(x_* - \varepsilon, \infty)$ contains all but a finite number of such terms. If $x^* = x_* = x$ then $(x - \varepsilon, x + \varepsilon)$ contains all but a finite number of terms of (x_n). But this is just the assertion that $x_n \to x$.

Now suppose that $x_n \to x$. For any $\varepsilon > 0$, there is an integer N such that

$$n \geq N \Rightarrow x_n < x + \varepsilon$$
$$\Rightarrow y_n \leq x + \varepsilon.$$

Hence $x^* \leq x + \varepsilon$. Since $\varepsilon > 0$ is arbitrary, we obtain

$$x^* \leq x.$$

Working with the sequence $(-x_n)$, whose limit is $-x$, the same procedure gives

$$(-x_n)^* \leq (-x),$$

or $x_* \geq x$. Since $x_* \leq x^*$, we must conclude that $x^* = x_* = x$.

(iv) With $\varepsilon = 1$ there is an integer N_1 such that $x_n \in (-\infty, x^* + 1)$ for all $n \geq N_1$. Since $(x^* - 1, \infty)$ contains an infinite number of the terms x_n, we can choose $n_1 > N_1$ such that $x_{n_1} \in (x^* - 1, \infty)$. It then follows that $x_{n_1} \in (x^* - 1, x^* + 1)$. If $\varepsilon = 1/2$, we can similarly find $n_2 > n_1$ such that $x_{n_2} \in (x^* - 1/2, x^* + 1/2)$. Suppose now that $n_1 < n_2 < \cdots < n_k$ have been chosen so that

$$x_{n_j} \in (x^* - 1/j, x^* + 1/j), \ 1 \leq j \leq k.$$

Since $(-\infty, x^* + (k+1)^{-1})$ contains all but a finite number of elements of (x_n), and $(x^* - (k+1)^{-1}, \infty)$ contains an infinite number of those elements, there is an $n_{k+1} > n_k$ such that

$$x_{n_{k+1}} \in (x^* - (k+1)^{-1}, x^* + (k+1)^{-1}).$$

By induction on k we thus form a subsequence (x_{n_k}) such that $|x_{n_k} - x^*| < 1/k$. Such a subsequence clearly converges to x^*.

By using the sequence $(-x_n)$ we can follow a similar procedure to construct a subsequence of (x_n) which converges to x_*.

(v) Suppose (x_{n_k}) is a subsequence of (x_n) which converges to x. Since $n_k \geq k$ for all $k \in \mathbb{N}$, we must have

$$y_k = \sup\{x_j : j \geq k\} \geq x_{n_k} \quad \text{for all } k \in \mathbb{N}.$$

Letting $k \to \infty$, we obtain $x^* \geq x$. The inequality $x_* \leq x$ is obtained by a similar procedure applied to the sequence $(-x_n)$. □

Exercises 3.6

1. If $x_n \leq y_n$ for all $n \in \mathbb{N}$, prove that

$$\limsup x_n \leq \limsup y_n$$
$$\liminf x_n \leq \liminf y_n$$
$$\liminf x_n \leq \limsup y_n.$$

Give an example where $\limsup x_n > \liminf y_n$.

2. Let (x_n) be a sequence. For any $\varepsilon > 0$, prove that

   (a) $x_n \in (\liminf x_n - \varepsilon, \infty)$ for all but a finite number of terms of the sequence,

   (b) $x_n \in (-\infty, \liminf x_n + \varepsilon)$ for an infinite number of terms of the sequence.

3. Determine the upper and lower limits for each of the following sequences:

   (a) $(-1)^n n$.
   (b) $(1, 1/2, 1, 1/3, 1, 1/4, \cdots)$.
   (c) $(0, 1, 0, 1, 2, 0, 1, 2, 3, 0, 1, 2, 3, 4, \cdots)$.

4. For any two sequences (x_n) and (y_n), prove that $\limsup(x_n + y_n) \leq \limsup x_n + \limsup y_n$ and $\liminf(x_n + y_n) \geq \liminf x_n + \liminf y_n$.

5. For any sequence (x_n) and $c \geq 0$, prove that
$$\limsup cx_n = c \limsup x_n$$
and $\liminf cx_n = c \liminf x_n$.

6. Given a positive sequence (x_n), what is the relation of $\limsup x_n$ and $\liminf x_n$ to $\limsup(1/x_n)$ and $\liminf(1/x_n)$?

3.7 Open and Closed Sets

This section introduces some topological notions in $\mathbb{R}$ which will further clarify its properties.

Definition 3.9
A set $A \subseteq \mathbb{R}$ is called *open* if A is a neighborhood of every point in A, that is, if for every $x \in A$ there is a positive number ε such that $(x - \varepsilon, x + \varepsilon) \subseteq A$.

Consider the following examples:

(a) Every open interval is an open set (Exercise 3.1.7).

(b) For any real number x the set $\mathbb{R}\setminus\{x\}$ is open (see Exercise 3.1.8).

(c) $[a, b)$ is not open since it is not a neighborhood of a.

(d) $\mathbb{Z}$ is not open (why?).

(e) $\mathbb{Q}$ is not open (why?).

Theorem 3.17
(i) $\mathbb{R}$ and $\varnothing$ are both open.
(ii) Any union of open sets in $\mathbb{R}$ is open.
(iii) Any finite intersection of open sets in $\mathbb{R}$ is open.

Proof

(i) $\mathbb{R}$ is clearly open. Since $\varnothing$ has no elements, Definition 3.9 is automatically satisfied.

(ii) Let G_λ be an open set for every $\lambda \in \Lambda$, where Λ is an arbitrary index set. If $x \in \bigcup_{\lambda \in \Lambda} G_\lambda$ then there is a $\lambda_0 \in \Lambda$ such that $x \in G_{\lambda_0}$. Since G_{λ_0} is open, there is a number $\varepsilon > 0$ such that
$$(x - \varepsilon, x + \varepsilon) \subseteq G_{\lambda_0}.$$
This implies $(x - \varepsilon, x + \varepsilon) \subseteq \bigcup_{\lambda \in \Lambda} G_\lambda$, hence the union $\bigcup_{\lambda \in \Lambda} G_\lambda$ is open.

(iii) Suppose $x \in \bigcap_{i=1}^{n} G_i$. This implies $x \in G_i$ for every $1 \leq i \leq n$. Since G_i is open, there is an $\varepsilon_i > 0$ such that
$$(x - \varepsilon_i, x + \varepsilon_i) \subseteq G_i.$$

Let $\varepsilon = \min\{\varepsilon_1, \varepsilon_2, \cdots, \varepsilon_n\}$. ε is clearly a positive number and
$$(x - \varepsilon, x + \varepsilon) \subseteq (x - \varepsilon_i, x + \varepsilon_i) \subseteq G_i \text{ for all } i = 1, 2, \cdots, n.$$
$$\Rightarrow (x - \varepsilon, x + \varepsilon) \subseteq \bigcap_{i=1}^{n} G_i.$$

Thus $\bigcap_{i=1}^{n} G_i$ is an open set. □

Remarks 3.5

1. Theorem 3.17 states that the open sets in $\mathbb{R}$ are closed under the operations of arbitrary unions and finite intersections, and that they include the universal set $\mathbb{R}$ and the empty set $\varnothing$. A collection of such sets is called a *topology* on $\mathbb{R}$. A topology on a set is used to define a system of neighborhoods in that set.

2. The set $G_n = (-1/n, 1/n)$ is clearly open for every $n \in \mathbb{N}$. Yet $\bigcap_{n=1}^{\infty} G_n = \{0\}$ is not open, hence an arbitrary intersection of open sets is not necessarily open.

The next theorem gives a useful characterization of an open set.

Theorem 3.18
A subset of $\mathbb{R}$ is open if, and only if, it is the union of a countable number of disjoint open intervals.

Proof. Let G be any set of real numbers. If G is a countable union of open intervals, then, by Theorem 3.16, it is open.

Now assume G is open. For each point $x \in G$, let $\mathcal{I}_x$ be the set of open intervals which contain x and which lie in G, that is,
$$\mathcal{I}_x = \{I : x \in I \subseteq G\}.$$

Since G is open $\mathcal{I}_x$ is not empty. Let
$$J_x = \cup\{I : I \in \mathcal{I}_x\},$$

and note that
(i) $x \in J_x \subseteq G$, hence $G = \bigcup_{x \in G} J_x$.

(ii) J_x is an open interval. More specifically, if A_x is the set of left end-points of the intervals in $\mathcal{I}_x$ and B_x is the set of right end-points, then it can be shown that $J_x = (a_x, b_x)$, where

$$a_x = \inf A_x, \ b_x = \sup B_x.$$

(iii) For any pair of points $x, y \in G$, either $J_x = J_y$ or $J_x \cap J_y = \varnothing$. To see that, we shall assume that $J_x \cap J_y \neq \varnothing$ and show that $J_x = J_y$. Since the intervals J_x and J_y intersect, their union $J_x \cup J_y$ is an open interval which contains y, hence it belongs to $\mathcal{I}_y$. Thus

$$J_x \subseteq J_x \cup J_y \subseteq J_y.$$

Similarly, $J_y \subseteq J_x$, and we conclude that $J_x = J_y$.
(iv) To show that the set $\{J_x : x \in G\}$ is countable, we use the density of $\mathbb{Q}$ in $\mathbb{R}$ to choose a rational number q_x in every J_x, and note that the function $J_x \mapsto q_x$ is injective, by (iii). Since $\mathbb{Q}$ is countable so is $\{J_x : x \in G\}$. □

Definition 3.10 A set $A \subseteq \mathbb{R}$ is *closed* if its complement A^c is open.

Here are some examples:

(a) The intervals $[a, b]$, $[a, \infty)$, $(-\infty, b]$ are all closed.

(b) For any real number x, the set $\{x\}$ is closed.

(c) $[a, b)$ is not closed.

(d) $\mathbb{Q}$ is not closed.

(e) $\mathbb{Z}$ is closed.

The next theorem is a direct consequence of Theorem 3.17 and De Morgan's laws.

Theorem 3.19
(i) $\mathbb{R}$ and $\varnothing$ are closed.
(ii) Any finite union of closed sets in $\mathbb{R}$ is closed.
(iii) Any intersection of closed sets in $\mathbb{R}$ is closed.

Theorem 3.20
For any set $F \subseteq \mathbb{R}$, the following statements are equivalent:
(i) F is closed.
(ii) $\hat{F} \subseteq F$.
(iii) If (x_n) is a convergent sequence of elements in F, then $\lim x_n \in F$.

Proof

(i)⇒(ii): Suppose F is closed. Let $x \notin F$. F^c being open, there is an $\varepsilon > 0$ such that $(x - \varepsilon, x + \varepsilon) \subseteq F^c$. Hence the interval $(x - \varepsilon, x + \varepsilon)$ contains no points of F, so $x \notin \hat{F}$. Hence $F^c \subseteq \hat{F}^c$, or $\hat{F} \subseteq F$.

(ii)⇒(iii): Suppose $\hat{F} \subseteq F$ and let $x_n \in F$ for every $n \in \mathbb{N}$, with $x_n \to x$. If $\{x_n : n \in \mathbb{N}\}$ is finite, it has a constant subsequence (x), and x is clearly in F. If, on the other hand, $\{x_n : n \in \mathbb{N}\}$ is infinite then, by Theorem 3.10, $x \in \hat{F}$ and hence $x \in F$.

(iii)⇒(i): Suppose F includes the limits of all its convergent sequences. If F is not closed then F^c is not open, hence there is a point $x \notin F$ such that
$$(x - \varepsilon, x + \varepsilon) \cap F \neq \varnothing \quad \text{for all } \varepsilon > 0.$$
Hence, for every $\varepsilon_n = 1/n$, $n \in \mathbb{N}$, there is an $x_n \in F$ such that $|x_n - x| < 1/n$. The sequence (x_n) is therefore in F but its limit x is not in F, in contradiction to our hypothesis. Thus F is closed. □

In view of Theorem 3.18, the definitions of open and closed sets imply that, in order to obtain all the closed sets in $\mathbb{R}$, it suffices to remove from $\mathbb{R}$ a countable collection of disjoint, open intervals. In the following example we do something of this sort, and we obtain a set with surprising properties, known as the *Cantor ternary set*.

Example 3.19 Let $F_0 = [0, 1]$, and remove from F_0 its middle third $(1/3, 2/3)$ to obtain
$$F_1 = \left[0, \frac{1}{3}\right] \cup \left[\frac{2}{3}, 1\right].$$
If we remove the middle third of each sub-interval, we obtain
$$F_2 = \left[0, \frac{1}{9}\right] \cup \left[\frac{2}{9}, \frac{1}{3}\right] \cup \left[\frac{2}{3}, \frac{7}{9}\right] \cup \left[\frac{8}{9}, 1\right].$$
Continuing in this fashion, we obtain a sequence of closed intervals $(F_0, F_1, F_3, \cdots)$. Now we define *Cantor's set* as the intersection
$$F = \bigcap_{n=0}^{\infty} F_n.$$

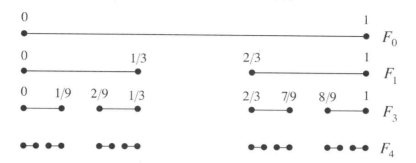

Figure 3.5

Note the following observations:

1. F is closed since every F_n is closed (Theorem 3.18).
2. It is clear that F includes the end-points $0, 1/3, 2/3, 1/9, 2/9, \cdots$ of the deleted intervals, but the question is: does it contain anything else? The surprising answer is that, not only does F contain many more points, but it is in fact equivalent to $[0, 1]$ and is therefore not countable. To prove that, we recall from section 2.5 that any number $x \in [0, 1]$ has a ternary expansion

$$x = 0.t_1 t_2 t_3 \cdots$$

which may also be expressed as

$$x = \sum_{i=1}^{\infty} \frac{t_i}{3^i}, \quad t_i \in \{0, 1, 2\}.$$

The number $x \in [0, 1]$ also has a binary expansion given by

$$x = 0.s_1 s_2 s_3 \cdots = \sum_{i=1}^{\infty} \frac{s_i}{2^i}, \quad s_i \in \{0, 1\}.$$

The Cantor set F is in fact the set of points for which a ternary expansion can be chosen so as not to contain the number 1, that is,

$$x = \sum_{i=1}^{\infty} \frac{t_i}{3^i} \in F \Leftrightarrow t_i \in \{0, 2\} \text{ for all } i \in \mathbb{N}.$$

Now we define the function

$$\Psi : [0, 1] \to F$$

as follows
$$\Psi\left(\sum_{i=1}^{\infty}\frac{s_i}{2^i}\right) = \sum_{i=1}^{\infty}\frac{2s_i}{3^i}.$$
Noting that Ψ is injective, we conclude that $F \sim [0,1]$.

3. The total length of the deleted intervals is given by
$$\begin{aligned}L &= \frac{1}{3} + \frac{2}{3^2} + \frac{2^2}{3^3} + \frac{2^3}{3^4} + \cdots \\ &= \frac{1}{3}\left[1 + \frac{2}{3} + \left(\frac{2}{3}\right)^2 + \left(\frac{2}{3}\right)^3 + \cdots\right] \\ &= \frac{1}{3}\frac{1}{1-2/3} \\ &= 1.\end{aligned}$$

Since the interval $[0,1]$ is also of length 1, this means that the set F has "length" 0, which would seem to to be inconsistent with the equivalence of F and $[0,1]$. This is another example of the pitfalls of accepting intuitive ideas as facts. We shall have occasion to return to the Cantor set later in the book, when we discuss *measure theory*.

EXERCISES 3.7

1. Prove the statements (a) to (e) following Definition 3.10.

2. If $A \subseteq \mathbb{R}$ is open and bounded, prove that $\sup A \notin A$ and $\inf A \notin A$.

3. If $A \subseteq \mathbb{R}$ is closed and bounded, prove that $\sup A \in A$ and $\inf A \in A$.

4. Give an example of a collection of closed sets whose union is not closed.

5. Prove that $A \subseteq \mathbb{R}$ is closed if, and only if, every Cauchy sequence in A has a limit in A.

6. Let $A \subseteq B \subseteq \mathbb{R}$. A is said to be *open in* B if, for every $x \in B$, there is a positive number ε such that
$$(x-\varepsilon, x+\varepsilon) \cap B \subseteq A.$$

(a) Prove that A is open in B if, and only if, there is an open set G such that $A = G \cap B$.

Give appropriate definitions of the following statements:

   i. A is a neighborhood of x in B.
   ii. A is closed in B.

7. For any subset A of $\mathbb{R}$, we define the *interior* of A, denoted by $A°$, as the union of all open subsets of A. Clearly $A°$ is the "largest" open set in A. The elements of $A°$ are called the *interior points* of A. Prove the following statements:

   (a) Every interior point of A has a neighborhood contained in A.
   (b) A is open if and only if $A = A°$.
   (c) $(A \cup B)° \supseteq A° \cup B°$, $(A \cap B)° = A° \cap B°$.

   Give an example of two sets A and B where $(A \cup B)° \neq A° \cup B°$.

8. The *closure* of any set $A \subseteq \mathbb{R}$, denoted by $\bar{A}$, is the intersection of all closed sets which contain A. Thus $\bar{A}$ is the "smallest" closed set which contains A. Prove the following statements:

   (a) A is closed if and only if $A = \bar{A}$.
   (b) $x \in \bar{A}$ if and only if $(x - \varepsilon, x + \varepsilon) \cap A \neq \emptyset$ for any $\varepsilon > 0$.
   (c) If $x \in \bar{A}$, then there is a sequence of elements of A which converges to x.
   (d) $\overline{A \cap B} \subseteq \bar{A} \cap \bar{B}$, $\overline{A \cup B} = \bar{A} \cup \bar{B}$.

   Give an example of two sets A and B where $\overline{A \cap B} \neq \bar{A} \cap \bar{B}$.
   (e) $\bar{A} = A \cup \hat{A}$.

9. The *boundary* of the set A, denoted by ∂A, is defined as
$$\partial A = \bar{A} \cap \overline{A^c}.$$
Prove that $A° \cap \partial A = \emptyset$ and $\bar{A} = A° \cup \partial A$. Consequently, a closed set contains its boundary points, while an open set contains none of its boundary points.

10. Prove that
$$\bar{A} = [(A^c)°]^c,$$
$$A° = (\overline{A^c})^c.$$

11. Give an example of sets A and B such that $\overline{A^\circ} \neq \bar{A}$, and $(\bar{B})^\circ \neq B$.

12. Determine the interior, closure, and boundary of each of the following sets:

    (a) $\mathbb{N}$
    (b) $\mathbb{Z}$
    (c) $\mathbb{R}$
    (d) $\mathbb{R}\setminus\{0\}$
    (e) $[0, \infty)$
    (f) $\mathbb{Q}$
    (g) $\mathbb{Q}^c$
    (h) $[a, b)$
    (i) $(2, 3] \cup (4, 5)$
    (j) $\{1/n : n \in \mathbb{N}\}$.

13. Which of the sets in Exercise 3.7.12 is open and which are closed?

14. What is the interior and the boundary of the Cantor set?

4

Infinite Series

An infinite series is a special kind of sequence. Its importance in this study will become more apparent in the last four chapters of the book, where we introduce the Riemann integral, analytic functions, and Lebesgue's theory of measure and integration. An infinite series, being a sequence, is subject to the convergence properties of sequences, as presented in chapter 3. This chapter is concerned with deriving the corresponding convergence properties of infinite series, including the so-called *convergence tests*.

4.1 Basic Properties

Suppose we have a sequence (x_n) of real numbers. We can then form another sequence

$$S_1 = x_1$$
$$S_2 = x_1 + x_2$$
$$S_3 = x_1 + x_2 + x_3$$
$$\vdots$$
$$S_n = x_1 + x_2 + \cdots + x_n = \sum_{k=1}^{n} x_k,$$

called the *sequence of partial sums* of (x_n). The sequence (S_n) is the *infinite series* generated by (x_n). If (S_n) is convergent, the infinite series is said to be *convergent*, otherwise it is *divergent*. When the series is convergent, $\lim S_n$ is called the *sum* of the series (S_n). We shall use the symbol

$$\sum_{n=1}^{\infty} x_n = x_1 + x_2 + x_3 + \cdots$$

to denote this sum. The real numbers x_n are called the *terms* of the series. The first term in the series need not be x_1, but may be x_0 or x_m, for some integer m, in which case we write $\sum_{n=0}^{\infty} x_n$ or $\sum_{n=m}^{\infty} x_n$, and the convergence is clearly not affected by the choice of m. Hence, for the sake of convenience, we shall use the notation $\sum x_n$ to refer to the infinite series, especially when the initial value of the index n is understood or is irrelevant. But, when there is no likelihood of confusion, a convergent series (S_n) will also be denoted by its sum $\sum_{n=m}^{\infty} x_n = \lim S_n$. This kind of flexibility in our use of notation is not uncommon, for we occasionally denote the function $x \mapsto f(x)$ by $f(x)$, the *value of* f at x, rather than by f.

Example 4.1 A typical example of an infinite series is the so-called *geometric series* $\sum_{n=0}^{\infty} a^n$, where a is a real number. Here we have $x_n = a^n$, and the sequence of partial sums is

$$S_n = 1 + a + a^2 + \cdots + a^n = \begin{cases} \dfrac{1-a^{n+1}}{1-a}, & a \neq 1 \\ n+1, & a = 1. \end{cases}$$

If $a = 1$ then $S_n = n+1$, which diverges (Example 3.4). If $a \neq 1$ then $\lim S_n$ exists if and only if $\lim a^n$ exists, that is if and only if $|a| < 1$ (see Exercise 3.2.12), in which case $\lim a^n = 0$, and

$$\sum_{k=0}^{\infty} a^k = \lim S_n = \frac{1}{1-a}.$$

Now we use the convergence properties of sequences, as presented in section 3.2, to deduce corresponding results for series.

Theorem 4.1
(i) If the series $\sum x_n$ and $\sum y_n$ are both convergent, then the series $\sum (x_n + y_n)$ is also convergent, and

$$\sum_{n=1}^{\infty} (x_n + y_n) = \sum_{n=1}^{\infty} x_n + \sum_{n=1}^{\infty} y_n.$$

(ii) If the series $\sum x_n$ is convergent and c is any real number, then $\sum c x_n$ is convergent, and

$$\sum_{n=1}^{\infty} c x_n = c \sum_{n=1}^{\infty} x_n.$$

Proof. This is a direct application of Theorem 3.4.

Remark 4.1 From Theorem 3.4 we can also conclude that the convergence of $\sum x_n$ and $\sum y_n$ implies the existence of the product $\sum x_n \sum y_n$, but it does not imply the convergence of $\sum x_n y_n$, and we shall presently give examples to illustrate this point.

The next theorem is known as the *Cauchy criterion for series.*

Theorem 4.2 (Cauchy's Criterion)
The series $\sum x_n$ is convergent if and only if, for every $\varepsilon > 0$ there is a positive integer $N = N(\varepsilon)$ such that

$$n > m \geq N \Rightarrow |S_n - S_m| = |x_{m+1} + \cdots + x_n| < \varepsilon. \qquad (4.1)$$

Proof. This is a direct application of the Cauchy criterion to the sequence $S_n = \sum_{k=1}^{n} x_k$.

Corollary 4.2 *If the series $\sum x_n$ is convergent, then $\lim x_n = 0$.*

Proof. Setting $n = m + 1$ in (4.1) yields the desired result. □

According to Corollary 4.2, $\lim x_n = 0$ is a necessary condition for the convergence of $\sum x_n$, but it is not a sufficient one, as the following example indicates.

Example 4.2 Although $\lim 1/\sqrt{n} = 0$ we shall now show that the series $\sum 1/\sqrt{n}$ is divergent.

$$S_n = 1 + \frac{1}{\sqrt{2}} + \cdots + \frac{1}{\sqrt{n}}$$
$$\geq \frac{1}{\sqrt{n}} + \frac{1}{\sqrt{n}} + \cdots + \frac{1}{\sqrt{n}}$$
$$= \frac{n}{\sqrt{n}}$$
$$= \sqrt{n} \to \infty.$$

Thus Corollary 4.2 is really a divergence test, rather than a convergence test, since the fact that $\lim x_n = 0$ does not tell us anything about the convergence or divergence of the series $\sum x_n$. But if $\lim x_n \neq 0$, then we conclude that the series is divergent.

Example 4.3 The series $\sum 1/n$ is known as the *harmonic series*, and it also satisfies the condition $\lim 1/n = 0$. Here we have

$$S_{2n} - S_n = \sum_{k=n+1}^{2n} \frac{1}{k}$$
$$= \frac{1}{n+1} + \frac{1}{n+2} + \cdots + \frac{1}{2n}$$
$$\geq \frac{1}{2n} + \frac{1}{2n} + \cdots + \frac{1}{2n}$$
$$= \frac{n}{2n} = \frac{1}{2},$$

which implies that the sequence (S_n) is not a Cauchy sequence and, by the Cauchy criterion, cannot converge. Since the sequence is increasing, we must conclude that
$$\sum_{n=1}^{\infty} \frac{1}{n} = \infty.$$

Needless to say, a series may diverge in other ways, such as $\sum(-1)^n$, whose n-th sum
$$S_n = \sum_{k=1}^{n}(-1)^k = -1 + 1 - 1 + \cdots + (-1)^n$$
is -1 if n is odd, and 0 if n is even, and hence approaches no limit as $n \to \infty$.

Definition 4.1 The series $\sum x_n$ is said to be *absolutely convergent* if the series $\sum |x_n|$ is convergent.

Theorem 4.3
If a series is absolutely convergent, then it is convergent.

Proof. Suppose $\sum |x_n|$ converges. We shall use Cauchy's criterion to show that $\sum x_n$ converges. Let ε be an arbitrary positive number. By Theorem 4.2 there a positive integer N such that
$$n > m \geq N \Rightarrow |x_{m+1}| + \cdots + |x_n| < \varepsilon$$
$$\Rightarrow |x_{m+1} + \cdots + x_n| < \varepsilon,$$
which means $\sum x_n$ satisfies the Cauchy criterion for series, and is therefore convergent. □

If the terms of a series do not change sign, then, clearly, there is no difference between convergence and absolute convergence. The difference appears when a subsequence of the terms is positive and another subsequence is negative. Here is a well known example of a convergent series which is not absolutely convergent.

Example 4.4 The series $\sum_{n=1}^{\infty}(-1)^{n+1}/n$ is not absolutely convergent, as we found in Example 4.3. Now we shall prove that this series is convergent. First let n be an even integer, say $n = 2m$. The sequence
$$S_{2m} = \sum_{k=1}^{2m} \frac{(-1)^{k+1}}{k}$$
$$= \left(1 - \frac{1}{2}\right) + \left(\frac{1}{3} - \frac{1}{4}\right) + \cdots + \left(\frac{1}{2m-1} - \frac{1}{2m}\right)$$
$$= \frac{1}{2 \cdot 1} + \frac{1}{4 \cdot 3} + \frac{1}{6 \cdot 5} + \cdots + \frac{1}{2m(2m-1)}$$

is positive and increasing with m, and since

$$S_{2m} = 1 - \left(\frac{1}{2} - \frac{1}{3}\right) - \left(\frac{1}{4} - \frac{1}{5}\right) - \cdots - \frac{1}{2m} < 1,$$

it is bounded above. Hence the sequence (S_{2m}) converges to some limit S. Now let $n = 2m+1$ be an odd integer. Since

$$S_{2m+1} = S_{2m} + \frac{1}{2m+1}$$

also converges to S, we can use the result of Example 3.5 to conclude that $\lim S_n$ exists.

Definition 4.2 The series $\sum x_n$ is said to be *conditionally convergent* if $\sum x_n$ is convergent and $\sum |x_n|$ is divergent.

Thus the series given in Example 4.4 is conditionally convergent.

Definition 4.3 The series $\sum y_n$ is called a *rearrangement* of the series $\sum x_n$ if there is a bijection $f : \mathbb{N} \to \mathbb{N}$ such $y_n = x_{f(n)}$ for all $n \in \mathbb{N}$.

Example 4.5 The series

$$1 - \frac{1}{2} + \frac{1}{3} - \frac{1}{4} + \frac{1}{5} - \frac{1}{6} + \cdots$$

of Example 4.4 may be rearranged in a number of ways, such as

$$1 + \frac{1}{3} - \frac{1}{2} + \frac{1}{5} + \frac{1}{7} - \frac{1}{4} + \frac{1}{9} + \frac{1}{11} - \frac{1}{6} + \cdots,$$

or

$$1 - \frac{1}{2} - \frac{1}{4} + \frac{1}{3} - \frac{1}{6} - \frac{1}{8} + \frac{1}{5} - \frac{1}{10} - \frac{1}{12} + \frac{1}{7} - \cdots,$$

or any other series in which the terms $1, -1/2, 1/3, -1/4, \cdots$ are added in a certain order.

Rearranging a convergent series would be expected to affect its convergence properties. In what way this happens will depend on the type of convergence of the original series, as we shall now see.

If the series $\sum x_n$ is conditionally convergent, then its positive terms form an unending subsequence of (x_n), otherwise the series would converge absolutely. For the same reason, the negative terms form an infinite subsequence of (x_n). Suppose (y_n) is the subsequence of positive terms and (z_n) is the subsequence of negative terms. It is not difficult to prove that the series $\sum y_n$ and $\sum z_n$ are both divergent (Exercise 4.1.5). We shall now see how a rearrangement of $\sum x_n$ can produce a divergent series.

Since $\sum_{k=1}^{n} y_k \to \infty$ there is a positive integer m_1 such that

$$T_1 = \sum_{k=1}^{m_1} y_k + z_1 \geq 1.$$

Similarly, there is an integer $m_2 > m_1$ such that

$$T_2 = T_1 + \sum_{m_1+1}^{m_2} y_k + z_2 = \sum_{k=1}^{m_2} y_k + z_1 + z_2 \geq 2,$$

and so on, for every $n \in \mathbb{N}$ there is an $m_n \in \mathbb{N}$ such that

$$T_n = \sum_{k=1}^{m_n} y_k + \sum_{k=1}^{n} z_k \geq n.$$

Clearly, $T_n \to \infty$, which means the rearrangement of $\sum x_n$ defined by the sequence

$$(y_1, \cdots, y_{m_1}, z_1, y_{m_1+1}, \cdots, y_{m_2}, z_2, \cdots)$$

diverges to ∞.

There is, in fact, a theorem (due to Riemann), which states that, for any real number c, there is a rearrangement of $\sum x_n$ which converges to c. The proof of this theorem relies on the observation that we can form a finite sum of the positive terms y_k (in the same order in which they appear in (x_n)) to obtain the smallest sum $Y_1 = \sum_{k=1}^{n_1} y_k$ which exceeds c, then add the least number of negative terms z_k (again in the order in which they appear in (x_n)) to obtain the largest sum $Y_2 = \sum_{k=1}^{n_2} y_k + \sum_{k=1}^{n_2} z_k$ which is less than c. Proceeding in this fashion, we obtain a sequence $Y_1, Y_2, Y_3, \cdots$ which converges to c, since $\lim y_n = \lim z_n = 0$. Thus we have proved

Theorem 4.4
Given any conditionally convergent series and any $c \in \bar{\mathbb{R}}$, there is a rearrangement of the series which converges to c.

An absolutely convergent series, on the other hand, is not affected by any rearrangement as the next theorem shows. Note, in this connection, that a series is also a rearrangement of itself in which the bijection in Definition 4.3 is $f(n) = n$.

Theorem 4.5

If $\sum x_n$ is absolutely convergent, then all its rearrangements also converge absolutely to the same limit.

Proof. Let $\sum x_n$ be absolutely convergent, and suppose $\sum y_n$ is a rearrangement of $\sum x_n$. Define

$$S_n = \sum_{k=1}^{n} x_n, \; S = \lim S_n, \; T_n = \sum_{k=1}^{n} y_k,$$

and let $\varepsilon > 0$. By the absolute convergence of $\sum x_n$, there is an integer N such that

$$|S - S_N| = \sum_{k=N+1}^{\infty} |x_k| = |x_{N+1}| + |x_{N+2}| + \cdots < \frac{\varepsilon}{2}.$$

Now we choose the integer M so that all the terms $x_1, x_2, \cdots, x_N$ appear in the first M terms of the rearranged series, that is, within the finite sequence $(y_1, y_2, \cdots, y_M)$. Hence these terms do not contribute to the difference $T_m - S_N$, where $m \geq M$, and we obtain

$$m \geq M \Rightarrow |T_m - S_N| \leq |x_{N+1}| + |x_{N+2}| + \cdots < \frac{\varepsilon}{2}$$
$$\Rightarrow |S - T_m| \leq |S - S_N| + |S_N - T_m| < \varepsilon.$$

Thus $\lim T_n = S$.

To show that the rearranged series $\sum y_n$ is absolutely convergent, we note that $\sum |y_n|$ is a rearrangement of the absolutely convergent series $\sum |x_n|$ and, by the argument above, converges to the same limit. □

Exercises 4.1

1. Let $\sum y_n$ be the series which results after all the zero terms are omitted from the series $\sum x_n$. Prove that $\sum x_n$ is convergent if, and only if, $\sum y_n$ is convergent, and that $\sum y_n = \sum x_n$ if either series is convergent.

2. Prove that changing a finite number of terms of a series does not affect its convergence.

3. If $x_n \geq 0$ for all $n \in \mathbb{N}$, prove that the series $\sum x_n$ converges if, and only if, the sequence $S_n = \sum_{k=1}^{n} x_n$ is bounded above.

4. If the series $\sum x_n$ converges, prove that the series
$$(x_1 + x_2) + (x_3 + x_4) + (x_5 + x_6) + \cdots$$
also converges. Give an example to show that the converse is false.

5. Let $\sum x_n$ be conditionally convergent, with positive terms given by the sequence (y_n) and negative terms by (z_n), both arranged in the order in which they appear in (x_n).

   (a) Prove that both $\sum y_n$ and $\sum z_n$ diverge.
   (b) Define a rearrangement of $\sum x_n$ which has no limit in $\bar{\mathbb{R}}$.

6. Prove that
$$\sum_{n=0}^{\infty} \frac{1}{(n+a)(n+1+a)} = \frac{1}{a} \quad \text{for all } a > 0.$$

7. If $x_n \geq 0$ for all n, prove that the convergence of $\sum x_n$ implies the convergence of $\sum x_n^2$, but that the converse is false.

8. Let $x_n > 0$ for all n and $y_n = \frac{1}{n}(x_1 + x_2 + \cdots + x_n)$. Prove that the series $\sum y_n$ is divergent.

9. Let
$$1 + \frac{1}{3} - \frac{1}{2} + \frac{1}{5} + \frac{1}{7} - \frac{1}{4} + \frac{1}{9} + \frac{1}{11} - \frac{1}{6} + \cdots$$
be a rearrangement of the conditionally convergent series
$$\sum_{n=1}^{\infty} \frac{(-1)^{n+1}}{n}.$$
Prove that this rearrangement is also convergent, but to a different limit from that of the original series.

10. If the series $\sum x_n$ is absolutely convergent, prove that
$$\sum_{n=1}^{\infty} x_n = (x_1 + x_3 + x_5 + \cdots) + (x_2 + x_4 + x_6 + \cdots).$$

4.2 Convergence Tests

This section covers the more important tests for the convergence of infinite series, excluding the so-called *integral test* which has to await the treatment of the Riemann (improper) integral in chapter 8. In general, these tests provide sufficient, but not necessary, conditions for convergence. This is in stark contrast to the Cauchy criterion (Theorem 4.2) which, though difficult to apply in practice, provides a necessary and sufficient condition for convergence, and remains one of the most effective theoretical means for proving the results of this section. The first of these is called the *comparison test*.

Theorem 4.6 (Comparison Test)
If (x_n) and (y_n) are positive sequences that satisfy the inequality $x_n \leq y_n$ for all $n \geq N$, for some fixed positive integer N, then the convergence of $\sum y_n$ implies the convergence of $\sum x_n$.

Proof. Let $\varepsilon > 0$. Since $\sum y_n$ is convergent there is a positive integer M such that

$$\sum_{k=m+1}^{n} y_k = y_{m+1} + \cdots + y_n < \varepsilon \text{ for all } n > m \geq M.$$

By choosing $K = \max\{M, N\}$ we obtain

$$n > m \geq K \implies \sum_{k=m+1}^{n} x_k \leq \sum_{k=m+1}^{n} y_k < \varepsilon.$$

By the Cauchy criterion, $\sum x_n$ is convergent. □

Corollary 4.6 *If $0 \leq x_n \leq y_n$ for all $n \geq N$ and the series $\sum x_n$ is divergent, then $\sum y_n$ is also divergent.*

Theorem 4.7 (Limit Comparison Test)
Suppose (x_n) and (y_n) are positive sequences such that $\lim (x_n/y_n)$ exists.
(i) If $\lim (x_n/y_n) \neq 0$, the series $\sum x_n$ and $\sum y_n$ either both converge or they both diverge.
(ii) If $\lim (x_n/y_n) = 0$ and the series $\sum y_n$ converges, then $\sum x_n$ also converges.

Proof
In case (i) we know that $\lim x_n/y_n$ must be a positive number. According to Theorems 3.2 and 3.3, the sequence (x_n/y_n) is bounded away

from 0, that is, there are two positive constants M and K and a positive integer N such that

$$n \geq N \Rightarrow M \leq \frac{x_n}{y_n} \leq K$$
$$\Rightarrow My_n \leq x_n \leq Ky_n.$$

Using the comparison test, we now conclude that the convergence of one of the two series implies the convergence of the other.

In case (ii) there is a positive integer N such that

$$n \geq N \Rightarrow 0 \leq \frac{x_n}{y_n} \leq 1$$
$$\Rightarrow 0 \leq x_n \leq y_n,$$

so, again by the comparison test, $\sum x_n$ converges if $\sum y_n$ converges. □

Example 4.6 Since $\dfrac{1}{n} \leq \dfrac{1}{n^p}$ for all $p \leq 1$, $n \in \mathbb{N}$, and since $\sum \dfrac{1}{n}$ is divergent (Example 4.3), the series $\sum \dfrac{1}{n^p}$ diverges for all $p \leq 1$ by the comparison test.

Example 4.7 Let

$$x_n = \frac{1}{n} - \frac{1}{n+1} = \frac{1}{n(n+1)}.$$

Then

$$S_n = \sum_{k=i}^{n} x_k = 1 - \frac{1}{2} + \frac{1}{2} - \frac{1}{3} + \cdots + \frac{1}{n} - \frac{1}{n+1}$$
$$= 1 - \frac{1}{n+1}.$$

Hence

$$\sum_{k=1}^{\infty} \frac{1}{n(n+1)} = \lim\left(1 - \frac{1}{n+1}\right) = 1.$$

To test the series $\sum 1/n^2$, note that the limit comparison test yields

$$\lim \frac{1}{n(n+1)} \div \frac{1}{n^2} = \lim \frac{1}{1+1/n} = 1,$$

hence $\sum 1/n^2$ is convergent. Since

$$\frac{1}{n^p} \leq \frac{1}{n^2} \quad \text{for all } p \geq 2,$$

it follows that $\sum 1/n^p$ converges for all $p \geq 2$. The series $\sum 1/n^p$ diverges for all $p \leq 1$, as we saw in Example 4.6, so it remains to determine its behaviour when $1 < p < 2$. The next example shows that we have convergence over these values as well.

Example 4.8 The series $\sum 1/n^p$ converges for all $p > 1$.

Proof. For any $n \in \mathbb{N}$, let

$$S_n = \sum_{k=1}^{n} \frac{1}{k^p}, \quad T_n = S_{2^n}.$$

Then

$$\begin{aligned} T_{n+1} - T_n &= \sum_{k=2^n+1}^{2^{n+1}} \frac{1}{k^p} \\ &= \frac{1}{(2^n+1)^p} + \frac{1}{(2^n+2)^p} + \cdots + \frac{1}{(2^{n+1})^p} \\ &< \frac{1}{2^{np}} + \frac{1}{2^{np}} + \cdots + \frac{1}{2^{np}} \\ &= \frac{2^n}{2^{np}} \\ &= \frac{1}{2^{n(p-1)}}. \end{aligned}$$

Hence

$$\begin{aligned} T_{n+1} &= \sum_{k=0}^{n} (T_{k+1} - T_k) + T_0 \\ &< \sum_{k=0}^{n} \frac{1}{2^{k(p-1)}} + T_0 \\ &< \frac{1}{1 - (1/2)^{p-1}} + T_0 = K, \end{aligned}$$

where we used the formula derived in Example 4.1 for the sum of the geometric series $\sum_{k=0}^{\infty} (1/2^{p-1})^k$. Thus the positive sequence (T_n) is bounded above by the constant K. Since the sequence (2^m) is increasing and unbounded, for any $n \in \mathbb{N}$ there is an $m \in \mathbb{N}$ such that $2^m > n$; and since (S_n) is an increasing sequence, we have

$$S_n \leq S_{2^m} = T_m \leq K,$$

which implies (S_n) is bounded and therefore convergent, by Theorem 3.7. □

The next theorem, known as the *root test,* is one of the fundamental convergence tests, since it makes no assumptions regarding the sequence of terms x_n.

Theorem 4.8 (Root Test)
Given a series $\sum x_n$, let
$$r = \limsup \sqrt[n]{|x_n|}.$$
Then
(i) $\sum x_n$ is absolutely convergent if $r < 1$.
(ii) $\sum x_n$ is divergent if $r > 1$.
(iii) The test is inconclusive if $r = 1$.

Proof
(i) If $r < 1$, choose a positive number $c \in (r, 1)$. Let $\varepsilon = c - r$. By Theorem 3.16(i), there is a positive integer N such that
$$n \geq N \Rightarrow \sqrt[n]{|x_n|} < r + \varepsilon = c$$
$$\Rightarrow |x_n| < c^n.$$
Since the geometric series $\sum c^n$ converges, $\sum |x_n|$ converges by comparison, and therefore $\sum x_n$ is absolutely convergent.

(ii) By Theorem 3.16(iv), there is a subsequence (x_{n_k}) of (x_n) such that
$$\sqrt[n_k]{|x_{n_k}|} \to r.$$
Since $r > 1$ there is a positive integer N such that
$$k \geq N \Rightarrow \sqrt[n_k]{|x_{n_k}|} > 1$$
$$\Rightarrow |x_{n_k}| > 1,$$
and we conclude that the condition $x_n \to 0$, which is a necessary condition for convergence, is not satisfied.

(iii) In Example 4.3 we found that $\sum 1/n$ was divergent, and in Example 4.7 we established the convergence of $\sum 1/n^2$. But in both cases $r = \lim \sqrt[n]{1/n^p} = 1$. □

Theorem 4.9
For any positive sequence (x_n),
$$\limsup \sqrt[n]{x_n} \leq \limsup \frac{x_{n+1}}{x_n}$$
$$\liminf \frac{x_{n+1}}{x_n} \leq \liminf \sqrt[n]{x_n}.$$

Proof. To prove the first inequality, let

$$L = \limsup \frac{x_{n+1}}{x_n}.$$

If $L = \infty$ there is nothing to prove. If $0 \leq L < \infty$ then, for any $M > L$, there is a positive integer N such that

$$\frac{x_{n+1}}{x_n} \leq M \quad \text{for all } n \geq N.$$

In particular, for any $k \in \mathbb{N}$,

$$\begin{aligned} x_{N+k} &\leq M x_{N+k-1} \\ &\leq M^2 x_{N+k-2} \\ &\leq \cdots \\ &\leq M^k x_N. \end{aligned}$$

Hence

$$\begin{aligned} n > N \Rightarrow x_n &\leq M^{n-N} x_N \\ \Rightarrow \sqrt[n]{x_n} &\leq M \sqrt[n]{x_N M^{-N}}. \end{aligned}$$

From Example 3.8, $\lim \sqrt[n]{x_N M^{-N}} = 1$, and from Exercise 3.6.1 and Theorem 3.16(iii) we obtain

$$\limsup \sqrt[n]{x_n} \leq M.$$

Since this is true for any $M > L$, it follows that $\limsup \sqrt[n]{x_n} \leq L$. The second inequality is proved by a similar procedure. $\square$

Corollary 4.9.1 *Given the series $\sum x_n$, define*

$$\limsup \left| \frac{x_{n+1}}{x_n} \right| = L, \quad \liminf \left| \frac{x_{n+1}}{x_n} \right| = \ell.$$

Then
(i) $\sum x_n$ converges absolutely if $L < 1$,
(ii) $\sum x_n$ diverges if $\ell > 1$,
(iii) The test is inconclusive if $\ell \leq 1 \leq L$.

Remark 4.2 Since omitting all zero terms in a sequence does not affect its convergence or its value (Exercise 4.1.1), the ratio test can be applied to the non-zero terms of the series only.

Proof. (i) and (ii) follow directly from Theorems 4.8 and 4.9. To prove (iii), we again consider the divergent series $\sum 1/n$ and the convergent series $\sum 1/n^2$. In both cases we have $L = \ell = 1$. □

Corollary 4.9.2 (Ratio Test) *If*

$$\lim \left| \frac{x_{n+1}}{x_n} \right| = \lambda$$

exists, then

$$\lim \sqrt[n]{x_n} = \lambda,$$

and the series $\sum x_n$ is absolutely convergent if $\lambda < 1$ and divergent if $\lambda > 1$. If $\lambda = 1$ the test fails.

Proof. From Theorem 4.9,

$$\limsup \sqrt[n]{|x_n|} \leq \lim \left| \frac{x_{n+1}}{x_n} \right|$$
$$\leq \liminf \sqrt[n]{|x_n|}.$$

Therefore $\limsup \sqrt[n]{|x_n|} = \liminf \sqrt[n]{|x_n|} = \lim \sqrt[n]{|x_n|} = \lambda$. □

Example 4.9 The root and ratio tests applied to the series *p-series* $\sum 1/n^p$ yield

$$\lim_{n \to \infty} \sqrt[n]{\frac{1}{n^p}} = 1$$

and

$$\lim_{n \to \infty} \left(\frac{n+1}{n} \right)^p = 1,$$

so both tests fail to provide any information on the convergence of this series. But we already know, from Examples 4.6 and 4.8, that $\sum 1/n^p$ diverges for $p \leq 1$ and converges for $p > 1$.

Example 4.10 To test $\sum 1/n!$ we use the ratio test which, in this case, is easier to apply than the root test:

$$\lim \frac{1}{(n+1)!} \div \frac{1}{n!} = \lim \frac{1}{n+1} = 0.$$

Thus the series is convergent.

Example 4.11 Let the series $\sum x_n$ be defined by

$$x_1 = \frac{1}{3}, \ x_2 = \frac{1}{2}, \ x_3 = \frac{1}{3^2}, \ x_4 = \frac{1}{2^2}, \ \cdots,$$

that is,
$$x_{2k-1} = \frac{1}{3^k}, \quad x_{2k} = \frac{1}{2^k}, \quad k \in \mathbb{N}.$$

Here we see that
$$\limsup \sqrt[n]{x_n} = \lim \sqrt[n]{\frac{1}{2^{n/2}}} = \frac{1}{\sqrt{2}},$$
$$\limsup \frac{x_{n+1}}{x_n} = \lim \frac{1/2^n}{1/3^n} = \infty,$$
$$\liminf \frac{x_{n+1}}{x_n} = \lim \frac{1/3^n}{1/2^{n-1}} = 0,$$

hence the ratio test gives no information, while the root test indicates convergence. This is consistent with the result of Theorem 4.9, and implies that the root test is more general than the ratio test.

All the convergence tests presented so far are in fact tests for absolute convergence, and cannot be used to test conditional convergence. Hence the significance of Theorem 4.10, known as the *alternating series test*, is that it may be used to test the conditional convergence of some absolutely divergent series.

Definition 4.4 The sequence (x_n) is called *alternating* if the sign of x_n is different from that of x_{n+1} for every n. The resulting series $\sum x_n$ is called an *alternating series*.

Theorem 4.10 (Alternating Series Test)
Let (x_n) be a positive, decreasing sequence whose limit is 0. Then the alternating series $\sum_{n=1}^{\infty} (-1)^{n+1} x_n$ is convergent, and

$$\left| \sum_{k=n+1}^{\infty} (-1)^{k+1} x_k \right| \leq x_{n+1}. \tag{4.2}$$

Proof. First we shall prove that the sequence of partial sums S_n converges, using ideas from Example 4.4.

$$S_{2n} = (x_1 - x_2) + (x_3 - x_4) + \cdots + (x_{2n-1} - x_{2n}).$$

Since $x_k - x_{k+1} \geq 0$ for all k, the sequence S_{2n} is increasing. It is also bounded above because

$$S_{2n} = x_1 - (x_2 - x_3) - \cdots - (x_{2n-2} - x_{2n-1}) - x_{2n} \leq x_1 \tag{4.3}$$

128   Elements of Real Analysis

for all $n \in \mathbb{N}$, hence $\lim S_{2n}$ exists; call it S. On the other hand,

$$|S_{2n+1} - S| = |S_{2n} + x_{2n+1} - S|$$
$$\leq |S_{2n} - S| + |x_{2n+1}|.$$

In the limit as $n \to \infty$, we obtain $\lim S_{2n+1} = S$. Therefore $S_n \to S$.

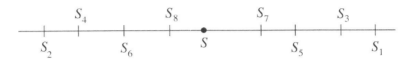

Figure 4.1

The relation (4.2) is equivalent to

$$|S - S_n| \leq x_{n+1},$$

that is, the error involved in approximating S by S_n does not exceed x_{n+1}. Now

$$S - S_n = (-1)^{n+2}(x_{n+1} - x_{n+2} + x_{n+3} - \cdots)$$
$$|S - S_n| = x_{n+1} - x_{n+2} + x_{n+3} - \cdots . \tag{4.4}$$

The series (4.4) is of the same alternating type as $S = x_1 - x_2 + x_3 - \cdots$, the only difference being that it starts with x_{n+1} instead of x_1. By (4.3) we have

$$S = \lim S_{2n} \leq x_1,$$

and the corresponding upper bound on the series (4.4) is therefore x_{n+1}. Thus $|S - S_n| \leq x_{n+1}$. □

Example 4.12 Since $1/n^p \to 0$ for all $p > 0$, the series

$$\sum \frac{(-1)^n}{n^p}$$

converges for all positive values of p. In view of Example 4.6 we may therefore conclude that this series converges conditionally for all p such that $0 < p \leq 1$. Now, assuming that $0 < p \leq 1/2$ and $0 < q \leq 1/2$, and defining

$$x_n = \frac{(-1)^n}{n^p}, \quad y_n = \frac{(-1)^n}{n^q},$$

we see that $\sum x_n$ and $\sum y_n$ are both convergent, but that $\sum x_n y_n$ is divergent.

In the following exercises, $\log n$ is the logarithm of n to the base e (see Example 3.12), or $\log_e n$ in the terminology of Section 2.4. It is the power to which e must be raised to give n. The logarithmic function will be defined more carefully later on, but, for the time being, only the elementary properties of the function are needed, such as $\log 1 = 0$ and $\log n$ increases monotonically to ∞ as $n \to \infty$.

Exercises 4.2

1. Test the following series for convergence:

    (a) $\sum_{n=1}^{\infty} \dfrac{1}{(n+1)(n+2)}$

    (b) $\sum_{n=1}^{\infty} (-1)^n \dfrac{\sqrt{n}}{n^2}$

    (c) $\sum_{n=1}^{\infty} (-1)^n \dfrac{\log n}{n}$

    (d) $\sum_{n=1}^{\infty} (-1)^n \dfrac{1}{\sqrt[3]{n^2+1}}$

    (e) $\sum_{n=1}^{\infty} \dfrac{n^2}{n!}$

    (f) $\sum_{n=2}^{\infty} \dfrac{1}{(\log n)^n}$

    (g) $\sum_{n=1}^{\infty} \dfrac{n!}{n^n}$

    (h) $\sum_{n=1}^{\infty} \dfrac{2^n n!}{n^n}$

    (i) $\sum_{n=1}^{\infty} \dfrac{3^n n!}{n^n}$

2. Prove that the convergence of $\sum a_n^2$ and $\sum b_n^2$ implies the convergence of $\sum a_n b_n$.

3. Prove that the convergence of $\sum a_n^2$ implies the convergence of $\sum a_n/n$.

4. Let the sequence (x_n) be positive and decreasing. If $\sum x_n$ converges prove that $\lim n x_n = 0$.

5. If the series $\sum x_n$ diverges show that $\sum \dfrac{x_n}{1+x_n}$ also diverges.

6. Calculate $\liminf \sqrt[n]{x_n}$ for the series defined in Example 4.11.

7. With reference to Theorem 4.7, if $\lim \dfrac{x_n}{y_n} = \infty$, what can you say about the convergence or divergence of the two series (x_n) and (y_n)?

8. Consider the series
$$\frac{1}{5} + \frac{1}{7} + \frac{1}{5^2} + \frac{1}{7^2} + \frac{1}{5^3} + \frac{1}{7^3} + \cdots.$$
Show that the series converges by the root test while the ratio test gives no information.

5

Limit of a Function

Having defined the limit of a sequence, we now turn our attention to the limit of a function at a point. In some sense this is a more general notion, as we shall presently discover, and one which will be central to all our future work.

5.1 Limit of a Function

Let c be a real number and f a real function defined on $D \subseteq \mathbb{R}$. We shall try to explain the meaning of the statement: "the limit of f at c is ℓ". Generally speaking, this means that $f(x)$, the value of f at x, will be "close to ℓ" if x is "close enough to c". Another, more dynamic, interpretation is that the numbers $f(x)$ 'approach ℓ" if we allow x to "approach c in any fashion". But we need to indicate precisely what we mean by the descriptive terms "close", "close enough", and "approach".

Consider the first interpretation. As in our discussion of the convergence of a sequence, the statement that "$f(x)$ is close to ℓ" is mathematically meaningful only if there is a prescribed measure of "closeness" $\varepsilon > 0$, and $f(x)$ is within an ε-neighborhood of ℓ, that is, $|f(x) - \ell| < \varepsilon$. Similarly the "closeness" of x to c is measured by another positive number δ and expressed by the inequality $|x - c| < \delta$. Thus, "the limit of f at c is ℓ" will mean that, given $\varepsilon > 0$, there is a $\delta > 0$ such that $f(x)$ lies in an ε-neighborhood of ℓ whenever $x \in D$ lies in a δ-neighborhood of c, or

$$x \in D, \ |x - c| < \delta \Rightarrow |f(x) - \ell| < \varepsilon. \tag{5.1}$$

Note that if there is a positive number δ such that $\{x \in D : 0 < |x - c| < \delta\} = \varnothing$, that is, if c is not a cluster point of D, then the implication (5.1) is (vacuously) satisfied for any $c \notin D$, in which case any real number ℓ will be a limit of f at c. For that reason it makes sense, in defining the limit of f at c, to stipulate that c be a *cluster point* of the domain of f. Furthermore, we are only interested in the behaviour of f *near* c, and not *at* c, so we shall also stipulate that $x \neq c$. Thus we arrive at the following definition:

Definition 5.1 Let $f : D \to \mathbb{R}$ and $c \in \hat{D}$. We say the *limit* of f at c is ℓ if, for every $\varepsilon > 0$, there is a $\delta > 0$ such that

$$x \in D, 0 < |x - c| < \delta \Rightarrow |f(x) - \ell| < \varepsilon,$$

and express this symbolically by writing

$$\lim_{x \to c} f(x) = \ell,$$

or

$$f(x) \to \ell \text{ as } x \to c.$$

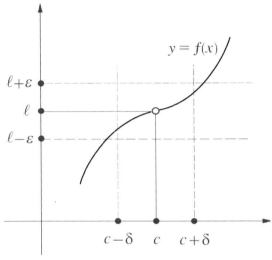

Figure 5.1

Example 5.1
$$\lim_{x \to 2}(3x - 5) = 1.$$

Proof. Given any $\varepsilon > 0$, we seek a $\delta > 0$ such that

$$0 < |x - 2| < \delta \Rightarrow |(3x - 5) - 1| < \varepsilon. \tag{5.2}$$

Note that

$$|(3x - 5) - 1| = |3x - 6| < \varepsilon$$
$$\Leftrightarrow 3|x - 2| < \varepsilon$$
$$\Leftrightarrow |x - 2| < \frac{\varepsilon}{3}.$$

If we take $|x-2| < \varepsilon/3$ we are guaranteed that $|(3x-5)-1| < \varepsilon$, hence we choose $\delta = \varepsilon/3$ in (5.2). Note that any other *smaller* positive value of δ will work. Note also that if we define another function which coincides with the polynomial $3x-5$ except at $x = 2$, say

$$f(x) = \begin{cases} 3x-5, & x \neq 2 \\ 7, & x = 2, \end{cases}$$

then $\lim_{x \to 2} f(x)$ would still be 1. In other words, the limit of a function at a point is not affected by how it is defined, or whether it is even defined, at that point.

Example 5.2 Let

$$f(x) = \begin{cases} x^2, & x \neq 2 \\ 0, & x = 2. \end{cases}$$

Then $\lim_{x \to 2} f(x) = 4$.

Proof. Let ε be any positive number. We wish to find a positive number δ such that

$$0 < |x-2| < \delta \Rightarrow |f(x) - 4| < \varepsilon.$$

With $x \neq 2$, we have $|f(x) - 4| = |x^2 - 4| = |x-2||x+2|$, which can be made less than ε if we can make

$$|x-2||x+2| < \varepsilon.$$

To that end, we first assume that $|x-2| < 1$, which implies

$$1 < x < 3$$
$$\Rightarrow 3 < x+2 < 5$$
$$\Rightarrow |x+2| < 5.$$

Consequently, with $x \neq 2$ and $|x-2| < 1$, we can make $|f(x) - 4| < \varepsilon$ by taking $5|x-2| < \varepsilon$, that is,

$$|x-2| < \varepsilon/5.$$

If we now choose $\delta = \min\{1, \varepsilon/5\}$, then

$$0 < |x-2| < \delta \Rightarrow |x+2| < 5, \ |x-2| < \varepsilon/5$$
$$\Rightarrow |x^2 - 4| = |x+2||x-2| < (5)(\varepsilon/5) = \varepsilon.$$

In both examples above we can verify that $\lim_{x \to c} f(x) = \ell$ by calculating f at values of x near c, tabulating the results, and noting that, as x gets closer to c, $f(x)$ gets closer to ℓ. This intuitive approach, however, will be of no use in the next example, where only the definition will serve us.

Example 5.3 Suppose $f : (0,1] \to \mathbb{R}$ is defined by

$$f(x) = \begin{cases} 0, & x \notin \mathbb{Q} \\ 1/q, & x = p/q \in \mathbb{Q}, \end{cases}$$

where the fraction p/q is in simple form, that is, p and q have no common factors. We shall prove that

$$\lim_{x \to c} f(x) = 0 \text{ for all } c \in (0,1].$$

Let $\varepsilon > 0$ be given. By the theorem of Archimedes (Theorem 2.7) we can choose $N \in \mathbb{N}$ so that $1/N < \varepsilon$. Now there is only a finite number of rational numbers $p/q \in (0,1]$ such that $q < N$, since the denominator q can only be one of the numbers $1, 2, \cdots, N-1$ and, for each such denominator i, the numerator p is one of the numbers $1, 2, \cdots, i$. Let A be the set of such rational numbers which are distinct from c, and define

$$\delta = \min\{|a - c| : a \in A\}.$$

Since A is finite, δ is positive, and

$$|a - c| \geq \delta \text{ for all } a \in A.$$

Suppose now that $x \in (0,1]$ and $0 < |x - c| < \delta$. There are two possibilities: either $x \notin \mathbb{Q}$, in which case $f(x) = 0$ and $|f(x) - 0| = 0 < \varepsilon$; or $x \in \mathbb{Q}$, in which case $x \notin A$ and hence $x = p/q$ (in simple form) with $q \geq N$. In this second case we have

$$f(x) = \frac{1}{q} \leq \frac{1}{N} < \varepsilon,$$

and again we obtain $|f(x) - 0| < \varepsilon$. Thus we have proved that

$$x \in (0,1], 0 < |x - c| < \delta \Rightarrow |f(x) - 0| < \varepsilon,$$

which just means that $\lim_{x \to c} f(x) = 0$.

Remarks 5.1

1. As mentioned in Example 5.1, if we can find a $\delta > 0$ which makes $|f(x) - \ell| < \varepsilon$, any $\delta' \in (0, \delta)$ will also work.

2. To prove that $\lim_{x \to c} f(x) = \ell$, it suffices to show that, for every $\varepsilon > 0$, there is a $\delta > 0$ such that

$$x \in D, 0 < |x - c| < \delta \Rightarrow |f(x) - \ell| < a\varepsilon,$$

where a is a positive constant which does not depend on x or δ. That is because $\varepsilon' = \varepsilon/a$ is positive whenever ε is positive, hence there is a $\delta' > 0$ such that

$$x \in D, 0 < |x - c| < \delta' \Rightarrow |f(x) - \ell| < a\varepsilon' = \varepsilon.$$

3. In the language of neighborhoods, the statement that $\lim_{x \to c} f(x) = \ell$ means that, for every neighborhood V of ℓ, there is a neighborhood U of c such that

$$x \in U \cap D, x \neq c \Rightarrow f(x) \in V.$$

In Chapter 3 we defined the limit of a convergent sequence. Since any sequence is a function defined on $\mathbb{N}$, it is legitimate to ask at this point whether that definition has any relation to Definition 5.1. To see the connection, recall that in the extended real numbers $\bar{\mathbb{R}} = \mathbb{R} \cup \{-\infty, +\infty\}$ we defined a neighborhood of $+\infty$ to be any subset of $\bar{\mathbb{R}}$ containing an interval $(N, +\infty]$ where $N \in \mathbb{R}$. Since every interval of the type $(N, +\infty]$ contains an infinite number of positive integers, this definition makes $+\infty$ a cluster point of $\mathbb{N}$. In fact, according to Example 3.15, $+\infty$ is the *only* cluster point of $\mathbb{N}$. As pointed out in Section 3.3, $+\infty$ will be abbreviated to ∞.

Now suppose (x_n) is a sequence in $\mathbb{R}$, and let

$$f : \mathbb{N} \to \mathbb{R}, \; f(n) = x_n.$$

According to Definition 5.1 and Remark 3 1.3, if

$$\lim_{n \to \infty} f(n) = \ell$$

then, for every neighborhood V of the point $\ell \in \mathbb{R}$, there is a neighborhood U of ∞ such that

$$n \in U \cap \mathbb{N} \Rightarrow f(n) \in V.$$

Hence, given $\varepsilon > 0$, there is a positive integer N such that

$$n \in [N, \infty) \Rightarrow f(n) \in (\ell - \varepsilon, \ell + \varepsilon),$$

which may be expressed as

$$n \geq N \Rightarrow |f(n) - \ell| < \varepsilon,$$

or

$$n \geq N \Rightarrow |x_n - \ell| < \varepsilon.$$

By Definition 3.2, this is just the statement that $\lim x_n = \ell$. Similarly, we can also show that, if $\lim x_n = \ell$ according to Definition 3.2, then $\lim_{n \to \infty} f(n) = \ell$ according to Definition 5.1. Hence Definition 3.2 of the limit of a sequence becomes a special case of Definition 5.1 where $c = \infty$.

But the connection between the limit of a sequence and that of a function runs deeper. Recall the dynamic picture which we used at the beginning of this section to motivate Definition 5.1, where the numbers $f(x)$ approach ℓ as x approaches c in any fashion. A most natural way for x to approach c would be through a *sequence* (x_n) in D with limit c, and then we would expect the sequence of images $f(x_n)$ to converge to ℓ. The qualification *in any fashion* simply means that it does not matter what sequence (x_n) is chosen, as long as $x_n \to c$. Adding the fine points that $c \in \hat{D}$ and that $x_n \neq c$, we arrive at the following result.

Theorem 5.1
Let $f : D \to \mathbb{R}$ and $c \in \hat{D}$. The following statements are equivalent:
(i) $\lim_{x \to c} f(x) = \ell$
(ii) For every sequence (x_n) in D such that $x_n \neq c$ for any $n \in \mathbb{N}$ and $x_n \to c$, the sequence $(f(x_n))$ converges to ℓ.

Proof
(i)$\Rightarrow$(ii): With $\lim_{x \to c} f(x) = \ell$ and (x_n) any sequence which satisfies the conditions in (ii), let ε be a positive number. By Definition 5.1 there is a $\delta > 0$ such that

$$x \in D, 0 < |x - c| < \delta \Rightarrow |f(x) - \ell| < \varepsilon. \tag{5.3}$$

Since $x_n \to c$ there is a positive integer N such that

$$n \geq N \Rightarrow |x_n - c| < \delta, \tag{5.4}$$

and since $x_n \neq c$ we conclude from (5.3) that

$$|f(x_n) - \ell| < \varepsilon \text{ for all } n \geq N,$$

i.e., $f(x_n) \to \ell$.

(ii)$\Rightarrow$(i): Suppose $\lim_{x \to c} f(x) \neq \ell$. We shall show that the statement (ii) is false by exhibiting a sequence (x_n) in D which satisfies the conditions in (ii), but for which $f(x_n) \not\to \ell$. The statement $\lim_{x \to c} f(x) \neq \ell$ means there is an $\varepsilon > 0$ such that, for every $\delta > 0$, we can find a point $x \in D$ which satisfies

$$0 < |x - c| < \delta \text{ and } |f(x) - \ell| \geq \varepsilon.$$

Taking $\delta = 1/n$, we see that for every $n \in \mathbb{N}$ there is an $x_n \in D$ such that

$$0 < |x_n - c| < \frac{1}{n} \text{ and } |f(x_n) - \ell| \geq \varepsilon.$$

Thus (x_n) satisfies the conditions in (ii) but $f(x_n) \not\to \ell$. □

Remarks 5.2

1. If $c \in D$ and $\ell = f(c)$, then the condition $x_n \neq c$ becomes unnecessary for the equivalence of (i) and (ii).

2. The significance of Theorem 5.1 is that it provides a criterion for proving that a function has no limit at c, as indicated below.

Corollary 5.1 *Let $f : D \to \mathbb{R}$ and $c \in \hat{D}$, and suppose S is the set of sequences (x_n) in D which satisfy $x_n \neq c$ and $x_n \to c$. If (i) there is a sequence (x_n) in S such that the sequence $(f(x_n))$ diverges, or (ii) there are two sequences (x_n) and (y_n) in S such that $\lim f(x_n) \neq \lim f(y_n)$, then $\lim_{x \to c} f(x)$ does not exist.*

Example 5.4 The function $f : \mathbb{R} \setminus \{0\} \to \mathbb{R}$ defined by the equation

$$f(x) = \frac{1}{x^2}$$

has no limit at 0. To see that let $x_n = 1/n$, and note that $x_n \neq 0$ for any n and $x_n \to 0$, but that $(f(x_n)) = (n^2)$ diverges. By Corollary 5.1(i), $\lim_{x \to 0} 1/x^2$ does not exist.

Example 5.5 The function sgn defined on $\mathbb{R}$ by

$$\text{sgn } x = \begin{cases} 1, & x > 0 \\ 0, & x = 0 \\ -1, & x < 0 \end{cases}$$

has no limit at 0. To see that, let $x_n = 1/n$ and $y_n = -1/n$, then $x_n \neq 0$ and $x_n \to 0$ and similarly $0 \neq y_n \to 0$, but sgn $x_n \to 1$ while sgn $y_n \to -1$. By Corollary 5.1(ii) $\lim_{x \to 0} \text{sgn } x$ does not exist.

Example 5.6 Let $f : \mathbb{R} \to \mathbb{R}$ be defined by

$$f(x) = \begin{cases} 1, & x \in \mathbb{Q} \\ 0, & x \notin \mathbb{Q}. \end{cases}$$

The function f has no limit at any $c \in \mathbb{R}$, since the density of $\mathbb{Q}$ and $\mathbb{Q}^c$ allows us to choose a sequence $x_n \in \mathbb{Q}$ such that $0 < |x_n - c| < 1/n$, and another sequence $y_n \in \mathbb{Q}^c$ such that $0 < |y_n - c| < 1/n$ for all $n \in \mathbb{N}$, whose images under f are $f(x_n) = 1$ and $f(y_n) = 0$.

Now we present a slightly stronger version of Theorem 5.1, which makes condition (ii) in Corollary 5.1 superfluous.

Theorem 5.2
Let $f:D \to \mathbb{R}$ and $c \in \hat{D}$. Then the following statements are equivalent:
(i) $\lim_{x \to c} f(x)$ exists
(ii) For every sequence (x_n) in D such that $x_n \neq c$ and $x_n \to c$, the sequence $(f(x_n))$ converges.

Proof
(i)$\Rightarrow$(ii): If $\lim_{x \to c} f(x) = \ell$ then statement (ii) follows directly from Theorem 5.1.
(ii)$\Rightarrow$(i): Let S be the set of sequences in D which converge to c but whose terms differ from c. Since c is a cluster point of D, S is not empty. Suppose (x_n) is a sequence in S such that $(f(x_n))$ converges to some limit ℓ. If (y_n) is another sequence in S whose image under f, $(f(y_n))$, converges to ℓ', we have to show that $\ell' = \ell$. To that end, consider the sequence $(z_n) = (x_1, y_1, x_2, y_2, \cdots)$. (z_n) is clearly in S (see Example 3.5), hence $(f(z_n))$ converges. By Theorem 3.13 the two subsequences $(f(z_{2k-1})) = (f(x_k))$ and $(f(z_{2k})) = (f(y_k))$ must have the same limit, that is, $\ell' = \ell$. Thus all the sequences in S converge to ℓ. By Theorem 5.1 we therefore conclude that $\lim_{x \to c} f(x) = \ell$. $\square$

As we saw in the case of a sequence, we would expect the limit of a function, at a point where it exists, to be unique; and we invite the reader to prove this by making the appropriate modifications to the proof of Theorem 3.1.

Theorem 5.3
Let $f : D \to \mathbb{R}$ and $c \in \hat{D}$. If $\lim_{x \to c} f(x)$ exists, it is unique.

EXERCISES 5.1

1. Determine which of the following statements correctly defines
$$\lim_{x \to c} f(x) = \ell.$$

   (a) For every $\delta > 0$ there is an $\varepsilon > 0$ such that
   $$x \in D, 0 < |x - c| < \varepsilon \Rightarrow |f(x) - \ell| < \delta.$$

   (b) For every $\varepsilon > 0$ there is a $\delta > 0$ such that
   $$x \in D, 0 < |x - c| < \delta \Rightarrow |f(x) - \ell| \leq \varepsilon.$$

   (c) For every $\delta > 0$ there is an $\varepsilon > 0$ such that
   $$x \in D, 0 < |x - c| < \delta \Rightarrow |f(x) - \ell| < \varepsilon.$$

   (d) For every $\varepsilon > 0$ there is a $\delta > 0$ such that
   $$x \in D, 0 < |x - c| \leq \delta \Rightarrow |f(x) - \ell| < 7\varepsilon.$$

   (e) For every $\varepsilon > 0$ there is a $\delta > 0$ such that
   $$x \in D, |f(x) - \ell| < \varepsilon \Rightarrow 0 < |x - c| < \delta.$$

   (f) For every $\varepsilon > 0$ there is a $\delta > 0$ such that
   $$x \in D, 0 < |x - c| < \delta/5 \Rightarrow |f(x) - \ell| < \varepsilon.$$

2. Using Definition 5.1, prove that $\lim_{x \to c} f(x) = \ell$ in each of the following cases:

   (a) $f(x) = x^3$, $\ell = c^3$.
   (b) $f(x) = 1/x$, $x \neq 0$, $c = 1$, $\ell = 1$.
   (c) $f(x) = \sqrt{x}$, $x \geq 0$, $c = 2$, $\ell = \sqrt{2}$.

3. Use Definition 5.1 or Theorem 5.1 to prove

   (a) $\lim_{x \to 1} \dfrac{3x}{1+x^2} = \dfrac{3}{2}$.

(b) $\lim_{x \to 1} \left(x^5 + \dfrac{1}{x} \right) = 2.$

(c) $\lim_{x \to 2} \dfrac{x^3 - 8}{x - 2} = 12.$

4. Prove that the following limits do not exist:

   (a) $\lim_{x \to 0} \dfrac{x}{|x|}.$

   (b) $\lim_{x \to -2} \dfrac{x}{x^2 - 4}.$

   (c) $\lim_{x \to 0} (x^2 + \operatorname{sgn} x).$

5. Let $f : \mathbb{R} \to \mathbb{R}$ be defined by
$$f(x) = \begin{cases} x, & x \in \mathbb{Q} \\ -x, & x \notin \mathbb{Q}. \end{cases}$$
Prove that $\lim_{x \to c} f(x)$ exists only when $c = 0$.

6. Let $f : \mathbb{R} \to \mathbb{R}$ and $\lim_{x \to 0} f(x) = \ell$. If $g : \mathbb{R} \to \mathbb{R}$ is defined by $g(x) = f(ax)$, where $a \neq 0$, prove that $\lim_{x \to 0} g(x) = \ell$.

7. Let $f : D \to \mathbb{R}$ and $c \in \hat{D}$. If $\lim_{x \to c} [f(x)]^2 = 0$, prove that $\lim_{x \to c} f(x) = 0$. Give an example of a function f for which $\lim_{x \to c} [f(x)]^2$ exists but $\lim_{x \to c} f(x)$ does not exist.

8. Suppose $A_n \subseteq [0, 1]$ is a finite set for every $n \in \mathbb{N}$, and $A_n \cap A_m = \emptyset$ whenever $n \neq m$. Define $f : [0, 1] \to \mathbb{R}$ by
$$f(x) = \begin{cases} 1/n, & x \in A_n \\ 0, & x \notin \cup_{n=1}^{\infty} A_n. \end{cases}$$
Prove that $\lim_{x \to c} f(x) = 0$ for every $c \in [0, 1]$.

9. Suppose that $f, g : D \to \mathbb{R}$ and that $c \in \hat{D}$. If $f(x) = g(x)$ for every $x \neq c$ in a neighborhood of c, and $\lim_{x \to c} f(x) = \ell$, show that $\lim_{x \to c} g(x) = \ell$.

5.2 Basic Theorems

Here we consider limits in the light of the algebraic and order properties of $\mathbb{R}$.

Theorem 5.4
Let $f : D \to \mathbb{R}$ and $c \in \hat{D}$. If f has a limit at c, then f is bounded in a neighborhood of c, that is, there is a neighborhood U of c and a positive number M such that
$$|f(x)| \leq M \quad \text{for all } x \in U \cap D.$$

Proof. Suppose $\lim_{x \to c} f(x) = \ell$. By Definition 5.1 there is a $\delta > 0$ such that
$$x \in D, 0 < |x - c| < \delta \Rightarrow |f(x) - \ell| < 1.$$
Setting $U = (c - \delta, c + \delta)$ we see that
$$x \in U \cap D, x \neq c \Rightarrow |f(x) - \ell| < 1$$
$$\Rightarrow |f(x)| < |\ell| + 1.$$
If $c \in D$ take $M = \max\{|f(c)|, |\ell| + 1\}$, otherwise set $M = |\ell| + 1$. □

Theorem 5.5
Let $f : D \to \mathbb{R}$ and $c \in \hat{D}$. If $\lim_{x \to c} f(x) = \ell$, where $\ell \neq 0$, then there is a neighborhood U of c and a positive number M such that
$$|f(x)| > M \quad \text{for all } x \in U \cap D \setminus \{c\}.$$

Proof. The proof closely follows that of Theorem 3.3 for sequences.

Theorem 5.6
Let $f, g : D \to \mathbb{R}$ and $c \in \hat{D}$. If $\lim_{x \to c} f(x) = \ell$ and $\lim_{x \to c} g(x) = m$, then
(i) $\lim_{x \to c}[f(x) + g(x)] = \ell + m$,
(ii) $\lim_{x \to c}[f(x) \cdot g(x)] = \ell m$,
(iii) $\lim_{x \to c} f(x)/g(x) = \ell/m$ provided $m \neq 0$.

Proof. This theorem corresponds to Theorem 3.4 for sequences, and its proof can be constructed along the lines of that earlier proof, an exercise worth undertaking for its own sake. Here we shall use Theorems 5.1 and 3.4 to prove part (iii) of the theorem.

If $m \neq 0$ then we know from Theorem 5.5 that there is a neighborhood U of c such that
$$|g(x)| > 0 \quad \text{for all } x \in U \cap D \setminus \{c\},$$

so that f/g is well defined on $U \cap D \setminus \{c\}$. Suppose now that (x_n) is any sequence in D such that $x_n \neq c$ for all n and that $x_n \to c$. By Theorem 5.1 we know that $f(x_n) \to \ell$ and $g(x_n) \to m$, and Theorem 3.4 implies that $f(x_n)/g(x_n) \to \ell/m$. Hence, by Theorem 5.1,

$$\lim_{x \to c} \frac{f(x)}{g(x)} = \frac{\ell}{m}. \quad \square$$

Note that parts (i) and (ii) of Theorem 5.6 imply

$$\lim_{x \to c} kf(x) = k\ell \text{ for any real number } k,$$
$$\lim_{x \to c}[f(x) - g(x)] = \ell - m.$$

Theorem 5.7
Suppose the two functions $f, g : D \to \mathbb{R}$ have limits at $c \in \hat{D}$. If there is a neighborhood U of c such that

$$f(x) \leq g(x) \text{ for all } x \in U \cap D \setminus \{c\}$$

then

$$\lim_{x \to c} f(x) \leq \lim_{x \to c} g(x).$$

Proof. This is left to the reader (see Theorem 3.5).

Finally, we have a "squeezing" theorem for functions corresponding to Theorem 3.6, which can be proved along similar lines.

Theorem 5.8
Let $f, g, h : D \to \mathbb{R}$ and $c \in \hat{D}$. If there is a neighborhood U of c such that

$$f(x) \leq g(x) \leq h(x) \text{ for all } x \in U \cap D \setminus \{c\}$$

and $\lim_{x \to c} f(x) = \lim_{x \to c} h(x) = \ell$, then $\lim_{x \to c} g(x) = \ell$.

Example 5.7 If p is a polynomial, then $\lim_{x \to c} p(x) = p(c)$ for any $c \in \mathbb{R}$. This follows directly from the observation that $\lim_{x \to c} x = c$ and the application of Theorem 5.6. Suppose $p(x) = a_0 + a_1 x + a_2 x^2 + \cdots + a_n x^n$. then, using parts (i) and (ii) of the theorem,

$$\lim_{x \to c} p(x) = a_0 + a_1 \lim_{x \to c} x + a_2 \lim_{x \to c}(x \cdot x) + \cdots + a_n \lim_{x \to c}(x \cdot x \cdots .x)$$
$$= a_0 + a_1 c + a_2 c^2 + \cdots + a_n c^n$$
$$= p(c).$$

For a rational function p/q, where p and q are polynomials such that $q(c) \neq 0$, part (iii) of Theorem 5.6 gives

$$\lim_{x \to c} \frac{p(x)}{q(x)} = \frac{p(c)}{q(c)}.$$

Now we consider some examples involving the trigonometric functions

$$\sin : \mathbb{R} \to [-1, 1],$$
$$\cos : \mathbb{R} \to [-1, 1],$$
$$\tan : \mathbb{R} \setminus \{\pm \pi/2, \pm 3\pi/2, \cdots\} \to \mathbb{R},$$

which we shall assume the reader is familiar with. These functions are defined through the coordinates of a point on the unit circle in the usual manner (see Figure 5.2), and their properties can be derived accordingly. Our purpose is to introduce more variety in our examples at this point, and not have to wait for more rigorous definitions before we discuss such functions.

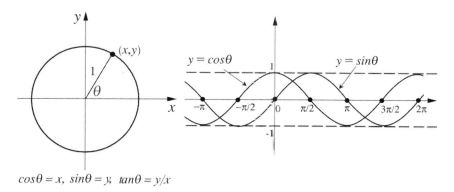

$\cos\theta = x$, $\sin\theta = y$, $\tan\theta = y/x$

Figure 5.2

Example 5.8

$$\lim_{\theta \to 0} \sin \theta = 0$$
$$\lim_{\theta \to 0} \cos \theta = 1$$

Proof. If $\theta \in [-\pi/2, \pi/2]$ then $|\theta|$ will be in the first quadrant of the coordinate plane, hence $\sin|\theta| \geq 0$. Referring to Figure 5.3 (where $0 < \theta < \pi/2$) we see that

$$\sin|\theta| = |BC| \leq |AB|, \qquad (5.5)$$

where $|AB|$ and $|BC|$ are the lengths of the sides AB and BC, respectively. Furthermore, $|\theta|$ is the length of the arc AB, hence

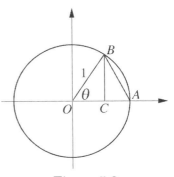

Figure 5.3

$$|AB| \leq |\theta|. \qquad (5.6)$$

Since $\sin\theta$ is an *odd* function, in the sense that $\sin(-\theta) = -\sin\theta$, we have $|\sin\theta| = \sin|\theta|$. Combining (5.5) and (5.6),

$$0 \leq |\sin\theta| \leq |\theta| \quad \text{for all } \theta \in [-\pi/2, \pi/2].$$

Taking the limit in this last inequality as $\theta \to 0$, and using Theorem 5.8, gives $\lim_{\theta \to 0} |\sin\theta| = 0$, hence $\lim_{\theta \to 0} \sin\theta = 0$ (see Exercise 5.2.6).

With $\theta \in [-\pi/2, \pi/2]$, $\cos\theta \geq 0$, and the relation $\sin^2\theta + \cos^2\theta = 1$ yields

$$\cos\theta = \sqrt{1 - \sin^2\theta} \to 1 \text{ as } \theta \to 0. \qquad \square$$

More generally, we have

$$\lim_{\theta \to \theta_0} \sin\theta = \sin\theta_0, \quad \lim_{\theta \to \theta_0} \cos\theta = \cos\theta_0 \quad \text{for any } \theta_0 \in \mathbb{R}.$$

To see that, we use a well-known trigonometric identity to write

$$\sin\theta = \sin(\theta_0 + (\theta - \theta_0))$$
$$= \sin\theta_0 \cos(\theta - \theta_0) + \cos\theta_0 \sin(\theta - \theta_0)$$
$$\to \sin\theta_0 \cdot 1 + \cos\theta_0 \cdot 0 = \sin\theta_0 \text{ as } \theta \to \theta_0.$$

Similarly, the second limit can be shown to hold.

Example 5.9 The function

$$f(x) = \sin\frac{1}{x}, \quad x \in \mathbb{R}\setminus\{0\},$$

has no limit at 0, for the two sequences

$$x_n = \frac{1}{n\pi}, \quad y_n = \frac{1}{2n\pi + \pi/2}$$

both tend to 0 as prescribed by Corollary 5.1, and

$$f(x_n) = \sin(n\pi) = 0 \text{ for all } n$$
$$f(y_n) = \sin(2n\pi + \pi/2) = 1 \text{ for all } n.$$

Thus $\lim f(x_n) = 0 \neq \lim f(y_n) = 1$, so that $\lim_{x \to 0} f(x)$ does not exist.

Example 5.10

$$\lim_{\theta \to 0} \frac{\sin \theta}{\theta} = 1.$$

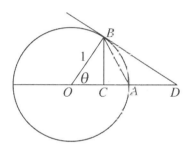

Figure 5.4

Proof. Take $\theta \in (-\pi/2, \pi/2)$ and $\theta \neq 0$ so that $|\theta|$ lies in the first quadrant. In Figure 5.4, where BD is tangent to the unit circle at B, we have

area of triangle $AOB \leq$ area of sector $AOB \leq$ area of triangle OBD,

that is,

$$\frac{1}{2}|BC| \leq \frac{1}{2}|\theta| \leq \frac{1}{2}|BD|,$$

or

$$\sin|\theta| \leq |\theta| \leq \tan|\theta| = \frac{\sin|\theta|}{\cos|\theta|}$$

$$\Rightarrow 1 \leq \frac{|\theta|}{|\sin \theta|} \leq \frac{1}{\cos|\theta|} \text{ for all } \theta \in \left(\frac{-\pi}{2}, \frac{\pi}{2}\right), \theta \neq 0. \quad (5.7)$$

Since cos is an even function, i.e., $\cos(-\theta) = \cos\theta$, we can write $\cos|\theta| = \cos\theta$; and since the sine function is odd, we have $|\theta|/|\sin\theta| = \theta/\sin\theta$. The inequality (5.7) therefore takes the form

$$1 \leq \frac{\theta}{\sin\theta} \leq \frac{1}{\cos\theta}, \quad 0 < |\theta| < \pi/2,$$

which allows us to conclude that $\lim_{\theta \to 0} \theta/\sin\theta = 1$, and hence

$$\lim_{\theta \to 0} \frac{\sin\theta}{\theta} = 1,$$

using Theorem 5.8. □

Example 5.11

$$\lim_{x \to 0} x \sin \frac{1}{x} = 0.$$

Proof. Since $\left|\sin \dfrac{1}{x}\right| \leq 1$ for all $x \neq 0$, we have

$$0 \leq \left|x \sin \frac{1}{x}\right| \leq |x|, \quad x \neq 0.$$

Therefore $\lim_{x \to 0} \left|x \sin \dfrac{1}{x}\right| = 0$ by Theorem 5.8. □

Example 5.12 To calculate the limit

$$\lim_{x \to 2} \frac{x^2 - 3x + 2}{x^2 - 4}$$

we cannot use Theorem 5.6(iii) directly, since the limit of the denominator at 2 is 0. But we know that, as long as $x \neq 2$, we can write

$$\frac{x^2 - 3x + 2}{x^2 - 4} = \frac{x - 1}{x + 2}, \quad x \neq 2,$$

$$\lim_{x \to 2} \frac{x^2 - 3x + 2}{x^2 - 4} = \lim_{x \to 2} \frac{x - 1}{x + 2} = \frac{1}{4}.$$

Exercises 5.2

1. Give examples of two functions f and g such that

    (a) $\lim_{x \to c} f(x)$ and $\lim_{x \to c} g(x)$ do not exist but $f+g$ has a limit at c.

    (b) $\lim_{x \to c} f(x)$ and $\lim_{x \to c} g(x)$ do not exist but $f \cdot g$ has a limit at c.

    (c) $\lim_{x \to c} g(x) = 0$ and f/g has a limit at c.

2. If $a \leq f(x) \leq b$ for all $x \in D_f$ and $\lim_{x \to c} f(x)$ exists, prove that $a \leq \lim_{x \to c} f(x) \leq b$. What can we say if $a < f(x) < b$ for all $x \in D_f$?

3. Let $f(x) \geq 0$ for all $x \in D_f$. If $\lim_{x \to c} f(x) = \ell$ prove that $\lim_{x \to c} \sqrt{f(x)} = \sqrt{\ell}$.

    Can you generalize this result to the n-th root of $f(x)$?

4. Determine the values of the following limits where they exist:

    (a) $\lim\limits_{x \to 3} \sqrt{\dfrac{2x^2 - 1}{x + 2}}$

    (b) $\lim\limits_{x \to -1} \dfrac{x^3 + x^2 - x - 1}{x^3 + 2x^2 + x}$

    (c) $\lim\limits_{x \to 0} \operatorname{sgn}\left(\sin \dfrac{1}{x}\right)$

    (d) $\lim\limits_{x \to 0} \dfrac{\sqrt{1 + 3x} - \sqrt{1 + x}}{3x - x^2}$

    (e) $\lim\limits_{x \to 0} \sqrt{x} \cdot \sin\left(\dfrac{1}{x}\right) \cdot \cos\left(\dfrac{1}{x}\right)$

5. If $\lim_{x \to c} f(x) = 0$ and g is a bounded function in some neighborhood of c, prove that $\lim_{x \to c} f(x)g(x) = 0$.

6. If $\lim_{x \to c} f(x) = \ell$, prove that $\lim_{x \to c} |f(x)| = |\ell|$. When is the converse also true?

7. If $\lim_{x \to 0} f(x)/x = \ell$, prove that $\lim_{x \to 0} f(ax)/x = a\ell$ for all $a \neq 0$. What happens if $a = 0$? Calculate $\lim_{x \to 0} (\sin 3x)/x$.

8. Recall the definition of $\exp_a$ given in Section 2.4. Prove that
$$\lim_{x \to 0} \exp_a(x) = 1.$$
Hint: Start with $a > 1$. Given $\varepsilon > 0$, choose $N \in \mathbb{N}$ such that $0 < a^{1/n} - 1 < \varepsilon$ for all $n \geqslant N$ in accordance with Example 3.8. Deduce that $\lim_{x \to 0+} \exp_a(x) = 1$.

9. Let $f : \mathbb{R} \to \mathbb{R}$ satisfy the relation $f(x + y) = f(x) + f(y)$ for all $x, y \in \mathbb{R}$. If f has a limit at some point in $\mathbb{R}$, prove that

   (a) f has a limit at every point in $\mathbb{R}$,
   (b) $\lim_{x \to 0} f(x) = 0$.

10. Let $f : \mathbb{R} \to \mathbb{R}$ satisfy the relation $f(x + y) = f(x)f(y)$ for all $x, y \in \mathbb{R}$. Prove the following:

    (a) $f(x) \geq 0$ for all $x \in \mathbb{R}$.
    (b) If $f(a) = 0$ for some $a \in \mathbb{R}$, then $f(x) = 0$ for all $x \in \mathbb{R}$.
    (c) If $f \neq 0$ then $f(0) = 1$.
    (d) If $f \neq 0$ and $\lim_{x \to a} f(x)$ exists for some $a \in \mathbb{R}$, then
    $$\lim_{x \to 0} f(x) = 1.$$

    The reader may recall that these are properties of the exponential function discussed in Section 2.4 and its exercises, together with Exercise 5.2.8.

11. If $\lim_{x \to c} f(x) = \ell$ and $\lim_{y \to \ell} g(y) = m$, show by a suitable example that it does not necessarily follow that $\lim_{x \to c} g \circ f(x) = m$.

5.3 Some Extensions of The Limit

In Examples 5.4, 5.5, and 5.9 we found that the following limits do not exist:

$$(i) \ \lim_{x \to 0} \operatorname{sgn} x, \quad (ii) \ \lim_{x \to 0} \frac{1}{x^2}, \quad (iii) \ \lim_{x \to 0} \sin \frac{1}{x}.$$

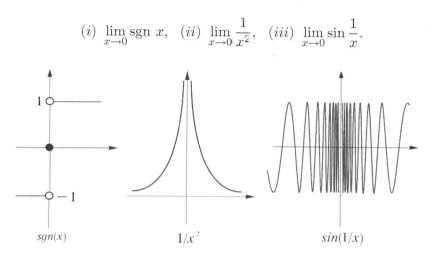

Figure 5.5

But Figure 5.5 shows that the behaviour of these functions is not the same in the neighborhood of 0. For one thing, the function $1/x^2$ is unbounded in such a neighborhood, whereas both sgn x and $\sin(1/x)$ are bounded (above by 1, below by -1). On the other hand, the function sgn x would have a limit at 0 if its domain were restricted from $(-\infty, \infty)$ to $(0, \infty)$ or $(-\infty, 0)$, but $\sin(1/x)$ fails to have a limit under such a restriction, as it continues to oscillate between -1 and 1 on both sides of 0.

Definition 5.2 Let $f : D \to \mathbb{R}$ and suppose c is a cluster point of $D \cap (c, \infty)$. The *right-hand limit* of f at c is defined as the limit at c of the restriction of f to $D \cap (c, \infty)$, and is denoted by $\lim_{x \to c^+} f(x)$ or $f(c^+)$. That is,

$$\lim_{x \to c^+} f(x) = \lim_{x \to c} f\big|_{D \cap (c,\infty)} (x).$$

If, on the other hand, c is a cluster point of $D \cap (-\infty, c)$, then the *left-hand limit* of f at c, denoted by $\lim_{x \to c^-} f(x)$ or $f(c^-)$, is the limit at c of $f\,|_{D\cap(-\infty,c)}$, i.e.,

$$\lim_{x\to c^-} f(x) = \lim_{x\to c} f\,|_{D\cap(-\infty,c)}(x).$$

Thus
$$\lim_{x\to c^+} f(x) = \ell$$
if, for every $\varepsilon > 0$, there is a $\delta > 0$ such that

$$x \in D, 0 < x - c < \delta \Rightarrow |f(x) - \ell| < \varepsilon.$$

Similarly,
$$\lim_{x\to c^-} f(x) = m$$
if, for every $\varepsilon > 0$, there is a $\delta > 0$ such that

$$x \in D, 0 < c - x < \delta \Rightarrow |f(x) - m| < \varepsilon.$$

Note that, since the definitions of the left-hand and right-hand limits are based on Definition 5.1 of the limit of a function, all the relevant theorems of Sections 5.1 and 5.2 apply to these one-sided limits.

Theorem 5.9
Let $f : D \to \mathbb{R}$ and suppose c is a cluster point of $D \cap (c, \infty)$ and $D \cap (-\infty, c)$. The limit of f at c exists if and only if $\lim_{x\to c^-} f(x)$ and $\lim_{x\to c^+} f(x)$ both exist and are equal, in which case

$$\lim_{x\to c^-} f(x) = \lim_{x\to c^+} f(x) = \lim_{x\to c} f(x).$$

Proof. If $\lim_{x\to c} f(x) = \ell$, it is not difficult to show that

$$\lim_{x\to c^-} f(x) = \lim_{x\to c^+} f(x) = \ell,$$

and we leave the details as an exercise.

Now assume that
$$\lim_{x\to c^-} f(x) = \lim_{x\to c^+} f(x) = \ell.$$

Given $\varepsilon > 0$, there are $\delta_1 > 0$ and $\delta_2 > 0$ such that

$$x \in D, 0 < x - c < \delta_1 \Rightarrow |f(x) - \ell| < \varepsilon$$
$$x \in D, 0 < c - x < \delta_2 \Rightarrow |f(x) - \ell| < \varepsilon.$$

Setting $\delta = \min\{\delta_1, \delta_2\}$, it is clear that $\delta > 0$. If $x \in D$ satisfies $0 < |x - c| < \delta$, then either $0 < x - c < \delta \leq \delta_1$ or $0 < c - x < \delta \leq \delta_2$. In either case we have
$$|f(x) - \ell| < \varepsilon.$$
Thus $\lim_{x \to c} f(x) = \ell$. $\square$

Example 5.13 Since
$$\lim_{x \to 0^+} \operatorname{sgn} x = 1, \quad \lim_{x \to 0^-} \operatorname{sgn} x = -1,$$
it follows that $\lim_{x \to 0} \operatorname{sgn} x$ does not exist.

Example 5.14 Let
$$f(x) = \begin{cases} \dfrac{x-4}{\sqrt{x}-2}, & x > 4 \\ 0, & x = 4 \\ 2x - 4, & x < 4. \end{cases}$$
Then
$$\lim_{x \to 4^+} f(x) = \lim_{x \to 4} \frac{x-4}{\sqrt{x}-2} = \lim_{x \to 4}(\sqrt{x}+2) = 4,$$
$$\lim_{x \to 4^-} f(x) = \lim_{x \to 4}(2x-4) = 4.$$
Therefore
$$\lim_{x \to 4} f(x) = 4.$$

Example 5.15 If $f : \mathbb{R}\setminus\{0\} \to \mathbb{R}$ is defined by
$$f(x) = \frac{1}{x},$$
then $f(0^+)$ does not exist because $f\big|_{(0,\infty)}$ is unbounded in any neighborhood of 0. For a similar reason, neither does $f(0^-)$.

Example 5.16 If
$$f(x) = \begin{cases} x, & x > 0 \\ \dfrac{1}{x}, & x < 0, \end{cases}$$
then $\lim_{x \to 0^+} f(x) = 0$, while $\lim_{x \to 0^-} f(x)$ does not exist.

In the extended real numbers $\bar{\mathbb{R}}$, we can extend the definition of the limit of a function as we did with the limit of a sequence. This is quite straightforward, as it ultimately relies on how the neighborhoods of $+\infty$ and $-\infty$ were defined in Section 3.3.

Definition 5.3 Let $f : D \to \mathbb{R}$ and $c \in \hat{D}$. The limit of f at c is said to be ∞, symbolically

$$\lim_{x \to c} f(x) = \infty,$$

if, given any $M \in \mathbb{R}$, there is a $\delta > 0$ such that

$$x \in D, 0 < |x - c| < \delta \Rightarrow f(x) > M.$$

Similarly,

$$\lim_{x \to c} f(x) = -\infty,$$

if, given any $M \in \mathbb{R}$, there is a $\delta > 0$ such that

$$x \in D, 0 < |x - c| < 0 \Rightarrow f(x) < M.$$

Recalling the definition of a cluster point, we can say that $\infty \in \hat{D}$ if $(M, \infty) \cap D \neq \varnothing$ for every $M \in \mathbb{R}$, and $-\infty \in \hat{D}$ if $(-\infty, M) \cap D \neq \varnothing$ for every $M \in \mathbb{R}$. Now we can define the limits of f at $\pm \infty$.

Definition 5.4 Let $f : D \to \mathbb{R}$, where $\infty \in \hat{D}$. The limit of f at ∞ is ℓ,

$$\lim_{x \to \infty} f(x) = \ell,$$

if, given any $\varepsilon > 0$, there is a real number M such that

$$x \in D, x > M \Rightarrow |f(x) - \ell| < \varepsilon.$$

On the other hand, when $-\infty \in \hat{D}$,

$$\lim_{x \to -\infty} f(x) = m$$

if, given any $\varepsilon > 0$, there is an $M \in \mathbb{R}$ such that

$$x \in D, x < M \Rightarrow |f(x) - m| < \varepsilon.$$

Example 5.17

$$\lim_{x \to 0} \frac{1}{x^2} = \infty.$$

To prove this, let M be any positive number. We wish to make $1/x^2 > M$, that is, $x^2 < 1/M$. It suffices therefore to take $|x| < 1/\sqrt{M}$, which is achieved by choosing δ in Definition 5.3 such that $0 < \delta \leq 1/\sqrt{M}$.

Example 5.18
$$\lim_{x \to \infty} \frac{1}{x^2} = 0.$$

To see this, let ε be any positive number. The inequality
$$\left|\frac{1}{x^2} - 0\right| = \frac{1}{x^2} < \varepsilon$$
is equivalent to $x^2 > 1/\varepsilon$. By choosing $M = \sqrt{1/\varepsilon}$, we have
$$x > M \Rightarrow \left|\frac{1}{x^2} - 0\right| < \varepsilon.$$

Similarly, it is straightforward to show that
$$\lim_{x \to -\infty} \frac{1}{x^2} = 0.$$

EXERCISES 5.3

1. Modify the statements of Theorems 5.1 to 5.8 so that they apply to one-sided limits, then prove the first and third theorems.

2. Let $f, g : D \to \mathbb{R}$ and $c \in \hat{D}$. If c has a neighborhood U where
$$f(x) \leq g(x) \quad \text{for all } x \in D \cap U/\{c\},$$
prove the following:

   (a) If $\lim_{x \to c} f(x) = \infty$, then $\lim_{x \to c} g(x) = \infty$.
   (b) If $\lim_{x \to c} g(x) = -\infty$, then $\lim_{x \to c} f(x) = -\infty$.

3. Given $f : D \to \mathbb{R}$ and $c \in \hat{D}$, define each of the following statements:

   (a) $\lim_{x \to c^+} f(x) = \infty$
   (b) $\lim_{x \to c^+} f(x) = -\infty$

(c) $\lim_{x \to c^-} f(x) = \infty$
(d) $\lim_{x \to c^-} f(x) = -\infty$.

4. Given $f : D \to \mathbb{R}$ and $\pm\infty \in \hat{D}$, define each of the following statements:

   (a) $\lim_{x \to \infty} f(x) = \infty$
   (b) $\lim_{x \to \infty} f(x) = -\infty$
   (c) $\lim_{x \to -\infty} f(x) = \infty$
   (d) $\lim_{x \to -\infty} f(x) = -\infty$.

5. Determine the right-hand and the left-hand limits of $f(x)$ at c where they exist:

   (a) $f(x) = \dfrac{x}{x-2}$, $c = 2$
   (b) $f(x) = \sqrt{x+1} - \sqrt{x}$, $c = \infty$
   (c) $f(x) = \dfrac{x}{\sqrt{x^2+2}}$, $c = -\infty$
   (d) $f(x) = \dfrac{3x^3 - 5x + 7}{2x^3 + 4x^2 - 1}$, $c = -\infty$
   (e) $f(x) = \dfrac{p(x)}{q(x)}$, $c = \infty$, where p and q are polynomials in x
   (f) $f(x) = x[x]$, $c = 0$
   (g) $f(x) = x[x]$, $c = 1$.

6. State and prove Theorem 5.1 for the case $\ell = \infty$.

7. State and prove Theorem 5.1 for the case $c = \infty$.

8. Let $f : (0, \infty) \to \mathbb{R}$. Prove that

$$\lim_{x \to 0^+} f\left(\frac{1}{x}\right) = \ell \Leftrightarrow \lim_{x \to \infty} f(x) = \ell.$$

9. Let U be a neighborhood of ∞, and suppose $f, g : U \to \mathbb{R}$. If $g(x) > 0$ for all $x \in U$, and $\lim_{x \to \infty} f(x)/g(x) = \ell > 0$, prove that

$$\lim_{x \to \infty} g(x) = \infty \Leftrightarrow \lim_{x \to \infty} f(x) = \infty.$$

What can we say if $\ell < 0$?

10. Let $\lim_{x \to c} f(x) = \ell$ and $\lim_{x \to c} g(x) = \infty$. If $\ell > 0$ prove that $\lim_{x \to c} f(x)g(x) = \infty$. What can we say if $\ell < 0$, and if $\ell = 0$?

11. If $f : D \to \mathbb{R}$ and $c \in \hat{D}$, define

$$\limsup_{x \to c} f(x) = \inf_{h>0} \sup\{f(x) : |x - c| < h\}$$

$$\liminf_{x \to c} f(x) = \sup_{h>0} \inf\{f(x) : |x - c| < h\}.$$

Show that $\lim_{x \to c} f(x) = \ell \in \bar{\mathbb{R}}$ if and only if $\limsup_{x \to c} f(x) = \liminf f(x) = \ell$.

5.4 Monotonic Functions

As defined in Section 2.2, a function is monotonic if it is either increasing or decreasing, and strictly monotonic if it is either strictly increasing or strictly decreasing. Clearly, f is (strictly) increasing if and only if $-f$ is (strictly) decreasing. Therefore, to study the properties of monotonic functions, we need only consider those properties that pertain to increasing functions.

Theorem 5.10

If $f : [a, b] \to \mathbb{R}$ is increasing, then for any $c \in (a, b)$ the limits $\lim_{x \to c^+} f(x)$ and $\lim_{x \to c^-} f(x)$ exist and satisfy the inequality

$$\lim_{x \to c^-} f(x) \leq f(c) \leq \lim_{x \to c^+} f(x).$$

At the end-points, $\lim_{x \to a^+} f(x)$ and $\lim_{x \to b^-} f(x)$ both exist and satisfy

$$f(a) \leq \lim_{x \to a^-} f(x) \leq \lim_{x \to b^+} f(x) \leq f(b).$$

Proof. Since f is increasing,

$$f(a) \leq f(x) \leq f(b) \text{ for all } x \in [a, b].$$

With $a < c < b$, the function f is bounded on the interval $[a, b]$, and hence on the intervals $[a, c]$ and $[c, b]$. Let

$$\ell = \inf\{f(x) : c < x \leq b\}, \ m = \sup\{f(x) : a \leq x < c\}.$$

First we shall prove

$$\lim_{x \to c^+} f(x) = \ell. \tag{5.8}$$

Given any $\varepsilon > 0$, the definition of ℓ implies that there is a point $x_0 \in (c, b]$ such that
$$\ell + \varepsilon > f(x_0). \tag{5.9}$$
Let $\delta = x_0 - c > 0$. Using the increasing property of f, we see from (5.9) that
$$\begin{aligned} 0 < x - c < \delta &\Rightarrow c < x < x_0 \\ &\Rightarrow f(x) \leq f(x_0) \\ &\Rightarrow f(x) < \ell + \varepsilon. \end{aligned} \tag{5.10}$$

We also conclude from the definition of ℓ that
$$f(x) \geq \ell \text{ for all } x \in (c, b]. \tag{5.11}$$

Now the relations (5.10) and (5.11) imply that
$$x \in [a, b], 0 < x - c < \delta \Rightarrow |\ell - f(x)| < \varepsilon,$$

which proves (5.8). With a few obvious modifications on this argument we can also prove that
$$\lim_{x \to c^-} f(x) = m.$$

Since f is increasing, $f(c)$ is a lower bound of $\{f(x) : x \in (c, b]\}$ and an upper bound of $\{f(x) : x \in [a, c)\}$, which implies that $f(c) \leq f(c^+)$ and $f(c) \geq f(c^-)$, hence $f(c^-) \leq f(c) \leq f(c^+)$. The results pertaining to the end-points a and b are left as an exercise. □

Remark 5.4 If f is decreasing on $[a, b]$, then $-f$ is increasing, and we obtain from Theorem 5.10 the following relations:

$$\lim_{x \to c^+} f(x) = \sup\{f(x) : x \in (c, b]\} \text{ for all } c \in (a, b)$$
$$\lim_{x \to c^-} f(x) = \inf\{f(x) : x \in [a, c)\} \text{ for all } c \in (a, b)$$
$$\lim_{x \to c^+} f(x) \leq f(c) \leq \lim_{x \to c^-} f(x).$$

The last theorem in this chapter indicates that a monotonic function has a limit at "almost every point" of its domain, in the sense that the set of points where it may fail to have a limit is countable.

Theorem 5.11

Let (a,b) be any open interval in $\mathbb{R}$. If the function $f : (a,b) \to \mathbb{R}$ is monotonic, the set of points $A \subseteq (a,b)$ where f has no limit is countable. Furthermore, for every $c \in (a,b)\setminus A$

$$\lim_{x \to c} f(x) = f(c).$$

Proof. We shall assume, without loss of generality, that f is increasing. By Theorem 5.9, f has a limit at c if and only if $f(c^+) = f(c^-)$, and, by Theorem 5.10, such a limit can only be $f(c)$. Hence

$$f(c^-) < f(c^+) \Leftrightarrow c \in A.$$

If $c \in A$ choose a rational number r_c such that

$$f(c^-) < r_c < f(c^+).$$

To show that the function $c \mapsto r_c$ is injective, let c and d be distinct points in A, say $c < d$, then

$$f(c^-) < r_c < f(c^+), \; f(d^-) < r_d < f(d^+). \tag{5.12}$$

Since f is increasing, we have $f(c^+) \leq f(t) \leq f(d^-)$ for any point $t \in (c,d)$, hence $r_c < r_d$ in view of (5.12). Thus there is a one-to-one correspondence between A and a subset of $\mathbb{Q}$, which proves A is countable. $\square$

Exercises 5.4

1. Prove that the inverse function of a strictly monotonic function is also strictly monotonic and of the same type, i.e., both increasing or both decreasing.

2. If f is increasing on (a,b) and unbounded above, prove that $f(b^-) = \infty$. What can we say about $f(b^-)$ if f were decreasing on (a,b) and unbounded below?

3. If f and g are increasing functions on D, prove that their sum $f + g$ is also increasing on D. What can you say about $f \cdot g$?

4. Let $f : \mathbb{R} \to \mathbb{R}$ satisfy the relation
$$f(x+y) = f(x) + f(y) \quad \text{for all } x, y \in \mathbb{R}.$$
If f is monotonic, prove that there is a real number α such that
$$f(x) = \alpha x \quad \text{for all } x \in \mathbb{R}.$$
Hint: First prove that $f(r) = \alpha r$ for all $r \in \mathbb{Q}$.

5. If $a > 1$, prove that $a^x \to 0$ as $x \to -\infty$, and $a^x \to \infty$ as $x \to \infty$. What can you say about $\lim_{x \to \pm\infty} a^x$ when $0 < a < 1$?

6. Let f be monotonically increasing on (a, b). If $c \in (a, b)$ prove that
$$\limsup_{x \to c} f(x) = f(c^+)$$
$$\liminf_{x \to c} f(x) = f(c^-).$$

6

Continuity

6.1 Continuous Functions

When we say that a function $f : D \to \mathbb{R}$ has a *limit* at the point c, it is tacitly assumed that c is a cluster point of D, which may or may not belong to D. In general, $\lim_{x \to c} f(x)$ has no relation to $f(c)$, the *value* of f at c. For example the function defined on $\mathbb{R}$ by

$$f(x) = \begin{cases} x^2, & x \neq 2 \\ 5, & x = 2 \end{cases}$$

has a limit at every point $c \in \mathbb{R}$, including the point 2, given by c^2. This coincides with the value of f at c, except when $x = 2$. At $x = 2$ we have $f(2) = 5 \neq \lim_{x \to 2} f(x) = 2^2 = 4$.

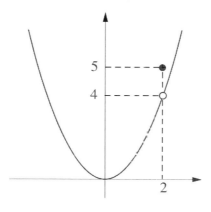

Figure 6.1

The result, as can be seen in Figure 6.1, is that the graph of $y = f(x)$ goes through an abrupt "jump" at $x = 2$ as it passes through the point $(2, 5)$. Clearly, if $f(2) = 4$, the point $(2, 4)$ would fill the vacant spot on the curve representing f, and the curve would thereby acquire a quality of "graphical continuity" which would otherwise be lacking. Our goal now is to characterize this graphical quality analytically.

Definition 6.1 Let $f : D \to \mathbb{R}$ and $c \in D$. The function f is said to be *continuous* at c if, given any $\varepsilon > 0$, there is a $\delta > 0$ such that

$$x \in D, |x - c| < \delta \Rightarrow |f(x) - f(c)| < \varepsilon. \tag{6.1}$$

This definition has an apparent resemblance to Definition 5.1 of the limit of a function at c. Here are some of the similarities and differences:

1. The limit of a function f is defined at a cluster point of its domain D (which may or may not lie in D), while continuity is defined only at a point in D (though it may be isolated in D).

2. Let $c \in D \cap \hat{D}$. If f is continuous at c, it is clear from Definition 6.1 that $\lim_{x \to c} f(x)$ exists and equals $f(c)$, since the condition $0 < |x - c| < \delta$ implies the condition $|x - c| < \delta$. On the other hand, if $\lim_{x \to c} f(x)$ exists, call it ℓ, then f is continuous at c provided $\ell = f(c)$, since $x = c$ in (6.1) implies $|f(c) - f(c)| = 0 < \varepsilon$. Thus we can say that, when $c \in D \cap \hat{D}$, the function f is continuous if and only if

$$\lim_{x \to c} f(x) = f(c). \tag{6.2}$$

3. If c is an isolated point in D, there is a $\delta > 0$ such that

$$(c - \delta, c + \delta) \cap D = \{c\},$$

in which case

$$f(x) - f(c) = 0 \quad \text{for all } x \in (c - \delta, c + \delta) \cap D.$$

With this choice of δ the implication (6.1) is satisfied for any positive number ε, which means f is continuous at every isolated point in its domain.

As was the case with Definition 5.1, we can also express Definition 6.1 in terms of neighborhoods as follows: $f : D \to \mathbb{R}$ is continuous at $c \in D$ if, for every neighborhood $(f(c) - \varepsilon, f(c) + \varepsilon)$ of the point $f(c)$, there is a neighborhood $(c - \delta, c + \delta)$ of the point c such that

$$x \in (c - \delta, c + \delta) \cap D \Rightarrow f(x) \in (f(c) - \varepsilon, f(c) + \varepsilon).$$

This is equivalent to saying: $f : D \to \mathbb{R}$ is continuous at $c \in D$ if, for every neighborhood V of $f(c)$, there is a neighborhood U of c such that

$f(U \cap D) \subseteq V$. Note how this last statement, brief as it is, captures the essence of Definition 6.1 in its full generality.

We shall say that f is *continuous on* $E \subseteq D$ if f is continuous at every point in E, and simply *continuous* if it is continuous on all D. The largest subset of D on which f is continuous will be referred to as the *domain of continuity of* f.

Using Theorem 5.1, continuity at a point may be expressed in terms of sequences.

Theorem 6.1
A function $f : D \to \mathbb{R}$ is continuous at $c \in D$ if, and only if, for every sequence (x_n) in D which converges to c, the sequence $(f(x_n))$ converges to $f(c)$.

Proof. If c is isolated in D, the result is obvious since f is then continuous at c, and every sequence (x_n) in D which converges to c must have a fixed tail. Consequently $f(x_n) \to f(c)$.

If $c \in \hat{D}$ then f is continuous at c if, and only if,

$$\lim_{x \to c} f(x) = f(c).$$

Hence, by Theorem 5.1 and Remark 5.2.1, $f(x_n) \to f(c)$ for any sequence (x_n) in D which converges to c. □

Corollary 6.1 *A function $f : D \to \mathbb{R}$ is not continuous at $c \in D$ if, and only if, there is a sequence (x_n) in D which converges to c, but whose sequence of images under f, $(f(x_n))$, does not converge to $f(c)$.*

Example 6.1 For any polynomial on $\mathbb{R}$,

$$f(x) = a_n x^n + \cdots + a_1 x + a_0,$$

where $n \in \mathbb{N}_0$, we have already seen in Example 5.7 that $\lim_{x \to c} f(x) = f(c)$ for all $c \in \mathbb{R}$. Hence every polynomial on $\mathbb{R}$ is continuous.

Example 6.2 The function $f : \mathbb{R} \to \mathbb{R}$ defined by

$$f(x) = \begin{cases} \dfrac{\sin x}{x}, & x \neq 0 \\ 0, & x = 0 \end{cases}$$

is continuous on $\mathbb{R} \setminus \{0\}$ since, by Theorem 5.6(iii) and Example 5.9, we have

$$\lim_{x \to c} f(x) = \lim_{x \to c} \frac{\sin x}{x}$$
$$= \frac{\sin c}{c}, \quad c \neq 0$$
$$= f(c).$$

At $c = 0$, using the result of Example 5.10,

$$\lim_{x \to 0} f(x) = \lim_{x \to 0} \frac{\sin x}{x} = 1.$$

Since $f(0) = 0 \neq \lim_{x \to 0} f(x)$ we conclude that f is not continuous at 0. Note that, had the function f been defined at $x = 0$ to be $f(0) = 1$, it would have been continuous at that point, and hence on $\mathbb{R}$.

Example 6.3 The function sgn x defined in Example 5.5 is continuous on $(0, \infty)$, where it has the constant value 1, and also on $(-\infty, 0)$ for a similar reason. At the cluster point $x = 0$ we saw in Example 5.13 that

$$\lim_{x \to 0+} \text{sgn } x = 1, \quad \lim_{x \to 0-} \text{sgn } x = -1,$$

hence $\lim_{x \to 0} \text{sgn } x$ does not exist, so the condition (6.2) is not satisfied. sgn x is therefore not continuous at $x = 0$.

Example 6.4 The function defined on $(0, \infty)$ by the rule

$$f(x) = \frac{1}{x}$$

is continuous, since

$$\lim_{x \to c} \frac{1}{x} = \frac{1}{c} \quad \text{for all } x > 0.$$

But this function cannot be defined at $x = 0$ so as to become continuous at that point, since $\lim_{x \to 0} 1/x$ does not exist.

Example 6.5 The function defined on $\mathbb{R} \setminus \{0\}$ by

$$f(x) = \sin \frac{1}{x}$$

is continuous, but it has no limit at $x = 0$ (see Example 5.9). Its domain of continuity cannot, therefore, be extended to $\mathbb{R}$. But the function

$$g(x) = x \sin \frac{1}{x}, \quad x \neq 0$$

has a limit at $x = 0$. In Example 5.11 it was shown that $\lim_{x \to 0} g(x) = 0$, so we can extend the definition of g to $\mathbb{R}$ as a continuous function by setting $g(0) = 0$, so that

$$g(x) = \begin{cases} x \sin \dfrac{1}{x}, & x \neq 0 \\ 0, & x = 0. \end{cases}$$

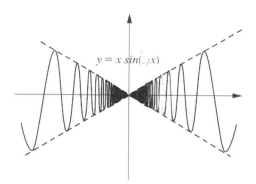

Figure 6.2

Example 6.6 The function $f : \mathbb{R} \to \mathbb{R}$ defined by

$$f(x) = \begin{cases} 1, & x \in \mathbb{Q} \\ 0, & x \notin \mathbb{Q} \end{cases}$$

has no limit at any point in $\mathbb{R}$ (example 5.6), so it is nowhere continuous.

Example 6.7 In Example 5.3 we found that the function defined on $(0, 1]$ by

$$f(x) = \begin{cases} 0, & x \notin \mathbb{Q} \cap (0, 1] \\ \dfrac{1}{q}, & x = \dfrac{p}{q} \in \mathbb{Q} \cap (0, 1], \end{cases}$$

where p/q is in simplest form, has the limit 0 at every point in $(0, 1]$. Consequently, f is continuous at every irrational point in $(0, 1]$, but not continuous at any rational point in $(0, 1]$. Can we define $f(0)$ so that f becomes continuous at $x = 0$?

From these examples we conclude that the discontinuity of the function $f : D \to \mathbb{R}$ at $c \in D$ may be due to different causes, but they all emanate from the failure to achieve the equality

$$\lim_{x \to c} f(x) = f(c), \tag{6.3}$$

164   Elements of Real Analysis

either because the limit of f at c does not exist, or because it is different from $f(c)$. This allows us to classify discontinuities into two main types:

1. If $\lim_{x \to c} f(x)$ exists but does not equal $f(c)$, f may be redefined at c so as to satisfy Equation (6.3), as pointed out in Example 6.2. This kind of discontinuity is said to be *removable* (see Figure 6.3).

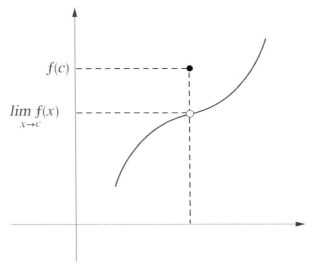

Figure 6.3

2. If, on the other hand, $\lim_{x \to c} f(x)$ does not exist, then the discontinuity at c is *non-removable*. In this case the one-sided limits $\lim_{x \to c^{\pm}} f(x)$ may or may not exist, and f may or may not be bounded in the neighborhood of c. Consequently, we can further classify non-removable discontinuities as follows:

   (a) If the right and the left-sided limits at c both exist but are not equal, the discontinuity is characterized by a "jump" in the value of the function at c, given by

   $$\lim_{x \to c^+} f(x) - \lim_{x \to c^-} f(x),$$

   as shown in Figure 6.4.

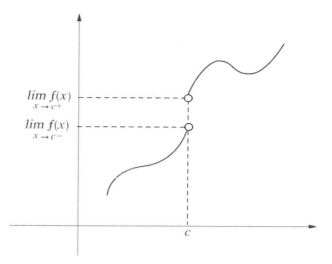

Figure 6.4

This type is called a *jump discontinuity*. Since $f(c^+)$ and $f(c^-)$ both exist, f is necessarily bounded in a neighborhood of c.

(b) If one (or both) of the limits $f(c^+)$ and $f(c^-)$ does not exist, then neither does $\lim_{x\to c} f(x)$. If, furthermore, there is a neighborhood U of c such that f is bounded on $D \cap U$, then the discontinuity of f at c is called *oscillatory*. The function $\sin(1/x)$, which is bounded on $(0, \infty)$ and has no limit at 0, provides a typical example of such a discontinuity (see Figure 6.5). In Examples 6.6 and 6.7 the discontinuities are also oscillatory.

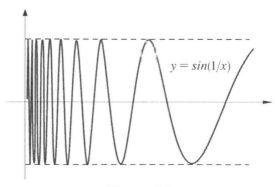

Figure 6.5

(c) If f is unbounded in every neighborhood of c, then, by Theorem 5.4, it cannot have a limit at c. f is then said to have an *infinite discontinuity* at c. Examples of infinite discontinuities at $x = 0$ are provided by the functions

$$f_1(x) = \frac{1}{x^2}, \quad f_2(x) = \frac{1}{x}, \quad f_3(x) = \frac{1}{x}\sin\frac{1}{x},$$

whose graphs are shown in Figure 6.6. Note that f_3, in addition to being unbounded in the neighborhood of 0, also exhibits oscillatory behaviour as $x \to 0$.

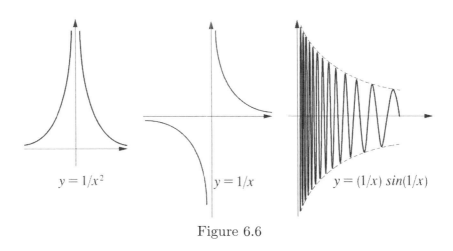

Figure 6.6

The points where a function $f : D \to \mathbb{R}$ is not continuous are often referred to as *singular points* of f. A singular point is therefore removable if f has a limit at that point, otherwise it is non-removable.

EXERCISES 6.1

1. Write the proof of Corollary 6.1.

2. Determine the domain of continuity for each of the following functions:

    (a) $f(x) = |x|$

(b) $g(x) = \begin{cases} \dfrac{|x|}{x}, & x \neq 0 \\ 1, & x = 0 \end{cases}$

(c) $h(x) = \begin{cases} \dfrac{1}{x^2} \sin x, & x \neq 0 \\ 0, & x = 0 \end{cases}$

3. Determine the points of discontinuity for the functions in Exercise 6.1.2 and their type.

4. Let $f : D \to \mathbb{R}$ be continuous at a point $c \in D \cap \hat{D}$. If $f(c) > 0$, prove that there is a positive number d and a neighborhood $(c - \delta, c + \delta)$ of c, where $\delta > 0$, such that $f(x) \geq d$ for all $x \in (c - \delta, c + \delta) \cap D$.

5. Let f be defined on $(-1, 1)$ and satisfy
$$|f(x)| \leq |x| \quad \text{for all } x \in (-1, 1).$$
Prove that f is continuous at $x = 0$.

6. Give an example of a function $f : D \to \mathbb{R}$ which is not continuous at $c \in D$, but whose absolute value
$$|f|(x) = |f(x)| \quad \text{for all } x \in D$$
is continuous at c.

7. Give an example of a function $f : D \to \mathbb{R}$ which is not continuous at any point in D, but whose absolute value $|f|$ is continuous on D.

8. If $f : D \to \mathbb{R}$ is continuous at $c \in D$, prove that $|f|$ is continuous at c.

9. Where possible, define each of the following functions at $x = 0$ so as to make it continuous at that point:

(a) $f(x) = \dfrac{1}{x} \sin 3x, \ x \neq 0$.

(b) $g(x) = \dfrac{1}{x} \sin x^2, \ x \neq 0$.

(c) $h(x) = \dfrac{1}{x} \sin \sqrt{x}, \ x > 0$.

(d) $k(x) = \dfrac{1}{\sqrt{|x|}} \sin x$, $x \neq 0$.

10. The function $f : \mathbb{R} \to \mathbb{Z}$ defined by $f(x) = [x]$ is called a *step function*. Determine the domain of continuity of each of the following functions:

    (a) $f(x) = [x]$.
    (b) $g(x) = x[x]$.
    (c) $h(x) = [\sin x]$.

11. If $f : D \to \mathbb{R}$ is continuous, prove that the set $\{x \in D : f(x) = 0\}$ is closed in D.

12. Let $f : D \to \mathbb{R}$ and g be the restriction of f to $E \subseteq D$.

    (a) If f is continuous at $c \in E$, prove that g is also continuous at c.
    (b) Give an example to show that the converse is false, that is, the continuity of g at c does not imply the continuity of f at c.

13. If the function $f : \mathbb{R} \to \mathbb{R}$ is continuous and $f(x) = 0$ for all $x \in \mathbb{Q}$, prove that $f(x) = 0$ for all $x \in \mathbb{R}$.

14. Suppose $f : D \to [0, \infty)$. If f is continuous on D, prove that the function $\sqrt{f}$, defined by $\sqrt{f}(x) = \sqrt{f(x)}$, is also continuous on D.

15. A polynomial q is said to have a *zero of degree* m at $x = c$, where $m \in \mathbb{N}$, if the limit
$$\lim_{x \to c} \frac{q(x)}{(x-c)^m}$$
exists and is not 0. If q has a zero at c of degree m and p is a polynomial, prove that the rational function p/q can be defined at c so as to be continuous at that point if, and only if, c is a zero of p of degree $k \geq m$.

16. Let $f : \mathbb{R} \to \mathbb{R}$ be defined by
$$f(x) = \begin{cases} x, & x \in \mathbb{Q} \\ 0, & \mathbb{R} \setminus \mathbb{Q}. \end{cases}$$
Where is f continuous, and where is it discontinuous?

17. Prove that a monotonic function on (a,b) can only have jump discontinuities. If the function f is defined at a and b so that f is monotonic on $[a,b]$, what type of discontinuities can f have at a and b?

18. Prove that the set of singular points of a monotonic function on a real interval is countable.

19. Let $f : \mathbb{R} \to \mathbb{R}$. What can you say about the cardinality of the set of points where f has a jump discontinuity?

6.2 Combinations of Continuous Functions

As we did in Section 5.2 with regard to limit operations, we now consider the conditions under which the continuity of two functions is transmitted to their sum, product, quotient and composition. Because of the close connection between limits and continuity, we should expect many parallel results to those of Section 5.2.

Theorem 6.2
If the functions $f : D \to \mathbb{R}$ and $g : D \to \mathbb{R}$ are continuous at $c \in D$, then
(i) $f + g$ and $f \cdot g$ are both continuous at c.
(ii) f/g is continuous at c provided $g(c) \neq 0$.

Proof. (i) If c is an isolated point in D then all the functions mentioned are continuous, so let $c \in \hat{D}$. Since f and g are continuous at c, we can use Theorem 5.6 to write

$$\begin{aligned}
\lim_{x \to c}(f+g)(x) &= \lim_{x \to c}[f(x) + g(x)] \\
&= \lim_{x \to c} f(x) + \lim_{x \to c} g(x) \\
&= f(c) + g(c) \\
&= (f+g)(c).
\end{aligned}$$

Therefore the function $f + g$ is continuous at c.
Similarly, we can prove the continuity of $f \cdot g$ and f/g where $g(c) \neq 0$.
□

As a special case of (i), kf is continuous for any constant k, hence $f - g = f + (-1)g$ is continuous wherever f and g are. Furthermore, if

f and g are continuous on D, then so is $f+g$, $f \cdot g$ and f/g, except of course where $g(x) = 0$ in the case of the quotient.

Example 6.8 Since every polynomial on $\mathbb{R}$ is a continuous function, a rational function p/q, where p and q are polynomials, is continuous on $\mathbb{R}$ except at the real zeros of q. Thus

$$f(x) = \frac{x^2 - 1}{x^2 - 4}, \quad x \neq \pm 2,$$

is continuous on $\mathbb{R}\setminus\{-2, 2\}$, and since $\lim_{x \to 2} f(x)$ and $\lim_{x \to -2} f(x)$ do not exist, f is not continuous at the points ± 2 no matter how f is defined there. Clearly the singular points ± 2 are of the infinite type. The rational function

$$g(x) = \frac{x^2 - 1}{x^2 + 1}, \quad x \in \mathbb{R},$$

is continuous on $\mathbb{R}$, since $x^2 + 1$ has no zeros in $\mathbb{R}$. On the other hand,

$$h(x) = \frac{x^2 - 1}{x - 1}, \quad x \neq 1,$$

is continuous on $\mathbb{R}\setminus\{1\}$, but since

$$\lim_{x \to 1} h(x) = \lim_{x \to 1} \frac{(x-1)(x+1)}{x-1}$$
$$= \lim_{x \to 1} (x + 1)$$
$$= 2,$$

the singular point $x = 1$ is removable. By defining $h(1) = 2$, the function h becomes continuous on $\mathbb{R}$.

Example 6.9 We saw in Example 5.8 that, for any $c \in \mathbb{R}$,

$$\lim_{x \to c} \sin x = \sin c$$
$$\lim_{x \to c} \cos x = \cos c,$$

which means these two functions are continuous on $\mathbb{R}$.

Now Theorem 6.2 allows us to conclude that the other trigonometric functions

$$\tan x = \frac{\sin x}{\cos x}, \quad x \neq \left(n + \frac{1}{2}\right)\pi, n \in \mathbb{Z},$$

$$\cot x = \frac{\cos x}{\sin x}, \quad x \neq n\pi, n \in \mathbb{Z},$$

$$\sec x = \frac{1}{\cos x}, \quad x \neq \left(n + \frac{1}{2}\right)\pi, n \in \mathbb{Z},$$

$$\csc x = \frac{1}{\sin x}, \quad x \neq n\pi, n \in \mathbb{Z},$$

are all continuous on their domains of definition. In each case the domain is $\mathbb{R}$ except for an infinite sequence of singular points. At each singular point the function has an infinite discontinuity.

Theorem 6.3
Let $f : D \to \mathbb{R}$, $g : E \to \mathbb{R}$, and $f(D) \subseteq E$. If f is continuous at $c \in D$ and g is continuous at $f(c)$, then the composite function $g \circ f : D \to \mathbb{R}$ is continuous at c.

Proof. We shall use Theorem 6.1 to prove the continuity of $g \circ f$. Let (x_n) be a sequence in D which converges to c. Since f is continuous at c, the sequence $(f(x_n))$ converges to $f(c)$, by Theorem 6.1. The same theorem implies, due to the continuity of g at $f(c)$, that $g(f(x_n)) \to g(f(c))$. In other words,

$$\lim g \circ f(x_n) = g \circ f(c). \quad \square$$

From this theorem it follows that if $f : D \to \mathbb{R}$ is continuous on D, and $g : E \to \mathbb{R}$ is continuous on E, where $f(D) \subseteq E$, then $g \circ f : D \to \mathbb{R}$ is continuous on D.

Example 6.10 Since

$$||x| - |c|| \leq |x - c| \quad \text{for all } x, c \in \mathbb{R},$$

the function $g(x) = |x|$ is continuous on $\mathbb{R}$. To see that, it suffices to take $\delta = \varepsilon$ in Definition 6.1. In view of Theorem 6.3, if $f : D \to \mathbb{R}$ is continuous, its absolute value $|f|$ is also continuous on D, for $|f| = g \circ f$.

Example 6.11 In Example 3.9 we saw that if (x_n) is a non-negative sequence which converges to x, then $\sqrt{x_n} \to \sqrt{x}$. Therefore, by Theorem 6.1, the function $g(x) = \sqrt{x}$ is continuous on $[0, \infty)$. Now, if $f : D \to [0, \infty)$ is continuous, the function $\sqrt{f} = g \circ f$ is also continuous on D.

Example 6.12 For any polynomial

$$p(x) = a_n x^n + \cdots + a_1 x + a_0$$

on $\mathbb{R}$, Theorem 6.3 implies that the functions

$$(p \circ \sin)(x) = a_n (\sin x)^n + \cdots + a_1 \sin x + a_0,$$
$$(\sin \circ p)(x) = \sin(a_n x^n + \cdots + a_1 x + a_0),$$

are both continuous on $\mathbb{R}$. If

$$f(x) = \frac{x}{x-1}, \quad x \neq 1,$$

then the composition

$$(f \circ \sin)(x) = \frac{\sin x}{\sin x - 1}$$

is continuous on its domain $\mathbb{R} \setminus \{\pi/2 + 2n\pi : n \in \mathbb{Z}\}$. But

$$(\sin \circ f)(x) = \sin\left(\frac{x}{x-1}\right)$$

is defined and continuous on $\mathbb{R} \setminus \{1\}$.

EXERCISES 6.2

1. Determine whether each of the following functions is continuous at $x = 0$:

    (a) $f(x) = \sqrt[3]{x}, \quad x \in \mathbb{R}$.
    (b) $g(x) = \sqrt{x + \sqrt{x}}, \quad x \geq 0$.
    (c) $h(x) = \begin{cases} \dfrac{|\sin x|}{x}, & x \neq 0 \\ 1, & x = 0. \end{cases}$

2. If $f : D \to \mathbb{R}$ is continuous, prove that the function f^n defined by

    $$f^n(x) = (f(x))^n, \quad x \in D, n \in \mathbb{N},$$

    is also continuous. Give an example to show that the converse is false.

3. Suppose $f, g : D \to \mathbb{R}$ are both continuous. Prove that the functions $f \vee g$ and $f \wedge g$ defined on D by

$$(f \vee g)(x) = f(x) \vee g(x) = \max\{f(x), g(x)\},$$
$$(f \wedge g)(x) = f(x) \wedge g(x) = \min\{f(x), g(x)\},$$

are continuous. Hint: Show that $f \vee g = \frac{1}{2}(f + g + |f - g|)$ and $f \wedge g = \frac{1}{2}(f + g - |f - g|)$.

4. Give an example of two functions, one of which is not continuous at c, whose product is continuous at c.

5. Give an example of two functions, one of which is not continuous at c, whose composition is continuous at c.

6. Determine the domain of continuity of the function $f(x) = x - [x]$ for all $x \in \mathbb{R}$.

7. Suppose $f : \mathbb{R} \to \mathbb{R}$ satisfies $f(x+y) = f(x) + f(y)$ for all $x, y \in \mathbb{R}$. If f has a limit at some point in $\mathbb{R}$, use the results of Exercise 5.2.9 to prove that f is continuous on $\mathbb{R}$.

8. Under the hypothesis of Exercise 5.2.7, if $f(1) = \alpha$ prove that $f(x) = \alpha x$ for all $x \in \mathbb{R}$. Hint: Start with $x \in \mathbb{Q}$.

9. Suppose $f : \mathbb{R} \to \mathbb{R}$ satisfies $f(x+y) = f(x)f(y)$ for all $x, y \in \mathbb{R}$. If f has a limit at some point in $\mathbb{R}$, use the results of Exercise 5.2.10 to show that f is continuous on $\mathbb{R}$.

10. Under the hypothesis of Exercise 6.2.9, if $f(1) = a$, prove that $f(x) = a^x$ for all $x \in \mathbb{R}$. Hint: Start with $x \in \mathbb{Q}$.

6.3 Continuity on an Interval

Intervals are the graphically connected subsets of the real line. Thus, to study the properties of continuity for a function of a real variable, it is only natural to start with a function defined on an interval, for then the implications and significance of continuity are more apparent.

In Theorem 5.4 we defined what it means for a function to be bounded on a neighborhood of a point. More generally, we have the following definition.

Definition 6.2 The function $f : D \to \mathbb{R}$ is *bounded* on $E \subseteq D$ if there is a constant $M \geq 0$ such that

$$|f(x)| \leq M \quad \text{for all } x \in E.$$

When f is bounded on D, it is said to be *bounded*.

Clearly, $f : D \to \mathbb{R}$ is bounded if and only if its range $f(D)$ is a bounded set. Thus $\sin x$ is bounded on $\mathbb{R}$ (by any number $M \geq 1$), but $f(x) = x^2$ is not bounded on $\mathbb{R}$.

Theorem 6.4
Let $I = [a,b]$ be a closed and bounded interval in $\mathbb{R}$. If the function $f : I \to \mathbb{R}$ is continuous, it is bounded.

Proof. Assuming f is is not bounded on I, there is, for every $n \in \mathbb{N}$, a point $x_n \in I$ such that $|f(x_n)| > n$. Now the sequence (x_n) is bounded, being in I. By the Bolzano-Weierstrass theorem, it has a subsequence (x_{n_k}) which converges to some limit, say x. Since I is closed, $x \in I$ by Theorem 3.20. Therefore $f(x_{n_k}) \to f(x)$, by the continuity of f. Being convergent, the sequence $(f(x_{n_k}))$ is necessarily bounded, by Theorem 3.2, which contradicts the assumption that $|f(x_{n_k})| > n_k$ for every $k \in \mathbb{N}$. □

Definition 6.3 The function $f : D \to \mathbb{R}$ is said to have a *minimum* on D if there is a point $x_1 \in D$ such that

$$f(x) \geq f(x_1) \quad \text{for all } x \in D,$$

and a *maximum* on D if there is a point $x_2 \in D$ such that

$$f(x) \leq f(x_2) \quad \text{for all } x \in D.$$

When they exist, $f(x_1)$ and $f(x_2)$ are called the *minimum value* and the *maximum value*, respectively, of f. An *extremum value* of f is either a minimum or a maximum value of the function.

When f has a minimum at x_1 and a maximum at x_2, we clearly have

$$f(x_1) \leq f(x) \leq f(x_2) \quad \text{for all } x \in D,$$

or $f(D) \subseteq [f(x_1), f(x_2)]$. Referring to Definition 2.6, we see that

$$f(x_1) \leq \inf f(D), \quad f(x_2) \geq \sup f(D).$$

But since $f(x_1), f(x_2) \in f(D)$, we conclude that

$$f(x_1) = \inf f(D), \quad f(x_2) = \sup f(D).$$

Hence, when it exists, an extremum value of a function is unique. But the function may take on that value at more than one point, such as the sine function on $\mathbb{R}$, which assumes its maximum value 1 at $\pi/2 + 2n\pi$ for all $n \in \mathbb{Z}$, and its minimum value -1 at $3\pi/2 + 2n\pi$ for every $n \in \mathbb{Z}$.

Not every function has an extremum, however, as the following examples show:

1. The function $f(x) = x^2$ defined on $\mathbb{R}$ has a minimum at $x = 0$ whose value is $f(0) = 0$, but no maximum. If this function is redefined at $x = 0$ so that $f(0) = 1$ (or any other positive number), it will no longer have a minimum.

2. $f(x) = x^2$ defined on $(0, 1)$ has neither a maximum nor a minimum. Note that $\sup f(D) = 1$ and $\inf f(D) = 0$, and both points are outside $f(D) = (0, 1)$. However, if the function is extended, using the same rule, to $D' = [0, 1]$ then $f(D') = [0, 1]$, and the function acquires a maximum at $x = 0$ and a minimum at $x = 1$.

3. The function $f : [0, 1] \to \mathbb{R}$ defined by

$$f(x) = \begin{cases} 1/x, & x \neq 0 \\ k, & x = 0 \end{cases}$$

has no maximum, regardless of the value of k, because $\sup f([0, 1]) = \infty$.

Thus, it would seem, the existence of an extremum for a function depends on its behaviour and its domain of definition.

Theorem 6.5
If I is a closed and bounded set, and the function $f : I \to \mathbb{R}$ is continuous, then f has a maximum and a minimum on I.

Proof. We know from Theorem 6.4 that $f(I)$ is bounded in $\mathbb{R}$, hence

$$M = \sup f(I), \quad m = \inf f(I)$$

exist, by the completeness of $\mathbb{R}$. It remains to prove that M and m lie in $f(I)$.

From the defining property of $\sup f(I)$, we know that, for every $n \in \mathbb{N}$, there is a point $x_n \in I$ such that

$$M - \frac{1}{n} < f(x_n) \leq M,$$

which implies

$$\lim f(x_n) = M.$$

Since (x_n) is a bounded sequence, being in I, it has a subsequence (x_{n_k}) which converges to some limit $c \in I$, I being closed. By Theorems 6.1 and 3.13,

$$f(c) = \lim f(x_{n_k}) = \lim f(x_n) = M.$$

Hence $M \in f(I)$.

On the other hand, $-f$ is also continuous on I and, by the same logic, there is a point $d \in I$ such that

$$(-f)(d) = \sup(-f)(I)$$
$$\Rightarrow -f(d) = -\inf f(I)$$
$$\Rightarrow f(d) = \inf f(I) = m. \square$$

Needless to say, there are functions which do not satisfy the conditions of Theorem 6.5 but which have maxima and minima, so these conditions are sufficient but not necessary for a function to achieve its extrema.

We should expect the graph of a continuous function on an interval to be a "connected" curve in the xy-plane, as pointed out in the beginning of this chapter. The next theorem, called the *intermediate value theorem*, characterizes this basic property in analytical terms.

Theorem 6.6 (Intermediate Value Property)
Let $f : [a, b] \to \mathbb{R}$ be continuous. If λ is a real number between $f(a)$ and $f(b)$, then there is a point $c \in (a, b)$ such that $f(c) = \lambda$.

Proof. Assume $f(a) < f(b)$, otherwise consider $-f$ instead of f. Let

$$S = \{x \in [a, b] : f(x) < \lambda\}.$$

The set S is clearly not empty, as it includes a, and is bounded above (by b). Therefore it has a supremum which lies in $[a, b]$. Let $c = \sup S$.

As we did in the proof of Theorem 6.5, we can form a sequence (x_n) in S which converges to c. Therefore $f(x_n) \to f(c)$ by the continuity of f. Since $f(x_n) < \lambda$ for every $n \in \mathbb{N}$, we can use Theorem 5.7 to write

$$f(c) \leq \lambda < f(b), \tag{6.4}$$

which implies $c < b$. Now we form a sequence (t_n) in $(c, b]$ which converges to c, such as

$$t_n = \min\left\{c + \frac{1}{n}, b\right\}.$$

Since $t_n \in [a, b]$ and $t_n > c = \sup S$, it follows that $t_n \notin S$ for any n. Hence $f(t_n) \geq \lambda$ for all $n \in \mathbb{N}$, and

$$f(c) = \lim f(t_n) \geq \lambda.$$

In view of (6.4) we must conclude that $f(c) = \lambda$. $\square$

It should be noted that the converse of this theorem is false. In other words, a function which satisfies the intermediate value property need not be continuous (see Exercise 6.3.8).

Corollary 6.6.1 *If I is an interval and $f : I \to \mathbb{R}$ is continuous, then $f(I)$ is an interval.*

Proof. Assuming $f(I)$ is a bounded set, let

$$\alpha = \inf f(I), \quad \beta = \sup f(I).$$

It is enough to show that

$$(\alpha, \beta) \subseteq f(I) \subseteq [\alpha, \beta].$$

The inclusion $f(I) \subseteq [\alpha, \beta]$ follows from the definition of α and β. To prove $(\alpha, \beta) \subseteq f(I)$, let $y \in (\alpha, \beta)$; hence $\alpha < y < \beta$. From the definition of the infimum and supremum, we know that there are points $x_1, x_2 \in I$ such that $\alpha \leq f(x_1) < y$ and $y < f(x_2) \leq \beta$. Therefore

$$f(x_1) < y < f(x_2).$$

By the Intermediate Value Theorem, applied to f on the closed interval with end-points x_1 and x_2, there is a point x between x_1 and x_2 (and hence in I) such that $y = f(x)$. This means $y \in f(I)$, hence $(\alpha, \beta) \subseteq f(I)$.

If $f(I)$ is bounded below and unbounded above, we have to show that $(\alpha, \infty) \subseteq f(I)$. This can be done by following the same procedure used above: Take any $y \in (\alpha, \infty)$ and show that there are $x_1, x_2 \in I$ such that $\alpha \leq f(x_1) < y < f(x_2) < \infty$. Now apply Theorem 6.6 to f on the closed interval determined by x_1 and x_2 to conclude that $y \in f(I)$.

The cases where $f(I)$ is unbounded below, or unbounded above and below, can be similarly treated. □

Corollary 6.6.2 *If f is continuous on a closed and bounded interval $[a,b]$, then $f([a,b])$ is a closed and bounded interval.*

Proof. By Theorem 6.5, f assumes its maximum M and minimum m on $[a,b]$, and we shall prove that

$$f([a,b]) = [m, M].$$

We clearly have $f([a,b]) \subseteq [m,M]$. By Corollary 6.6.1, $f([a,b])$ is an interval, and since it contains m and M, we must have $[m,M] \subseteq f([a,b])$. □

When $\lambda = 0$ in Theorem 6.6, we obtain the following result, which is often used to determine the zeros of a continuous function.

Corollary 6.6.3 *If the function $f : [a,b] \to \mathbb{R}$ is continuous, and its values at a and b have opposite signs, that is, either $f(a) < 0 < f(b)$ or $f(b) < 0 < f(a)$, then there is a point $x \in (a,b)$ where $f(x) = 0$.*

Example 6.13 Suppose we wish to approximate the irrational number $\sqrt{7}$. It is clearly a zero of the function

$$f(x) = x^2 - 7,$$

which is defined and continuous on $\mathbb{R}$. Since $f(2) = -3$ and $f(3) = 2$, f has a zero in the open interval $(2,3)$. As a first approximation of $\sqrt{7}$, we can take the mid-point of the interval $(2,3)$,

$$x_1 = \frac{2+3}{2} = 2.5.$$

Since $f(x_1) = -0.75$, f has a zero in $(2.5, 3)$. Let

$$x_2 = \frac{2.5 + 3}{2} = 2.75.$$

Now $f(x_2) = 0.5625$, hence f has a zero in $(2.5, 2.75)$. If we take

$$x_3 = \frac{2.5 + 2.75}{2} = 2.625$$

as an approximation of the positive root of the equation $x^2 - 7 = 0$, the error in the approximation will not exceed half the length of the interval $(2.5, 2.75)$, i.e.,

$$\frac{1}{2}(2.75 - 2.5) = \frac{1}{8}.$$

Hence $|2.625 - \sqrt{7}| < 1/8$, or $\sqrt{7} \approx 2.65$ up to an error of 0.125.

If we continue this process, we obtain a sequence (x_n) which satisfies

$$\left|x_n - \sqrt{7}\right| < \frac{1}{2^n}.$$

x_n clearly converges to $\sqrt{7}$, and provides an approximation of $\sqrt{7}$ to any degree of accuracy we desire.

Example 6.14 Let p be a real polynomial on $\mathbb{R}$ of degree n. If n is odd, the equation $p(x) = 0$ has at least one root in $\mathbb{R}$.

Proof. Suppose

$$p(x) = a_n x^n + \cdots + a_1 x + a_0, \quad x \in \mathbb{R},$$

and define the function $g : \mathbb{R}\setminus\{0\} \to \mathbb{R}$ by

$$g(x) = \frac{p(x)}{x^n} = a_n + \cdots + \frac{a_1}{x^{n-1}} + \frac{a_0}{x^n}.$$

Thus

$$\lim_{x \to \infty} g(x) = \lim_{x \to -\infty} g(x) = a_n \neq 0 \qquad (6.5)$$

since p has degree n.

If $a_n > 0$, we can use equation (6.5) and Definition 5.4 to conclude that there is a number $k > 0$ such that

$$g(x) \geq \frac{1}{2} a_n \text{ for all } |x| \geq k.$$

Since $p(x) = x^n g(x)$ and n is odd, we have

$$p(k) > 0, \ p(-k) < 0,$$

which implies, in view of Corollary 6.6.3, that $p(c) = 0$ for some $c \in (-k, k)$.

If $a_n < 0$, then
$$p(k) < 0, \quad p(-k) > 0$$
and we reach the same conclusion. □

Example 6.15 If $f : [0, 1] \to [0, 1]$ is a continuous function, prove that it has a *fixed point*, that is, a point $x_0 \in [0, 1]$ where $f(x_0) = x_0$.

Proof. Let $g(x) = f(x) - x$ on $[0, 1]$. We have to prove that the function g has a zero in $[0, 1]$. If $f(0) = 0$ we can take $x_0 = 0$, and if $f(1) = 1$ we take $x_0 = 1$. If $f(0) \neq 0$ and $f(1) \neq 1$, then
$$g(0) = f(0) > 0, \ g(1) = f(1) - 1 < 0.$$
Being the difference between two continuous functions on $[0, 1]$, g is continuous on $[0, 1]$ and, by Corollary 6.6.3, there is a point $x_0 \in (0, 1)$ where $g(x_0) = 0$ and hence $f(x_0) = x_0$. □

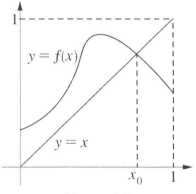

Figure 6.7

The geometric interpretation of this example is that the curve $y = f(x)$ of a continuous function $f : [0, 1] \to [0, 1]$ must intersect the line $y = x$ in the square $[0, 1] \times [0, 1]$, as shown in Figure 6.7.

Theorem 6.7
Let $f : I \to \mathbb{R}$, where I is an interval. If f is continuous and injective, then it is strictly monotonic.

Proof. We shall prove the theorem for the case when $I = [a, b]$, and leave the other cases as an exercise. Without loss of generality, we can assume $f(a) < f(b)$ and we have to show that f is strictly increasing (otherwise we consider $-f$).

First we shall prove that
$$f(a) < f(x) < f(b) \quad \text{for all } x \in (a,b). \tag{6.6}$$
Since f is injective, $f(x) \neq f(a)$ and $f(x) \neq f(b)$, and we have to exclude the two possibilities
$$f(x) < f(a) < f(b), \tag{6.7}$$
$$f(a) < f(b) < f(x). \tag{6.8}$$
If the inequality (6.7) were valid, then, by the Intermediate Value Theorem, there would exist a point $c \in (x,b)$ where $f(c) = f(a)$, which contradicts the injective property of f. Similarly, inequality (6.8) implies that there is a $c \in (a,x)$ where $f(c) = f(b)$, again contradicting the injectivity of f. This proves (6.6).

To complete the proof, we have to show that, if $x, y \in (a,b)$ and $x < y$, then $f(x) < f(y)$. From (6.6) we have
$$f(a) < f(y) < f(b).$$
If $f(y) < f(x)$, then
$$f(a) < f(y) < f(x),$$
in which case there would have to be a point $c \in (a,x)$ where $f(c) = f(y)$, thereby contradicting the injectivity of f. Therefore $f(y) > f(x)$. □

The next result is a sort of converse to Theorem 6.7.

Lemma 6.1 *Let f be a monotonic function on the interval I. If $f(I)$, the range of I, is an interval, the function f is continuous.*

Proof. As before, we shall only consider the case where f is strictly increasing. Assuming f is not continuous at some point $c \in I$, we shall show that this leads to a contradiction. The proof applies to the case where c is an interior point of I, but the argument can easily be modified if c is an end-point.

In view of Theorem 5.10, the discontinuity of f at c can only be a jump discontinuity, where $f(c^-) < f(c^+)$. But this is not consistent with the assumption that $f(I)$ is an interval (see Figure 6.8) for the following reason:

Let $y \neq f(c)$ be any point in the open interval $(f(c^-), f(c^+))$. Choose two sequences in I, (u_n) and (v_n), such that
$$u_n < c < v_n \quad \text{for all } n \in \mathbb{N}$$
$$f(u_n) \to f(c^-), \ f(v_n) \to f(c^+).$$

Since f is strictly increasing, Theorem 3.3 implies that there is a number N such that

$$f(u_n) < y < f(v_n) \quad \text{for all } n \geq N. \tag{6.9}$$

Since $f(I)$ is an interval, y lies in $f(I)$, so there is a point $x \in I$ where $y = f(x)$. The relation (6.9) and the monotonicity of f yield

$$u_n < x < v_n \quad \text{for all } n \geq N,$$

which implies $x = c$, the common limit of u_n and v_n. Consequently $y = f(c)$, which contradicts our choice of y. $\square$

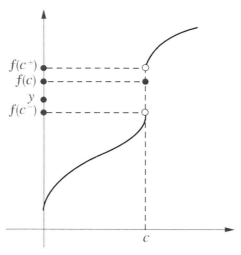

Figure 6.8

Theorem 6.8
Let $f : I \to \mathbb{R}$, where I is an interval. If f is continuous and injective, then f^{-1} is continuous and strictly monotonic.

Proof. By Theorem 6.7, f is strictly monotonic. To see that f^{-1} is also monotonic, suppose f is strictly increasing and $y_1, y_2 \in f(I)$ with $y_1 < y_2$. If $x_1 = f^{-1}(y_1)$ and $x_2 = f^{-1}(y_2)$, then $x_1 < x_2$; for the inequality $x_1 \geq x_2$ implies

$$y_1 = f(x_1) \geq f(x_2) = y_2,$$

since f is increasing. The continuity of $f^{-1} : f(I) \to I$ follows from Lemma 6.1, I and $f(I)$ being intervals. $\square$

Example 6.16 For every $n \in \mathbb{N}$, the function

$$f(x) = x^n$$

is continuous and strictly increasing on $[0, \infty)$. Its inverse function $g : [0, \infty) \to [0, \infty)$, defined by

$$g(x) = \sqrt[n]{x}$$

is therefore continuous and strictly increasing.

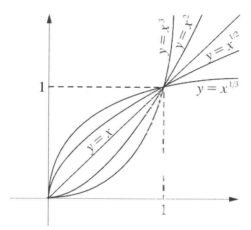

Figure 6.9

Example 6.17 Let f be defined on $[0, 1] \cup (2, 3]$ by

$$f(x) = \begin{cases} x, & 0 \leq x \leq 1 \\ x - 1, & 2 < x \leq 3. \end{cases}$$

It is a simple matter to verify that f is injective, and that its inverse function $g : [0, 2] \to [0, 1] \cup (2, 3]$ is defined by

$$g(x) = \begin{cases} x, & 0 \leq x \leq 1 \\ x + 1, & 1 < x \leq 2. \end{cases}$$

But, while f is continuous, g is not, as may be seen in Figure 6.10. Note here that the domain of f is not an interval. We shall return to this point in Theorem 6.16, where we take the domain of f to be closed and bounded.

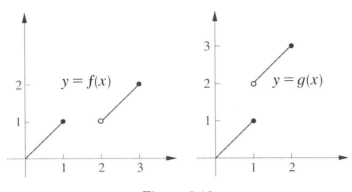

Figure 6.10

EXERCISES 6.3

1. Let the function $f : [a, b] \to \mathbb{R}$ be continuous and satisfy the condition: For every $x \in [a, b]$, there is a $y \in [a, b]$ such that $|f(y)| \leq \frac{1}{2}|f(x)|$. Prove that there is a $c \in [a, b]$ where $f(c) = 0$.

2. If the function $f : [a, b] \to \mathbb{R}$ is continuous and satisfies $f(x) > 0$ for all $x \in [a, b]$, prove that there is a positive number α such that $f(x) > \alpha$ for all $x \in [a, b]$.

3. Give an example of a real polynomial which has no real root.

4. Give an example of a function from $[0, 1]$ to $[0, 1]$ which has no fixed point.

5. Let the functions f and g be continuous on $[a, b]$ and $f(a) \geq g(a)$, $f(b) \leq g(b)$. Prove that there is a point $x_0 \in [a, b]$ where $f(x_0) = g(x_0)$.

6. Prove that the equation $\cos x = x$ has a solution in $(0, \pi/2)$.

7. Prove that the equation $x2^x = 1$ has a solution in $(0, 1)$.

8. Prove that the function

$$f(x) = \begin{cases} \sin \frac{1}{x}, & x \neq 0 \\ 0, & x = 0 \end{cases}$$

satisfies the intermediate value property on $\mathbb{R}$, but is not continuous at $x = 0$.

9. Determine a real interval of length 1 where the equation
$$xe^x = 1$$
has a solution.

10. Determine a real interval of length $1/2$ where the equation
$$x^3 - 6x + 2.5 = 0$$
has a solution. Approximate the solution to two significant decimal figures, using the averaging method of Example 6.13.

11. Let f be continuous on $[0, 1]$, and suppose $f(0) = f(1)$. Prove the existence of a point $c \in [0, 1/2]$ where
$$f(c) = f(c + 1/2).$$
Hint: Use the function $g(x) = f(x) - f(x + 1/2)$.

Assuming the temperature f is a continuous function of distance x along a great circle on the surface of the earth, such as the equator, this result has the following physical interpretation: There are two opposite points on the circle, i.e., collinear with the center of the earth, which have the same temperature.

12. Let $f : [0, \pi/2] \to \mathbb{R}$ be defined by
$$f(x) = \max\{x^2, \cos x\}.$$
Prove that the function f has a minimum at some point $c \in [0, \pi/2]$ which satisfies $\cos c = c^2$.

13. Let f be a continuous function on $\mathbb{R}$ such that $\lim_{x \to -\infty} f(x) = \lim_{x \to \infty} f(x) = 0$. Prove that f is bounded and that it has at least one extremum. Give an example to show that f may not have both a maximum and a minimum.

6.4 Uniform Continuity

Suppose $f : D \to \mathbb{R}$ is continuous. If $c \in D$, Definition 6.1 states that for every $\varepsilon > 0$ there is a $\delta > 0$ such that

$$x \in D, |x - c| < \delta \Rightarrow |f(x) - f(c)| < \varepsilon. \tag{6.10}$$

Clearly, the number δ depends on the choice of ε as long as we are dealing with a fixed point c. But, if we move to a different point $c' \in D$ while keeping ε fixed, the same δ may not satisfy the implication (6.10). Since f is continuous at c' as well, there is a $\delta' > 0$, not necessarily equal to δ, such that

$$x \in D, |x - c'| < \delta' \Rightarrow |f(x) - f(c')| < \varepsilon.$$

In other words, the number δ in (6.10) depends, in general, on c as well as ε, and we sometimes write $\delta = \delta(\varepsilon, c)$ to emphasize this dependence.

Take the function $f : (0, 1) \to \mathbb{R}$ defined by

$$f(x) = \frac{1}{x},$$

for example. Clearly this rational function is continuous. Given a positive number ε, we wish to choose a $\delta > 0$ such that the implication (6.10) holds at $c \in (0, 1)$. Note that

$$|f(x) - f(c)| = \left| \frac{1}{x} - \frac{1}{c} \right|, \quad x \in (0, 1)$$

$$= \frac{|x - c|}{xc}. \tag{6.11}$$

Suppose that

$$|x - c| < \frac{1}{2}c. \tag{6.12}$$

By a simple calculation, we would then have

$$\frac{1}{2}c < x < \frac{3}{2}c$$

$$\Rightarrow \frac{1}{x} < \frac{2}{c}$$

$$\Rightarrow \frac{1}{cx} < \frac{2}{c^2}. \tag{6.13}$$

Now equation (6.11) and inequalities (6.12) and (6.13) imply

$$|f(x) - f(c)| < \frac{2}{c^2}|x - c|. \tag{6.14}$$

To satisfy the inequality $|f(x) - f(c)| < \varepsilon$ it suffices to make the right-hand side of (6.14) less than ε, i.e.

$$\frac{2}{c^2}|x - c| < \varepsilon$$
$$\Rightarrow |x - c| < \frac{1}{2}c^2\varepsilon. \tag{6.15}$$

Clearly, the inequality $|f(x) - f(c)| < \varepsilon$ follows from the two inequalities (6.12) and (6.15). Hence (6.10) becomes valid if we choose

$$\delta = \min\{\frac{1}{2}c, \frac{1}{2}c^2\varepsilon\}.$$

This choice of δ clearly depends on c as well as ε.

It is natural to ask, at this point, whether we can choose a δ in (6.10) which will work for all points $c \in D$, in other words a δ which depends only on ε. This would certainly be the case if D were a finite set, say $\{c_1, c_2, \cdots, c_n\}$; for then the choice

$$\delta = \min\{\delta(\varepsilon, c_i) : i = 1, 2, \cdots, n\},$$

which is a positive number, would serve the purpose. It may seem that the choice

$$\delta = \inf\{\delta(\varepsilon, c) : c \in D\}$$

would also work in the general case, but the problem then is that the infimum of an infinite set of positive numbers may not be positive, and the choice $\delta = 0$ is inadmissible according to definition 6.1. In fact, for the function $f(x) = 1/x$ discussed above, where $\delta(\varepsilon, c) = \min\{c/2, c^2\varepsilon/2\}$, we have

$$\inf\{\delta(\varepsilon, c) : c \in (0, 1)\} = 0.$$

But the domain of continuity is not the only factor which determines δ. Consider the function $g : \mathbb{R} \to \mathbb{R}$ defined by

$$g(x) = x.$$

Here we see that $\delta = \varepsilon$ meets the requirement
$$|x - c| < \delta \Rightarrow |g(x) - g(c)| < \varepsilon$$
for any $c \in \mathbb{R}$. This would seem to indicate that the function $g(x) = x$ possesses a type of continuity on $\mathbb{R}$ (or any subset thereof) which the function $f(x) = 1/x$ lacks on $(0,1)$.

Definition 6.4 A function $f : D \to \mathbb{R}$ is said to be *uniformly continuous* on $E \subseteq D$ if, for every $\varepsilon > 0$, there is a $\delta = \delta(\varepsilon) > 0$ such that
$$x, t \in E, |x - t| < \delta \Rightarrow |f(x) - f(t)| < \varepsilon.$$
When f is uniformly continuous on D, it is simply said to be uniformly continuous.

Remarks 6.1

1. Uniform continuity is defined over a *set* and makes no useful sense at a single point.
2. A uniformly continuous function is continuous.

The next theorem provides a criterion for proving that a function is *not* uniformly continuous.

Theorem 6.9
A function $f : D \to \mathbb{R}$ is not uniformly continuous if, and only if, there exist two sequences (x_n) and (t_n) in D such that
(i) $|x_n - t_n| \to 0$
(ii) $|f(x_n) - f(t_n)| \not\to 0$.

Proof. If f is not uniformly continuous, then there is an $\varepsilon > 0$ such that, for all $\delta > 0$, we can find two points $x, t \in D$ that satisfy $|x - t| < \delta$ and $|f(x) - f(t)| \geq \varepsilon$. By taking
$$\delta = 1, \frac{1}{2}, \frac{1}{3}, \cdots,$$
we can form two sequences (x_n) and (t_n) in D such that
$$|x_n - t_n| < \frac{1}{n}, \ |f(x_n) - f(t_n)| \geq \varepsilon.$$

Conversely, suppose f is uniformly continuous and $(x_n), (t_n)$ are two sequences in D which satisfy (i). If ε is any positive number, the uniform continuity of f implies that there is a $\delta > 0$ such that
$$x, t \in D, |x - t| < \delta \Rightarrow |f(x) - f(t)| < \varepsilon.$$

Since $|x_n - t_n| \to 0$, there is an integer N such that

$$n \geq N \Rightarrow |x_n - t_n| < \delta \Rightarrow |f(x_n) - f(t_n)| < \varepsilon,$$

and therefore $|f(x_n) - f(t_n)| \to 0$. □

Corollary 6.9 *The function $f : D \to \mathbb{R}$ is uniformly continuous if and only if, for every pair of sequences (x_n) and (t_n) in D satisfying $|x_n - t_n| \to 0$, we have $|f(x_n) - f(t_n)| \to 0$.*

From a practical point of view, however, this variation on the statement of Theorem 6.9 is not as useful as the theorem, as it involves investigating the behaviour of $|f(x_n) - f(t_n)|$ for *all* sequences which satisfy $|x_n - t_n| \to 0$.

Example 6.18 To verify that the function $f(x) = 1/x$ is not uniformly continuous on $(0, 1)$, using Theorem 6.9, choose

$$x_n = \frac{1}{n}, \ t_n = \frac{1}{2n}.$$

Now $|x_n - t_n| = 1/2n \to 0$ while $|f(x_n) - f(t_n)| = n \nrightarrow 0$.

But this function is uniformly continuous on $[a, \infty)$, where a is any positive constant. To see that, let $\varepsilon > 0$ be arbitrary. By choosing δ to be the positive number

$$\delta = \min\left\{\frac{1}{2}a, \frac{1}{2}a^2\varepsilon\right\},$$

we obtain

$$x, t \in [a, \infty), |x - t| < \delta \Rightarrow |x - t| < \frac{1}{2}a < \frac{1}{2}t$$

$$\Rightarrow x > \frac{1}{2}t$$

$$\Rightarrow xt > \frac{1}{2}t^2 \geq \frac{1}{2}a^2$$

$$\Rightarrow \frac{1}{xt} < \frac{2}{a^2}$$

$$\Rightarrow \left|\frac{1}{x} - \frac{1}{t}\right| = \frac{|x-t|}{xt} < \frac{2|x-t|}{a^2} \leq \frac{1}{2}a^2\varepsilon \cdot \frac{2}{a^2} = \varepsilon.$$

The reader may wonder how we arrive at this choice of δ which makes things work. Earlier on in this section, we found that we can make $|f(x) - f(c)| < \varepsilon$ if we take $\delta(c, \varepsilon) = \min\{c/2, c^2\varepsilon/2\}$. By choosing

$\delta = \delta(a, \varepsilon)$ we are assured that δ is positive (which is not the case if $a = 0$) and that it works for all points c in $[a, \infty)$.

Example 6.19 The function $f(x) = x^2$ is continuous on $\mathbb{R}$, but not uniformly. Let
$$x_n = n, \quad t_n = n + \frac{1}{n}.$$
Clearly, $|x_n - t_n| = 1/n \to 0$, but
$$|f(x_n) - f(t_n)| = 2 + \frac{1}{n^2} \not\to 0.$$

At this point it is natural to ask: When is a continuous function on $D \subseteq \mathbb{R}$ uniformly continuous? In Example 6.18 the domain $(0, 1)$ was not closed, and in Example 6.19 it was not bounded; and in both cases the function failed to achieve uniform continuity. It turns out that these are precisely the qualities which determine the answer.

Theorem 6.10
Let $f : D \to \mathbb{R}$ be continuous. If D is closed and bounded, then f is uniformly continuous.

Proof. Suppose that f is continuous on the closed and bounded set D, but not uniformly continuous. We shall show that this leads to a contradiction.

Since f is not uniformly continuous, there is a positive number ε and two sequences, (x_n) and (t_n), in D such that
$$|x_n - t_n| < \frac{1}{n}, \quad |f(x_n) - f(t_n)| \geq \varepsilon. \qquad (6.16)$$

Since D is bounded, so is (x_n), and by Theorem 3.14 it has a convergent subsequence (x_{n_k}). Let x be the limit of (x_{n_k}). Since D is closed, $x \in D$. The corresponding subsequence (t_{n_k}) also converges to x, since
$$|t_{n_k} - x| \leq |t_{n_k} - x_{n_k}| + |x_{n_k} - x|$$
$$< \frac{1}{n_k} + |x_{n_k} - x| \to 0.$$

Now the continuity of f at x allows us to write
$$f(x_{n_k}) \to f(x), \quad f(t_{n_k}) \to f(x).$$
Hence $|f(x_{n_k}) - f(t_{n_k})| \to 0$, which is inconsistent with (6.16). $\square$

Needless to say, this theorem gives sufficient, not necessary, conditions for uniform continuity. The function $f(x) = x$, for example, is uniformly continuous on any subset of $\mathbb{R}$.

Example 6.20 We shall now prove that the function $f : [0, \infty) \to \mathbb{R}$ defined by
$$f(x) = \sqrt{x}$$
is uniformly continuous.

Since f is continuous on $[0, 2]$, it is uniformly continuous on this interval by Theorem 6.10. On the other hand, if $x, t \in [1, \infty)$, then
$$|f(x) - f(t)| = \left|\sqrt{x} - \sqrt{t}\right| = \frac{|x-t|}{\sqrt{x} + \sqrt{t}} \leq |x - t|,$$
which implies f is uniformly continuous on $[1, \infty)$.

Given any $\varepsilon > 0$, there is a $\delta_1(\varepsilon) > 0$ such that
$$x, t \in [0, 2], \ |x - t| < \delta_1 \Rightarrow |f(x) - f(t)| < \varepsilon.$$
There is also a $\delta_2(\varepsilon) > 0$ which satisfies
$$x, t \in [1, \infty), \ |x - t| < \delta_2 \Rightarrow |f(x) - f(t)| < \varepsilon.$$
Define
$$\delta(\varepsilon) = \min\{1, \delta_1(\varepsilon), \delta_2(\varepsilon)\},$$
and let $x, t \in [0, \infty)$ satisfy $|x - t| < \delta$. Then the points x, t belong to the interval $[0, 2]$ or to $[1, \infty)$, and in either case we obtain $|f(x) - f(t)| < \varepsilon$.

As noted earlier, a continuous function, such as $f(x) = 1/x$ on $(0, 1)$, need not have a continuous extension to the closure of its domain, in this case $[0, 1]$. The next theorem shows that this situation cannot arise if the function is uniformly continuous.

Theorem 6.11
Let $f : D \to \mathbb{R}$ be uniformly continuous. Then it has an extension to $\bar{D} = D \cup \hat{D}$ which is uniformly continuous.

Proof. Let $x \in \bar{D}$. Then there is a sequence in D such that $x_n \to x$. (x_n) is clearly a Cauchy sequence, and we shall now show that its image under f, $(f(x_n))$, is also a Cauchy sequence. Suppose $\varepsilon > 0$. By the uniform continuity of f on D, there is a $\delta > 0$ such that
$$x, t \in D, |x - t| < \delta \Rightarrow |f(x) - f(t)| < \varepsilon. \tag{6.17}$$

Since (x_n) is a Cauchy sequence, there is an integer N such that
$$m, n \geq N \Rightarrow |x_m - x_n| < \delta. \tag{6.18}$$
Now (6.17) and (6.18) together imply
$$m, n \geq N \Rightarrow |f(x_m) - f(x_n)| < \varepsilon,$$
which means $(f(x_n))$ is a Cauchy sequence, and therefore convergent, by Cauchy's criterion.

Let
$$\lim f(x_n) = g(x). \tag{6.19}$$
To make sure that equation (6.19) defines a function on $\bar{D}$, we have to check that $g(x)$ does not depend on the choice of sequence (x_n) that converges to x. Let (x'_n) be another sequence in D which converges to x. Since $|x'_n - x_n| \to 0$, it follows from Corollary 6.9 that $|f(x'_n) - f(x_n)| \to 0$, hence $f(x'_n) \to g(x)$.

g is clearly an extension of f to $\bar{D}$, and it remains to show that g is uniformly continuous. Let $\varepsilon > 0$. f being uniformly continuous, there is a $\delta > 0$ such that (6.17) is satisfied for any pair of points $x, t \in D$. Now take $x, t \in \bar{D}$ and assume $|x - t| < \delta/3$. There are two sequences (x_n) and (t_n) in D such that $x_n \to x$ and $t_n \to t$, and, by (6.19)
$$f(x_n) \to g(x), \; f(t_n) \to g(t).$$
Therefore there is an integer M such that
$$n \geq M \Rightarrow |x_n - x| < \frac{\delta}{3}, \; |t_n - t| < \frac{\delta}{3}$$
$$\Rightarrow |f(x_n) - g(x)| < \varepsilon, \; |f(t_n) - g(t)| < \varepsilon. \tag{6.20}$$
Consequently,
$$n \geq M \Rightarrow |x_n - t_n| \leq |x_n - x| + |x - t| + |t - t_n| < \delta.$$
By (6.17),
$$|f(x_n) - f(t_n)| < \varepsilon \text{ for all } n \geq M. \tag{6.21}$$
Now, in view of (6.20) and (6.21), if $n \geq M$, then
$$|g(x) - g(t)| \leq |g(x) - f(x_n)| + |f(x_n) - f(t_n)| + |f(t_n) - g(t)| < 3\varepsilon$$
for all $x, t \in \bar{D}$ such that $|x - t| < \delta/3$. This proves g is uniformly continuous on $\bar{D}$. $\square$

EXERCISES 6.4

1. Determine which functions are uniformly continuous:

   (a) $f(x) = 1/x^2$, $D_f = (0, \infty)$.
   (b) $g(x) = \sqrt[3]{x}$, $D_g = \mathbb{R}$.
   (c) $h(x) = x^{3/2}$, $D_h = [0, \infty)$.

2. Prove that the function $f(x) = \dfrac{1}{x^2 + 1}$ is uniformly continuous on $\mathbb{R}$.

3. Suppose the functions f and g are uniformly continuous on D. Prove that $f + g$ is also uniformly continuous on D. Show that $f \cdot g$ is not necessarily uniformly continuous by a suitable example.

4. If f and g are both uniformly continuous and bounded on D, prove that $f \cdot g$ is also uniformly continuous on D.

5. If $f : D \to E$ and $g : E \to \mathbb{R}$ are uniformly continuous, is $g \circ f : D \to \mathbb{R}$ also uniformly continuous?

6. Let
$$f(x) = \begin{cases} x \sin \dfrac{1}{x}, & x \neq 0 \\ 0, & x = 0. \end{cases}$$

   (a) Prove that f is uniformly continuous on any bounded subset of $\mathbb{R}$.
   (b) Is f uniformly continuous on $\mathbb{R}$?

7. In the proof of Theorem 6.11 it was shown that the image of a Cauchy sequence under a uniformly continuous function was a Cauchy sequence. Give an example to show that this is not necessarily the case if the function is merely continuous.

8. Prove that a uniformly continuous function on a bounded interval is bounded. Show by example that this is not true if f is merely continuous.

9. A function $f : D \to \mathbb{R}$ is said to satisfy a *Lipschitz condition* if there is a positive constant $k > 0$ such that
$$|f(x) - f(t)| \leq k\,|x - t| \quad \text{for all } x, t \in D.$$

Prove that a function which satisfies a Lipschitz condition is uniformly continuous. Using the function $f(x) = \sqrt{x}$ on $[0, \infty)$, prove that the converse is false.

10. Prove that the function $f(x) = x \sin x$ is not uniformly continuous on $\mathbb{R}$ (see Exercise 6.5.4).

11. A function $f : \mathbb{R} \to \mathbb{R}$ is called *periodic* if there exists a positive constant T such that
$$f(x + T) = f(x) \quad \text{for all } x \in \mathbb{R}.$$
Prove that every periodic and continuous function (on $\mathbb{R}$) is uniformly continuous.

6.5 Compact Sets and Continuity

In this section we define what it means for a set to be *compact*. This is one of the important concepts in mathematical analysis, not only in $\mathbb{R}$ but also in more general topological spaces. The notion of compactness will be used to reformulate some of the results of the last two sections, in order to see the properties of continuity in a more general context.

Suppose G_λ is an open subset of $\mathbb{R}$ for every $\lambda \in \Lambda$, where Λ is a general index set which may be finite or infinite, countable (such as $\mathbb{N}$) or uncountable (such as $\mathbb{R}$). If E is a subset of $\mathbb{R}$ such that
$$E \subseteq \bigcup_{\lambda \in \Lambda} G_\lambda,$$
the collection $\{G_\lambda : \lambda \in \Lambda\}$ is called an *open cover* (or *covering*) of E. When the index set Λ is finite the open cover $\{G_\lambda : \lambda \in \Lambda\}$ is called a *finite (open) cover*.

Example 6.21
(i) The collection $\{(0, n) : n \in \mathbb{N}\}$ constitutes an open cover of the open interval $(0, \infty)$.
(ii) The collection $\{(n - 1/n, n + 1/n) : n \in \mathbb{N}\}$ covers $\mathbb{N}$.
(iii) $\{(-n, n) : n \in \mathbb{N}\}$ is an open cover of $\mathbb{R}$.
(iv) $\{\mathbb{R}\}$ is an open (and finite) cover of any subset of $\mathbb{R}$.

Definition 6.5 $E \subseteq \mathbb{R}$ is *compact* if every open cover $\{G_\lambda : \lambda \in \Lambda\}$ of E contains a finite subcover $\{G_{\lambda_1}, G_{\lambda_2}, \cdots, G_{\lambda_n}\}$ of E.

Continuity   195

By this definition, E is compact if, for *every* open cover $\{G_\lambda : \lambda \in \Lambda\}$ of E, we can find a finite subset of Λ, say $\lambda_1, \lambda_2, \cdots, \lambda_n$, such that

$$E \subseteq G_{\lambda_1} \cup G_{\lambda_2} \cup \cdots \cup G_{\lambda_n}.$$

There are many ways a set can be covered, so it will not be easy to use this definition to test its compactness. Conversely, E is not compact if it has an open cover which contains no finite subcover of E. This is easier to check, as the following examples indicate:

Example 6.22 The set of real numbers $\mathbb{R}$ is not compact. To see that, we take the collection of open intervals $\{(n-1, n+1) : n \in \mathbb{Z}\}$ as a cover of $\mathbb{R}$. For every $x \in \mathbb{R}$, there is an integer n such that $x \in (n-1, n+1)$, hence

$$\mathbb{R} \subseteq \bigcup_{n \in \mathbb{Z}} (n-1, n+1).$$

But, since every integer n_0 lies in one, and only one, of the these open intervals, namely $(n_0 - 1, n_0 + 1)$, every interval in the infinite collection $\{(n-1, n+1) : n \in \mathbb{Z}\}$ is needed to cover $\mathbb{R}$. Thus no finite subcover will suffice to cover $\mathbb{R}$.

Another example of a cover of $\mathbb{R}$ which has no finite subcover is $\{(-n, n) : n \in \mathbb{N}\}$.

Example 6.23 Since

$$\bigcup_{n=2}^{\infty} \left(0, 1 - \frac{1}{n}\right) = (0, 1),$$

the collection of open intervals $\{(0, 1 - 1/n) : n = 2, 3, \cdots\}$ is an open cover of the interval $(0, 1)$. Here again we cannot extract a finite subcollection which will cover $(0, 1)$, for any such collection will be of the form

$$\{(0, 1 - 1/n_1), (0, 1 - 1/n_2), \cdots, (0, 1 - 1/n_k)\},$$

where $n_i \in \{2, 3, \cdots\}$; and if $m = \max\{n_1, n_2, \cdots, n_k\}$, then

$$\bigcup_{i=1}^{k} (0, 1 - 1/n_i) = (0, 1 - 1/m) \subset (0, 1).$$

Therefore $(0, 1)$ is not compact.

Example 6.24 Every finite set of real numbers $\{x_1, x_2, \cdots, x_n\}$ is compact. To prove that, assume $\{G_\lambda : \lambda \in \Lambda\}$ is an open cover of

$\{x_1, \cdots, x_n\}$. It then follows that, for every $k \in \{1, \cdots, n\}$, there is an open set G_{λ_k} in the collection $\{G_\lambda : \lambda \in \Lambda\}$ such that $x_k \in G_{\lambda_k}$. Hence the finite collection $\{G_{\lambda_1}, \cdots, G_{\lambda_n}\}$ covers $\{x_1, \cdots, x_n\}$, which must therefore be compact.

Example 6.24 demonstrates one of the few instances where we can apply Definition 6.5 directly to prove compactness. That is why the next theorem, known as the *Heine-Borel theorem*, is of crucial importance, as it completely characterizes the compact subsets of $\mathbb{R}$.

Theorem 6.12 (Heine-Borel)
A set $E \subseteq \mathbb{R}$ is compact if, and only if, it is closed and bounded.

Proof
(i) Suppose E is compact. We shall first prove that it is bounded. For every $n \in \mathbb{N}$, let $I_n = (-n, n)$. Clearly,

$$E \subseteq \bigcup_{n=1}^{\infty} I_n = \mathbb{R}.$$

Since E is compact, there is a positive integer N_1 such that

$$E \subseteq \bigcup_{n=1}^{N_1} (-n, n) = (-N_1, N_1),$$

which means E is bounded.

Now we shall prove that E is closed by proving that its complement E^c is open. Suppose $x \in E^c$. For every $n \in \mathbb{N}$, define

$$G_n = \left\{y \in \mathbb{R} : |y - x| > \frac{1}{n}\right\}$$
$$= \left[x - \frac{1}{n}, x + \frac{1}{n}\right]^c.$$

Being the complement of a closed interval, G_n is open for every $n \in \mathbb{N}$. Furthermore,

$$\bigcup_{n=1}^{\infty} G_n = \bigcup_{n=1}^{\infty} \left[x - \frac{1}{n}, x + \frac{1}{n}\right]^c$$
$$= \left\{\bigcap_{n=1}^{\infty} \left[x - \frac{1}{n}, x + \frac{1}{n}\right]\right\}^c$$
$$= \{x\}^c = \mathbb{R} \setminus \{x\}.$$

Since $x \notin E$ it follows that $E \subseteq \bigcup_{n=1}^{\infty} G_n$, and since E is compact, there is an $N_2 \in \mathbb{N}$ such that

$$E \subseteq \bigcup_{n=1}^{N_2} G_n = G_{N_2} = \left[x - \frac{1}{N_2}, x + \frac{1}{N_2} \right]^c$$

$$\Rightarrow \left[x - \frac{1}{N_2}, x + \frac{1}{N_2} \right] \subseteq E^c$$

$$\Rightarrow \left(x - \frac{1}{N_2}, x + \frac{1}{N_2} \right) \subseteq E^c.$$

This implies E^c is open, and hence E is closed.

(ii) Conversely, suppose E is closed and bounded, and let $\{G_\lambda : \lambda \in \Lambda\}$ be an open cover of E. We shall prove that a finite number of open sets in $\{G_\lambda : \lambda \in \Lambda\}$ suffices to cover E.

Assume, on the contrary, that no finite sub-collection of $\{G_\lambda : \lambda \in \Lambda\}$ covers E. Since E is bounded, there is a finite closed interval $I_0 = [a_0, b_0]$ which contains E. Divide this interval at its center into the two closed subintervals

$$I_0' = \left[a_0, \frac{a_0 + b_0}{2} \right], \quad I_0'' = \left[\frac{a_0 + b_0}{2}, b_0 \right].$$

At least one of the sets $E \cap I_0'$ and $E \cap I_0''$ cannot be covered by a finite number of open sets in $\{G_\lambda : \lambda \in \Lambda\}$, for if both could be so covered then the same would apply to

$$E = (E \cap I_0') \cup (E \cap I_0''),$$

which would violate our assumption. Suppose $I_1 = [a_1, b_1]$ is one of the subintervals I_0' or I_0'' whose intersection with E cannot be covered by a finite sub-collection of $\{G_\lambda : \lambda \in \Lambda\}$.

Again subdivide I_1 into two subintervals of equal length, I_1' and I_1'', and conclude that at least one of them, say $I_2 = [a_2, b_2]$ intersects E in a set which cannot be covered by a finite sub-collection of $\{G_\lambda : \lambda \in \Lambda\}$. Continuing in this manner, we end up with a sequence of closed intervals $I_n = [a_n, b_n]$ which satisfies

(a) $I_{n+1} \subseteq I_n$ for all $n \in \mathbb{N}$

(b) For every $n \in \mathbb{N}$, $E \cap I_n$ cannot be covered by a finite sub-collection of $\{G_\lambda : \lambda \in \Lambda\}$

(c) $l(I_n) = \frac{1}{2} l(I_{n-1}) = \frac{b_0 - a_0}{2^n}.$

By Cantor's Theorem 3.11 for nested intervals, the intersection $\bigcap I_n$ is not empty, and since

$$\inf\{l(I_n) : n \in \mathbb{N}\} = 0,$$

it contains exactly one point, say x. Since $E \cap I_n$ is not empty for every n, we can choose a sequence (x_n) in E such that $x_n \in I_n$ for every $n \in \mathbb{N}$. This implies

$$|x_n - x| \leq l(I_n) = \frac{b_0 - a_0}{2^n} \to 0,$$

hence $x = \lim x_n$. But E is closed, so, by Theorem 3.20, $x \in E$. Since $\{G_\lambda : \lambda \in \Lambda\}$ covers E, there is an open set G_{λ_0} in the cover such that $x \in G_{\lambda_0}$. There is therefore a positive number ε such that $(x-\varepsilon, x+\varepsilon) \subseteq G_{\lambda_0}$. Choose n so that

$$\frac{b_0 - a_0}{2^n} = l(I_n) < \varepsilon,$$

from which we conclude that $I_n \subseteq (x - \varepsilon, x + \varepsilon) \subseteq G_{\lambda_0}$, and hence $E \cap I_n \subseteq G_{\lambda_0}$. Thus a single element of the collection $\{G_\lambda : \lambda \in \Lambda\}$ covers $E \cap I_n$, thereby contradicting our choice of I_n, and proving that the assumption that no finite sub-collection of $\{G_\lambda : \lambda \in \Lambda\}$ covers E is false. $\square$

If compact sets in $\mathbb{R}$ are precisely the closed and bounded sets, why not define them as such and avoid Definition 6.5 altogether? The answer is twofold: The Heine-Borel theorem does not hold in all topological spaces, so we have to stick to Definition 6.5 in a general topological space. Secondly, even in $\mathbb{R}$, we sometimes need to resort to the characteristic property of a compact set as spelled out in the definition (see the proof of Theorem 6.14, for example). Using the theorem of Bolzano-Weierstrass, the next theorem characterizes compactness in terms of sequences.

Theorem 6.13
For any set $E \subseteq \mathbb{R}$, the following statements are equivalent:
(i) E is compact.
(ii) E is closed and bounded.
(iii) Every sequence in E has a subsequence which converges to a point in E.

Proof. The equivalence of (i) and (ii) is the content of Heine-Borel's theorem, so we need only prove the equivalence of (ii) and (iii). To prove that (ii) implies (iii), let (x_n) be a sequence in the closed and bounded set E. Being bounded, the sequence (x_n) has a convergent subsequence (x_{n_k}) by Theorem 3.14. Since E is closed, $\lim x_{n_k}$ lies in E by Theorem 3.20.

In the other direction, suppose statement (iii) is true. If E were unbounded, we would be able to form a sequence (x_n) in E such that

$$|x_n| > n \text{ for all } n \in \mathbb{N}.$$

But such a sequence would have no convergent subsequence, thereby violating (iii). Hence E is bounded. On the other hand, let (x_n) be a sequence in E such that $x_n \to x$. (iii) implies (x_n) has a subsequence which converges to a point in E. But such a subsequence can only converge to x. Hence $x \in E$ and, by Theorem 3.20, E is closed. □

Compactness has many implications for continuity, the first of which is given in the following theorem. This theorem is actually a restatement of Theorem 6.10, but the proof supplied here relies on the defining property of compactness.

Theorem 6.14
If D is a compact subset of $\mathbb{R}$ and the function $f : D \to \mathbb{R}$ is continuous, then f is uniformly continuous.

Proof. Let $\varepsilon > 0$ be given. Since f is continuous at every $x \in D$, there is a $\delta(x) > 0$ such that

$$t \in D, |t - x| < \delta(x) \Rightarrow |f(t) - f(x)| < \varepsilon. \tag{6.22}$$

The collection of open intervals

$$\left\{ \left(x - \frac{1}{2}\delta(x), x + \frac{1}{2}\delta(x) \right) : x \in D \right\}$$

clearly covers D. Since D is compact, there is a finite number of points $x_1, x_2, \cdots, x_n$ in D such that

$$D \subseteq \bigcup_{i=1}^{n} \left(x_i - \frac{1}{2}\delta(x_i), x_i + \frac{1}{2}\delta(x_i) \right). \tag{6.23}$$

Define

$$\delta = \frac{1}{2} \min\{\delta(x_1), \cdots, \delta(x_n)\},$$

which is positive, being the minimum of a finite set of positive numbers.

Now suppose that $x, t \in D$ and $|t - x| < \delta$. From (6.23) we conclude that

$$x \in \left(x_i - \frac{1}{2}\delta(x_i), x_i + \frac{1}{2}\delta(x_i)\right) \text{ for some } i \in \{1, \cdots, n\}$$
$$\Rightarrow |t - x_i| \leq |t - x| + |x - x_i| < \delta + \frac{1}{2}\delta(x_i) \leq \delta(x_i).$$

And from (6.22) we arrive at

$$|f(t) - f(x)| \leq |f(x) - f(x_i)| + |f(x_i) - f(t)| < 2\varepsilon. \quad \square$$

Some results of section 6.3 pertaining to the properties of a continuous function on a compact interval can now be extended to compact sets.

Theorem 6.15
If f is a continuous function on the compact set $D \subseteq \mathbb{R}$, then $f(D)$ is also compact in $\mathbb{R}$.

Proof. Let (y_n) be any sequence in $f(D)$. $f(D)$ is compact if we can find a convergent subsequence of (y_n) whose limit lies in $f(D)$. For every $n \in \mathbb{N}$ there is an $x_n \in D$ such that $y_n = f(x_n)$. Since (x_n) is a sequence in the compact set D, it has, according to Theorem 6.13, a subsequence (x_{n_k}) which converges to some point $x \in D$. Now the continuity of f implies

$$y_{n_k} = f(x_{n_k}) \to f(x) \in f(D). \quad \square$$

Corollary 6.15 *If f is a continuous function on the compact set D, then there are points $x_1, x_2 \in D$ such that*

$$f(x_1) \leq f(x) \leq f(x_2) \text{ for all } x \in D.$$

Proof. $f(D)$ is compact by Theorem 6.15, hence it is bounded and closed. Therefore $\inf f(D)$ and $\sup f(D)$ both exist and belong to $f(D)$, which means there are points $x_1, x_2 \in D$ such that $f(x_1) = \inf f(D)$ and $f(x_2) = \sup f(D)$. This implies

$$f(x_1) \leq f(x) \leq f(x_2) \text{ for all } x \in D. \quad \square$$

Theorem 6.16
If $D \subseteq \mathbb{R}$ is compact and $f : D \to \mathbb{R}$ is continuous and injective, then $f^{-1} : f(D) \to D$ is also continuous.

Proof. Suppose (y_n) is a sequence in $f(D)$ which converges to some $y \in f(D)$. It suffices to show that $f^{-1}(y_n) \to f^{-1}(y)$.

By definition of $f(D)$, there are points $x_n, x \in D$ such that $y_n = f(x_n)$ and $y = f(x)$, and we have to show that $x_n \to x$. Since D is bounded, the sequence (x_n) is bounded. By Theorem 3.15, it suffices to prove that all its convergent subsequences have the same limit x.

Let (x_{n_k}) be a convergent subsequence of (x_n) with limit x'. Since D is closed, $x' \in D$, hence

$$f(x_{n_k}) \to f(x')$$

by the continuity of f. But we already have

$$f(x_{n_k}) = y_{n_k} \to y = f(x).$$

Therefore $f(x') = f(x)$ by the uniqueness of the limit, and $x' = x$ by the injective property of f. $\square$

Theorem 6.15 states that the image under a continuous function of a compact set is compact. But this invariance under a continuous mapping breaks down if we consider the image of an open or a closed set, as may be seen from the following examples:

(i) If f is a constant function defined on the open interval $(0,1)$, it is obviously continuous and its range $f((0,1))$ is a single point, which is clearly not open.

(ii) Let $g : \mathbb{R} \to \mathbb{R}$ be defined by

$$g(x) = \frac{1}{1 + x^2}.$$

In this case $g(\mathbb{R}) = (0, 1]$, which is neither open nor closed, though $\mathbb{R}$ is both.

The situation is better if we look at the inverse of the function.

Theorem 6.17
A function $f : D \to \mathbb{R}$ is continuous if and only if, for every open set G there is an open set H such that $f^{-1}(G) = H \cap D$.

Proof

(i) Suppose f is continuous and $G \subseteq \mathbb{R}$ is open. If $G \cap f(D) = \emptyset$, then $f^{-1}(G) = \emptyset$ and we take H to be the open set $\emptyset$. Assume, therefore, that $c \in f^{-1}(G)$, so that $f(c) \in G$. Since G is open, there is a positive number ε such that

$$(f(c) - \varepsilon, f(c) + \varepsilon) \subseteq G. \tag{6.24}$$

By the continuity of f, there is a $\delta > 0$ such that

$$x \in (c - \delta, c + \delta) \cap D \Rightarrow f(x) \in (f(c) - \varepsilon, f(c) + \varepsilon). \tag{6.25}$$

Using the notation $U_c = (c - \delta, c + \delta)$, $V_c = (f(c) - \varepsilon, f(c) + \varepsilon)$, (6.24) and (6.25) are equivalent to

$$f(U_c \cap D) \subseteq V_c \subseteq G,$$

which implies

$$U_c \cap D \subseteq f^{-1}(G). \tag{6.26}$$

Let

$$H = \bigcup \{U_c : c \in f^{-1}(G)\},$$

which is clearly open, and, in view of (6.26), satisfies

$$H \cap D \subseteq f^{-1}(G).$$

But since every point $c \in f^{-1}(G)$ lies in $U_c \cap D$, we actually have

$$H \cap D = f^{-1}(G).$$

(ii) Conversely, suppose that for every open set G there is an open set H such that $f^{-1}(G) = H \cap D$. Take any $c \in D$ and let ε be any positive number. Since the interval

$$G = (f(c) - \varepsilon, f(c) + \varepsilon)$$

is open, there is an open set H which satisfies $H \cap D = f^{-1}(G)$. Since $f(c) \in G$,

$$c \in f^{-1}(G) \subseteq H;$$

and since H is open, there is a $\delta > 0$ such that

$$(c - \delta, c + \delta) \subseteq H,$$

from which we conclude that

$$
\begin{aligned}
x \in (c - \delta, c + \delta) \cap D &\Rightarrow x \in H \cap D \\
&\Rightarrow x \in f^{-1}(G) \\
&\Rightarrow f(x) \in G \\
&\Rightarrow f(x) \in (f(c) - \varepsilon, f(c) + \varepsilon).
\end{aligned}
$$

Thus f is continuous at c. But since c was an arbitrary point in D, f is continuous on D. $\square$

Definition 6.6 Let D be a subset of $\mathbb{R}$. A set $E \subseteq D$ is said to be *open (closed)* in D if there is an open (closed) set H such that $H \cap D = E$ (see exercise 3.7.6).

Needless to say, when a set is described as being open (closed) without qualification, it is tacitly assumed that it is open (closed) in $\mathbb{R}$. According to Definition 6.6, Theorem 6.17 simply states that, under a continuous function $f : D \to \mathbb{R}$, the inverse image of every open set (in $\mathbb{R}$) is an open set in D.

Corollary 6.17.1 *A function $f : \mathbb{R} \to \mathbb{R}$ is continuous if and only if the inverse image of every open set is open.*

Corollary 6.17.2 *A function $f : D \to \mathbb{R}$ is continuous if and only if, for every closed set F, there is a closed set K such that $f^{-1}(F) = K \cap D$, i.e., if and only if the inverse image of every closed set is closed in D.*

Proof. Assume f is continuous and F is closed. It follows that $G = F^c$ is open and, by Theorem 6.17, there is an open set H such that $f^{-1}(G) = H \cap D$. Now

$$
\begin{aligned}
f^{-1}(F) &= f^{-1}(G^c) \\
&= [f^{-1}(G)]^c \cap D \\
&= (H^c \cup D^c) \cap D \\
&= H^c \cap D.
\end{aligned}
$$

Setting $K = H^c$, which is closed, we obtain $f^{-1}(F) = K \cap D$.

The proof in the other direction is left as an exercise. $\square$

Exercises 6.5

1. Define an open cover of $(-1, 1]$ which has no finite subcover of $(-1, 1]$.

2. Let $E = \{1/n : n \in \mathbb{N}\}$. Use Definition 6.5 to prove that E is not compact, but that $E \cup \{0\}$ is compact.

3. Let K be a compact set and $F \subseteq K$. If F is closed in K, prove that F is compact.

4. If K_1 and K_2 are compact sets, prove that $K_1 \cup K_2$ and $K_1 \cap K_2$ are also compact.

5. If K_n is a compact set for every $n \in \mathbb{N}$, what can you say about $\bigcup_{n=1}^{\infty} K_n$ and $\bigcap_{n=1}^{\infty} K_n$?

6. If K is a compact set, prove that $\sup K$ and $\inf K$ both exist and lie in K.

7. For any $E \subseteq \mathbb{R}$ and $c \in \mathbb{R}$, the *distance* between the point c and the set E is defined as the non-negative number
$$d(c, E) = \inf\{|x - c| : x \in E\}.$$
If E is compact, prove that there is a point $a \in E$ such that $|a - c| = d(c, E)$.

8. Suppose the function f is continuous on the compact set D. Prove that the set $\{x : 0 \leq f(x) \leq 1\}$ is compact.

9. Let the function f be defined on the set $D = [0, 1] \cup [2, 3)$ by
$$f(x) = \begin{cases} x, & x \in [0, 1] \\ 4 - x, & x \in [2, 3). \end{cases}$$

   (a) Prove that f is continuous and injective.

   (b) Prove that $f^{-1} : [0, 2] \to D$ is not continuous at 1 by considering the sequence $f^{-1}(1 + (-1)^n/n)$.

   (c) Explain how this example is consistent with Theorems 6.8 and 6.16.

10. Define a non-continuous function whose inverse is continuous.

11. Use the definition of a compact set and Theorem 6.17 to prove Theorem 6.15.

12. Give an example to show that the inverse image under a continuous function of a compact set is not necessarily compact. What additional condition would make it compact?

7

Differentiation

The *derivative* of a function, as presented in standard courses in differential calculus, is usually tied up with the notion of the *slope* of the tangent to the curve which represents the function, or with the *rate of change* of the function. These interpretations are, in fact, the reason differentiation was introduced. Once analytic geometry was developed in the early part of the seventeenth century, many geometric and physical relations were expressed by algebraic relations. As it turned out, the slope of the geometric curve which represents the equation $y = f(x)$ and the rate of change of the physical quantity y with respect to x are one and the same, expressed by the concept of the derivative. *René Descartes* (1596-1650) is credited with the introduction of analytic geometry, which was shortly followed by the development of differential and integral calculus at the hands of *Isaac Newton* (1642-1727) and *Gottfried Wilhelm Leibnitz* (1646-1716). Thus, in the course of about forty years, from 1630 to 1670, mathematics went through a phase of remarkable expansion unparalleled in its entire recorded history.

In this chapter we shall be mainly interested in the derivative from an analytical point of view, as a development of the concepts which we have presented so far. Its physical interpretations, important as they are, lie outside the scope of this treatment. Integration will be presented along the lines of *Bernhard Riemann* (1826-1866) in Chapter 8.

7.1 The Derivative

Given a function $f : D \to \mathbb{R}$ and a point $c \in D$, the function

$$g(x) = \frac{f(x) - f(c)}{x - c}$$

is defined on $D \setminus \{c\}$. If c is a cluster point of D, we can discuss the limit of g at c,

$$\lim_{x \to c} \frac{f(x) - f(c)}{x - c},$$

in the context of Definition 5.1. If this limit exists, we call it the *derivative* of f at c. But the significance of the derivative comes out more naturally if c is not merely a point in $D \cap \hat{D}$, but an *interior point* of D; that is, if there is a positive number ε such that $(c - \varepsilon, c + \varepsilon) \subseteq D$.

It is therefore preferable to restrict the domain of f to an open set when discussing its derivative. In view of Theorem 3.18, any open subset of $\mathbb{R}$ is the union of a countable number of disjoint open intervals. Hence there is little loss of generality if we assume from the outset that D is simply an open interval (a, b). But since we sometimes need to consider right-hand and left-hand derivatives at the end points a and b, D is taken to be a real interval I, which may include one, or both, of its end points. Thus we arrive at the following definition.

Definition 7.1 Let $f : I \to \mathbb{R}$, where I is a real interval, and $c \in I$. The limit
$$\lim_{x \to c} \frac{f(x) - f(c)}{x - c},$$
if it exists, is called the *derivative* of f at c, and is denoted by $f'(c)$. The function f is said to be *differentiable at c* if $f'(c)$ exists. If $f'(c)$ exists at every $c \in I$, f is said to be *differentiable on I*, or simply *differentiable*, and its derivative is the function $f' : I \to \mathbb{R}$.

It is worth noting that x tends to c, in accordance with Definition 5.1, through points in I (where f is defined). Hence, when $I = [a, b]$, Definition 7.1 gives the derivatives at the end points as
$$f'(a) = \lim_{x \to a^+} \frac{f(x) - f(a)}{x - a},$$
$$f'(b) = \lim_{x \to b^-} \frac{f(x) - f(b)}{x - b}.$$

It is common in calculus to introduce the independent variable $y = f(x)$ and write $\left.\dfrac{dy}{dx}\right|_{x=c}$ for $f'(c)$.

Example 7.1 The derivative of the constant function
$$f(x) = k \quad \text{for all } x \in \mathbb{R}$$
is
$$f'(c) = \lim_{x \to c} \frac{f(x) - f(c)}{x - c} = \lim_{x \to c} \frac{k - k}{x - c} = 0$$
for all $c \in \mathbb{R}$.

The function
$$g(x) = x^n, \quad x \in \mathbb{R},$$
where $n \in \mathbb{N}$, is also differentiable on $\mathbb{R}$, and its derivative is given by
$$\begin{aligned} g'(c) &= \lim_{x \to c} \frac{x^n - c^n}{x - c} \\ &= \lim_{x \to c}(x^{n-1} + x^{n-2}c + \cdots + xc^{n-2} + c^{n-1}) \\ &= nc^{n-1} \quad \text{for all } c \in \mathbb{R}. \end{aligned}$$

Example 7.2 The derivative of the function
$$f(x) = |x|, \quad x \in \mathbb{R},$$
is the limit
$$f'(c) = \lim_{x \to c} \frac{|x| - |c|}{x - c}.$$

Let (x_n) be any sequence in $\mathbb{R}$ which converges to c. If $c > 0$, there is an $N_1 \in \mathbb{N}$ such that $x_n > 0$ for all $n \geq N_1$, in which case
$$\frac{|x_n| - |c|}{x_n - c} = \frac{x_n - c}{x_n - c} = 1 \quad \text{for all } n \geq N_1$$
$$\Rightarrow f'(c) = 1.$$

But if $c < 0$, then there is an $N_2 \in \mathbb{N}$ such that $x_n < 0$ for all $n \geq N_2$, and we we obtain
$$\frac{|x_n| - |c|}{x_n - c} = \frac{-x_n - (-c)}{x_n - c} = -1 \quad \text{for all } n \geq N_2$$
$$\Rightarrow f'(c) = -1.$$

If, on the other hand, $c = 0$ then the sequence $x_n = (-1)^n/n$ tends to 0, while
$$\frac{|x_n| - 0}{x_n - 0} = \frac{1/n}{(-1)^n/n} = (-1)^n$$
does not converge. This implies that $f'(0)$ does not exist.

Thus $f(x) = |x|$ is differentiable on $\mathbb{R}\setminus\{0\}$, where its derivative is the function
$$f'(x) = \begin{cases} 1, & x > 0 \\ -1, & x < 0. \end{cases}$$

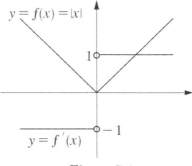

Figure 7.1

Example 7.3 For the function

$$f(x) = \sin x, \quad x \in \mathbb{R},$$

we can use the identity

$$\sin x - \sin c = 2\cos\left(\frac{x+c}{2}\right)\sin\left(\frac{x-c}{2}\right)$$

to obtain

$$\begin{aligned} f'(c) &= \lim_{x \to c} \frac{\sin x - \sin c}{x - c} \\ &= \lim_{x \to c} \frac{1}{x-c} 2\cos\left(\frac{x+c}{2}\right)\sin\left(\frac{x-c}{2}\right) \\ &= \lim_{x \to c} \cos\left(\frac{x+c}{2}\right) \cdot \lim_{x \to c} \sin\left(\frac{x-c}{2}\right) \div \left(\frac{x-c}{2}\right), \end{aligned}$$

where we have used Theorem 5.6 in the last equality. Since

$$\lim_{x \to c} \cos\left(\frac{x+c}{2}\right) = \cos c$$

by the continuity of the cosine function, and

$$\lim_{x \to c} \sin\left(\frac{x-c}{2}\right) \div \left(\frac{x-c}{2}\right) = 1,$$

based on the result of Example 5.10, we obtain

$$f'(c) = \cos c \text{ for all } c \in \mathbb{R}.$$

The relationship between continuity and differentiability is expressed by the following theorem.

Theorem 7.1
If the function $f : I \to \mathbb{R}$ is differentiable at $c \in I$, then it is continuous at c.

Proof. Assume $x \neq c$. Then
$$f(x) - f(c) = \frac{f(x) - f(c)}{x - c}(x - c).$$
Since both functions $\frac{f(x) - f(c)}{x - c}$ and $(x - c)$ have limits at c,
$$\lim_{x \to c} f(x) - f(c) = \lim_{x \to c} \frac{f(x) - f(c)}{x - c} \cdot \lim_{x \to c}(x - c)$$
$$= f'(c) \cdot 0 = 0$$
$$\Rightarrow \lim_{x \to c} f(x) = f(c). \qquad \square$$

Note that the converse of this theorem is false, as may be seen from Example 7.2 where $f(x) = |x|$ is continuous, but not differentiable, at $x = 0$.

When functions are combined under the operations of addition, multiplication or division, we can use Theorem 5.6 to express the derivative of the resulting function in terms of the original functions and their derivatives.

Theorem 7.2
If the functions f and g are defined on the interval I and differentiable at $c \in I$, then
(i) $f + g$ is differentiable at c and $(f + g)'(c) = f'(c) + g'(c)$,
(ii) fg is differentiable at c and $(fg)'(c) = f'(c)g(c) + f(c)g'(c)$,
(iii) if $g(c) \neq 0$, f/g is differentiable at c and
$$\left(\frac{f}{g}\right)'(c) = \frac{f'(c)g(c) - f(c)g'(c)}{g^2(c)}.$$

Proof
(i) This follows directly from Theorem 5.6.

(ii) Suppose $x \neq c$. Then
$$\frac{(fg)(x) - (fg)(c)}{x - c} = \frac{f(x)g(x) - f(c)g(x) + f(c)g(x) - f(c)g(c)}{x - c}$$
$$= \frac{f(x) - f(c)}{x - c} g(x) + f(c) \frac{g(x) - g(c)}{x - c}.$$

Since g is continuous at c (by Theorem 7.1), $g(x) \to g(c)$ as $x \to c$. And since f and g are differentiable at c,
$$\frac{f(x) - f(c)}{x - c} \to f'(c)$$
$$\frac{g(x) - g(c)}{x - c} \to g'(c).$$

Hence
$$\lim_{x \to c} \frac{(fg)(x) - (fg)(c)}{x - c} = f'(c)g(c) + f(c)g'(c).$$

(iii) If $g(c) \neq 0$ then, by the continuity of g, there is a neighborhood U of c where $g(x) \neq 0$. With $x \in U$ and $x \neq c$ we therefore have
$$\frac{(f/g)(x) - (f/g)(c)}{x - c} = \frac{f(x)g(c) - f(c)g(x)}{g(x)g(c)(x - c)}$$
$$= \frac{f(x)g(c) - f(c)g(c) + f(c)g(c) - f(c)g(x)}{g(x)g(c)(x - c)}$$
$$= \frac{1}{g(x)g(c)} \left[\frac{f(x) - f(c)}{x - c} g(c) - f(c) \frac{g(x) - g(c)}{x - c} \right].$$

Now the continuity of g at c yields
$$\lim_{x \to c} \frac{(f/g)(x) - (f/g)(c)}{x - c} = \frac{f'(c)g(c) - f(c)g'(c)}{g^2(c)}. \quad \square$$

If f is the constant function k in part (ii), then we obtain
$$(kg)'(c) = kg'(c)$$
for any constant number k, and if we replace g by $(-1)g$ in (i) then
$$(f - g)'(c) = f'(c) - g'(c).$$
Setting $g = f$ in (ii) yields
$$(f^2)'(c) = 2f(c)f'(c),$$

and using induction, we arrive at the more general result:

Corollary 7.2 If f is differentiable at c, then so is f^n for any $n \in \mathbb{N}$, and
$$(f^n)'(c) = nf^{n-1}(c)f'(c).$$

Since the derivative of the function $f(x) = x$ is
$$f'(c) = \lim_{x \to c} \frac{x-c}{x-c} = 1 \quad \text{for all } c \in \mathbb{R},$$
the derivative of $g(x) = x^n$ is therefore
$$g'(c) = nc^{n-1},$$
in agreement with Example 7.1.

Example 7.4 Using Theorem 7.2 and its corollary, we can differentiate any polynomial of degree n,
$$p(x) = a_n x^n + a_{n-1} x^{n-1} + \cdots + a_1 x + a_0, \quad x \in \mathbb{R},$$
at any point in $\mathbb{R}$ to obtain the polynomial of degree $n-1$
$$p'(x) = na_n x^{n-1} + (n-1)c_{n-1} x^{n-2} + \cdots + a_1.$$

Example 7.5 To differentiate the function
$$h(x) = x^{-n}, \quad x \neq 0, n \in \mathbb{N},$$
we apply rule (iii) of Theorem 7.2 with $f(x) = 1$ and $g(x) = x^n$. Thus
$$h'(x) = \frac{(0)(x^n) - (1)(nx^{n-1})}{x^{2n}} = \frac{-n}{x^{n+1}} = -nx^{-n-1}.$$

The next theorem gives the conditions under which the composition $g \circ f$ is differentiable and the form of its derivative.

Theorem 7.3
Let the functions f and g be defined on the real intervals I and J, respectively, such that $f(I) \subseteq J$. If f is differentiable at $c \in I$, and g is differentiable at $f(c)$, then the composed function
$$g \circ f : I \to \mathbb{R}$$

is differentiable at c, where its derivative is given by
$$(g \circ f)'(c) = g'(f(c)) \cdot f'(c).$$

Before we prove this theorem, it is worth pointing out a quick, but false, "proof" which begins with writing

$$\frac{(g \circ f)(x) - (g \circ f)(c)}{x - c} = \frac{g(f(x)) - g(f(c))}{f(x) - f(c)} \cdot \frac{f(x) - f(c)}{x - c},$$

and then taking the limit as $x \to c$. But this equation is valid only if $f(x) - f(c) \neq 0$. Though we know that $x - c \neq 0$, we cannot assume that $f(x) - f(c) \neq 0$. The following proof covers the case when $f(x) - f(c) = 0$.

Proof. Let $d = f(c)$ and define the function φ on J by

$$\varphi(y) = \begin{cases} \dfrac{g(y) - g(d)}{y - d}, & y \neq d \\ g'(d), & y = d. \end{cases}$$

Observe that
$$\lim_{y \to d} \varphi(y) = g'(d) = \varphi(d), \tag{7.1}$$

and that $g(y) - g(d) = (y - d)\varphi(y)$ for all $y \in J$, including $y = d$.
Now let $x \in I \setminus \{c\}$ and $y = f(x)$.

$$(g \circ f)(x) - (g \circ f)(c) = g(y) - g(d)$$
$$= (y - d)\varphi(y)$$

$$\Rightarrow \frac{(g \circ f)(x) - (g \circ f)(c)}{x - c} = \frac{y - d}{x - c}\varphi(y)$$
$$= \frac{f(x) - f(c)}{x - c}\varphi(y).$$

As $x \to c$,
$$\frac{f(x) - f(c)}{x - c} \to f'(c),$$
and $y \to d$, by the continuity of f, hence $\varphi(y) \to g'(d)$ by (7.1). Thus
$$\lim_{x \to c} \frac{(g \circ f)(x) - (g \circ f)(c)}{x - c} = g'(d) \cdot f'(c). \quad \square$$

Remarks 7.1

1. The rule for differentiating $g \circ f$, according to Theorem 7.3, is expressed by
$$(g \circ f)' = (g' \circ f) \cdot f'.$$
Another form of this result can be obtained by writing
$$y = f(x), \quad w = g(y)$$
$$\frac{dy}{dx} = f'(x), \quad \frac{dw}{dy} = g'(y).$$
This leads to the more familiar (and suggestive) expression
$$\frac{dw}{dx} = \frac{dw}{dy} \cdot \frac{dy}{dx},$$
known as the *chain rule*.

2. The rule
$$(f^n)' = nf^{n-1} \cdot f',$$
which we derived in Corollary 7.2, can also be obtained as a special case of Theorem 7.3. Suppose
$$g(y) = y^n, \quad y \in \mathbb{R}, \; n \in \mathbb{N}.$$
Then
$$g'(y) = ny^{n-1},$$
and
$$(g \circ f)(x) = f^n(x).$$
Applying the chain rule, we obtain
$$(f^n)'(x) = g'(f(x)) \cdot f'(x)$$
$$= nf^{n-1}(x) \cdot f'(x).$$

Example 7.6 To differentiate the function
$$f(x) = \cos x, \quad x \in \mathbb{R},$$

we apply the chain rule to the identity
$$\cos x = \sin\left(x + \frac{\pi}{2}\right).$$
Let $g(x) = x + \pi/2$ on $\mathbb{R}$, so that
$$\cos(x) = \sin(g(x))$$
$$\cos'(x) = \sin'(g(x)) \cdot g'(x).$$
From Example 7.3 we know that $\sin'(g(x)) = \cos(g(x)) = \cos(x+\pi/2)$, and since $g'(x) = 1$, we arrive at
$$\cos'(x) = \cos\left(x + \frac{\pi}{2}\right) = -\sin x, \quad x \in \mathbb{R}.$$

Example 7.7
(i) The function
$$f(x) = \begin{cases} x \sin \dfrac{1}{x}, & x \neq 0 \\ 0, & x = 0 \end{cases}$$
can be differentiated on $\mathbb{R}\setminus\{0\}$, using Theorems 7.2 and 7.3, to obtain
$$f'(x) = \sin\frac{1}{x} + x\cos\frac{1}{x}\left(-\frac{1}{x^2}\right)$$
$$= \sin\frac{1}{x} - \frac{1}{x}\cos\frac{1}{x}.$$
At $x = 0$ we cannot apply these theorems (why?), so we go back to the definition
$$f'(0) = \lim_{x \to 0} \frac{f(x) - f(0)}{x - 0} = \lim_{x \to 0} \sin\frac{1}{x}.$$
Since this limit does not exist, we conclude that f is not differentiable at $x = 0$.

(ii) If, on the other hand, the function we are dealing with is defined by
$$g(x) = \begin{cases} x^2 \sin \dfrac{1}{x}, & x \neq 0 \\ 0, & x = 0, \end{cases}$$
then
$$g'(x) = 2x\sin\frac{1}{x} - \cos\frac{1}{x}, \quad x \neq 0,$$

and

$$g'(0) = \lim_{x \to 0} \frac{g(x) - g(0)}{x - 0}$$
$$= \lim_{x \to 0} x \sin \frac{1}{x} = 0.$$

Hence g is differentiable on $\mathbb{R}$, including 0, where its derivative is 0.

It is worth noting in this example that both f and g are continuous at 0, where

$$\lim_{x \to 0} f(x) = \lim_{x \to 0} g(x) = 0.$$

But $g(x)$ tends to 0 faster than does $f(x)$, and that is why

$$\lim_{x \to 0} g(x)/x = g'(0)$$

exists while $\lim_{x \to 0} f(x)/x = f'(0)$ does not.

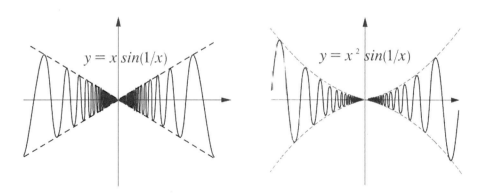

Figure 7.2

Note also that, though g is differentiable at $x = 0$, its derivative g' is not continuous at that point, for $\lim_{x \to 0} g'(x)$ does not exist. Thus the existence of the derivative does not guarantee its continuity.

Given an injective and differentiable function f, we can ask about the conditions under which $(f^{-1})'$ exists and the relationship between $(f^{-1})'$ and f'. Let us first assume that f is an injective, continuous function on an interval I, so that $f(I)$ is an interval. Then

$$f(f^{-1}(x)) = x \text{ for all } x \in f(I).$$

If f^{-1} is differentiable at c and f is differentiable at $b = f^{-1}(c)$,
$$f'(b) \cdot (f^{-1})'(c) = 1$$
by the chain rule. Based on this equation, it is clear that, for $(f^{-1})'(c)$ to exist, $f'(b)$ must not vanish, in which case
$$(f^{-1})'(c) = \frac{1}{f'(b)} = \frac{1}{f'(f^{-1}(c))}.$$
The question as to whether this condition is sufficient for the existence of $(f^{-1})'(c)$ is answered in the affirmative by the following theorem.

Theorem 7.4
Let f be an injective, continuous function on the interval I. If f is differentiable at $b \in I$, then f^{-1} is differentiable at $c = f(b)$ if and only if $f'(b) \neq 0$, in which case
$$(f^{-1})'(c) = \frac{1}{f'(b)}.$$

Proof. That $f'(b) \neq 0$ is a necessary condition for the existence of $(f^{-1})'(c)$ has already been shown. Suppose now that $f'(b) \neq 0$. To evaluate the limit
$$\lim \frac{f^{-1}(y) - f^{-1}(c)}{y - c},$$
we note that y is a point in $f(I) \setminus \{c\}$ and may therefore be expressed as $y = f(x)$, where $x = f^{-1}(y) \in I$. Since f^{-1} is injective, $x \neq b$ and we can write
$$\frac{f^{-1}(y) - f^{-1}(c)}{y - c} = \frac{x - b}{f(x) - f(b)}. \tag{7.2}$$
From Theorem 6.8 we know that f^{-1} is continuous, hence
$$y \to c \Rightarrow f^{-1}(y) \to f^{-1}(c)$$
$$\Rightarrow x \to b,$$
and the limit
$$\lim_{x \to b} \frac{f(x) - f(b)}{x - b}$$
exists and equals $f'(b)$. Since $f'(b) \neq 0$,
$$\lim_{x \to b} \frac{x - b}{f(x) - f(b)} = \frac{1}{f'(b)}.$$

The limit on the left-hand side of equation (7.2) therefore exists as $y \to c$, and gives $(f^{-1})'(c) = 1/f'(b)$. □

Example 7.8 The function
$$f(x) = \sin x, \ x \in \left[-\frac{\pi}{2}, \frac{\pi}{2}\right]$$
is injective and differentiable. Since
$$f'(x) = \cos x > 0 \ \text{ for all } x \in \left(-\frac{\pi}{2}, \frac{\pi}{2}\right),$$
the function $g : [-1, 1] \to [-\pi/2, \pi/2]$ defined by
$$g(x) = f^{-1}(x) = \arcsin x$$
is differentiable on $(-1, 1)$. Its derivative is
$$\begin{aligned} g'(x) &= \frac{1}{f'(g(x))} \\ &= \frac{1}{\cos(\arcsin x)} \\ &= \frac{1}{\sqrt{1 - \sin^2(\arcsin x)}} \\ &= \frac{1}{\sqrt{1 - x^2}}. \end{aligned}$$

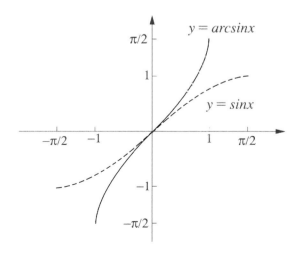

Figure 7.3

Note that g is not differentiable at either -1 or 1, since $f'(-\pi/2) = f'(\pi/2) = 0$.

Example 7.9 For any $n \in \mathbb{Z}$ we have already seen, in Examples 7.1 and 7.5, that the derivative of the function

$$f(x) = x^n, \quad x \in \mathbb{R}\setminus\{0\},$$

is

$$f'(x) = nx^{n-1}, \quad x \in \mathbb{R}\setminus\{0\}. \tag{7.3}$$

Now we shall extend this result from $\mathbb{Z}$ to $\mathbb{Q}$.

For any $n \in \mathbb{N}$ we define the function

$$f(x) = x^n, \quad x \in D,$$

where $D = \mathbb{R}$ if n in odd and $D = [0, \infty)$ if n is even. This function is strictly increasing and differentiable on D. Its inverse function

$$g(x) = x^{1/n}, \quad x \in D,$$

is also strictly increasing and differentiable on $D\setminus\{0\}$. Since $f'(x) = nx^{n-1}$ is positive on $D\setminus\{0\}$, we can use Theorem 7.4 to write

$$g'(x) = \frac{1}{f'(g(x))} = \frac{1}{n(x^{1/n})^{n-1}} = \frac{1}{n}x^{1/n-1}, \quad x \in D\setminus\{0\}.$$

For any $m \in \mathbb{Z}$, let

$$h(x) = g^m(x) = x^{m/n}, \quad x \in D\setminus\{0\}.$$

By the chain rule,

$$h'(x) = m(x^{1/n})^{m-1} \cdot g'(x)$$
$$= \frac{m}{n}x^{(m/n)-1}, \quad x \in D\setminus\{0\}.$$

Since every rational number can be represented by m/n, where $m \in \mathbb{Z}$ and $n \in \mathbb{N}$, we have therefore proved that the rule for differentiating the function $h(x) = x^r$ for any $r \in \mathbb{Q}$ is

$$h'(x) = rx^{r-1}, \quad x \in D\setminus\{0\},$$

which has the same form as (7.3). We can also show, using the exponential function, that this result actually holds for any real power of x (see Chapter 9).

Exercises 7.1

1. Determine the derivative of each of the following functions where it exists:

   (a) $f(x) = \tan x$, $x \in (-\pi/2, \pi/2)$
   (b) $g(x) = 1/\sqrt{x}$, $x > 0$
   (c) $h(x) = [x]$, $x \in \mathbb{R}$.

2. Let
$$f(x) = \begin{cases} x^2, & x \in \mathbb{Q} \\ 0, & x \in \mathbb{Q}^c. \end{cases}$$
Prove that f is differentiable at $x = 0$, and evaluate $f'(0)$.

3. If the function f satisfies $|f(x)| \le |x|^\alpha$, where $\alpha > 1$, prove that f is differentiable at $x = 0$. Compare this to the results of Examples 7.2 and 7.7.

4. Assuming f is differentiable at c, prove that

   (a) $f'(c) = \lim\limits_{h \to 0} \dfrac{f(c+h) - f(c)}{h}$
   (b) $f'(c) = \lim\limits_{h \to 0} \dfrac{f(c+h) - f(c-h)}{2h}$.

5. Let $f : \mathbb{R} \to \mathbb{R}$. The function f is *even* if $f(-x) = f(x)$ for all $x \in \mathbb{R}$, and *odd* if $f(-x) = -f(x)$ for all $x \in \mathbb{R}$. If f is differentiable, prove that f' is odd when f is even and even when f is odd.

6. For any $n \in \mathbb{N}$, the derivative of f of order n, denoted by $f^{(n)}$ is defined inductively as follows:
$$f^{(1)} = f'$$
$$f^{(2)} = f'' = (f')'$$
$$\vdots$$
$$f^{(n)} = (f^{(n-1)})'.$$

Determine $f^{(n)}(x)$ for each of the following functions, where $n \in \mathbb{N}$ and $x \in \mathbb{R}$:

(a) $f(x) = \sin x$
(b) $f(x) = \cos x$
(c) $f(x) = x^m$, $m \in \mathbb{N}$ and $0 \le m < n$
(d) $f(x) = x^n$
(e) $f(x) = x^m$, $m \in \mathbb{N}$ and $m > n$.

7. Let
$$f(x) = \begin{cases} x^n, & x \ge 0, n \in \mathbb{N} \\ 0, & x < 0. \end{cases}$$

(a) Prove that f is differentiable to order $n-1$ at $x = 0$ and compute $f^{(n-1)}(0)$, where $f^{(0)}$ is defined to be f.
(b) Prove that $f^{(n)}(0)$ does not exist.

8. Determine the derivatives of the following functions where they exist:

(a) $f(x) = \sqrt{|x|}$, $x \in \mathbb{R}$
(b) $g(x) = x|x|$, $x \in \mathbb{R}$
(c) $h(x) = |x^2 - 1|$, $x \in \mathbb{R}$.

9. Let $f(x) = \cos x$, where $x \in [0, \pi]$, and determine the derivative of $f^{-1}(x) = \arccos x$ on $[-1, 1]$.

10. Prove that a differentiable function at c satisfies a Lipschitz condition (see Exercise 6.4.9) in a neighborhood of c.

11. Use mathematical induction to prove *Leibnitz' rule* for the derivative of order n of the product of two functions:
$$(fg)^{(n)}(x) = \sum_{k=0}^{n} \binom{n}{k} f^{(n-k)}(x) \cdot g^{(k)}(x).$$

Compute the derivative of order 6 of the function $x^8 \sin x$.

12. Let n be an odd, positive integer and $f(x) = |x^n|$, where $x \in \mathbb{R}$. Determine $f^{(m)}(x)$ for any $m < n$ and show that $f^{(n)}(0)$ does not exist. What can we say about the case where n is even?

7.2 The Mean Value Theorem

In Chapter 5 we introduced the notion of the maximum and the minimum values of a function defined on a given domain. Such values are called *absolute extrema* of the function, in order to distinguish them from the *relative extrema* defined below.

Definition 7.2 The function $f : D \to \mathbb{R}$ is said to have a *local maximum* at $c \in D$ if there is a neighborhood $U = (c - \delta, c + \delta)$ of the point c such that
$$f(x) \leq f(c) \text{ for all } x \in U \cap D.$$

f has a *local minimum* at $c \in D$ if there is a neighborhood of $U = (c - \delta, c + \delta)$ such that
$$f(x) \geq f(c) \text{ for all } x \in U \cap D.$$

Referring to Definition 6.3, we see that every absolute maximum (minimum) is also a local maximum (minimum), but not vice-versa, as Figure 7.4 shows.

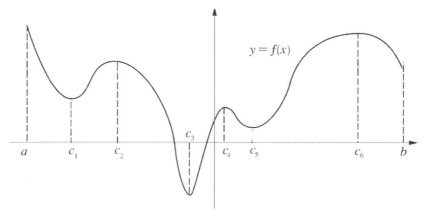

Figure 7.4

In this figure, the function $f : [a, b] \to \mathbb{R}$ takes its absolute maximum at the point a, and its local maxima at a, c_2, c_4 and c_6. Its absolute minimum is assumed at c_3, and its local minima at c_1, c_3, c_5 and b. Note that the local minimum value $f(c_1)$ is greater than the local maximum value $f(c_4)$.

Where a function assumes its extreme values, both absolute and local, is one of the fundamental questions in science, engineering, economics, and other disciplines. The following theorem, together with Theorem 6.5, provides an effective tool for tackling this problem.

Theorem 7.5
If the function f has an extremum on the open interval (a, b) at the point c, and if f is differentiable at c, then $f'(c) = 0$.

Proof. Suppose f has a maximum at c. Then
$$f(x) \leq f(c) \text{ for all } x \in (a, b),$$
and therefore
$$\frac{f(x) - f(c)}{x - c} \geq 0 \text{ for all } x \in (a, c).$$
Hence
$$\lim_{x \to c^-} \frac{f(x) - f(c)}{x - c} = f'(c) \geq 0. \tag{7.4}$$
On the other hand, if $x \in (c, b)$, then
$$\frac{f(x) - f(c)}{x - c} \leq 0$$
$$\Rightarrow \lim_{x \to c^+} \frac{f(x) - f(c)}{x - c} = f'(c) \leq 0, \tag{7.5}$$
and we conclude from (7.4) and (7.5) that $f'(c) = 0$.

If f has a minimum at c, then the sign of $f'(c)$ changes in both (7.4) and (7.5), and again we obtain $f'(c) = 0$. □

Remarks 7.2
1. $c \in D_f$ is called a *critical point* of the function f if $f'(c) = 0$ or $f'(c)$ does not exist. In view of Theorem 7.5, if f has an extremum on an open interval, the extremum is attained at a critical point in the interval.

2. If a function $f : D \to \mathbb{R}$ has a local extremum at a point c in $D°$, the interior of D, then c is necessarily a critical point of f. But if c lies on ∂D, the boundary of D, then c may not be a critical point of the function. The function $f(x) = x$, for example, has a local (and absolute) maximum on the interval $[0, 1]$ at $x = 1$, but $f'(1) = 1$.

3. In Theorem 7.5, the condition that c be a critical point of f is necessary, but not sufficient, for the function to have a local extremum

at c. 0 is a critical point of the function $f(x) = x^3$ since $f'(0) = 0$, but $f(x)$ has no maximum or minimum at $x = 0$ in any open interval containing 0.

Example 7.10
To determine the absolute extrema of the function $f(x) = 6x^{4/3} - 3x^{1/3}$ on the interval $[-1, 1]$, we first note that this function is continuous on the compact interval $[-1, 1]$, hence it attains its extrema in the interval. These values will be attained at points which lie either in the open interval $(-1, 1)$, or at the end-points $-1, 1$. In the former case, such points are necessarily critical points. Consequently, the function takes its extrema in the set $C \cup \{-1, 1\}$, where C is the set of critical points of f in $(-1, 1)$.

f is not differentiable at $x = 0$, and when $x \neq 0$,

$$f'(x) = 8x^{1/3} - x^{-2/3}$$
$$= \frac{8x - 1}{x^{2/3}},$$

which implies that $x = 0$ and $x = 1/8$ are the critical points in $(-1, 1)$. f can therefore assume its maxima and minima only at points in the set $\{0, 1/8, -1, 1\}$. Since

$$f(0) = 0, \quad f(1/8) = -9/8, \quad f(-1) = 9, \quad f(1) = 3,$$

we conclude that the maximum value of f on $[-1, 1]$ is 9, and is attained at $x = -1$, while its minimum is $-9/8$, which is attained at $x = 1/8$.

Thus the process of determining the absolute extrema of a function f on a given compact interval I is relatively straightforward, as long as the function is continuous and the set of critical points in the interior of I is finite. The process is not as simple when we wish to determine the *local* extrema of f. Though local extrema are also absolute extrema when viewed on a certain neighborhood of the critical point, it is not clear beforehand what that neighborhood is. Fortunately, there is another criterion for determining the type of extremum which depends on the sign of f' near the critical point, and which ultimately relies on one of the most important theorems of differential calculus, the *mean value theorem*. We shall start with a special case of that theorem, known as *Rolle's theorem*.

Theorem 7.6 (Rolle's Theorem)
If the function f is continuous on $[a,b]$, differentiable on (a,b), and satisfies the equality $f(a) = f(b)$, then there is a point c in (a,b) where $f'(c) = 0$.

Proof. The function defined by $g(x) = f(x) - f(a)$ is also continuous on $[a,b]$ and differentiable on (a,b), and it satisfies

$$g(a) = g(b) = 0. \qquad (7.6)$$

Since g is continuous on $[a,b]$ it achieves its maximum and minimum values on this interval, by Theorem 6.5. If both of these extreme values are 0, then g is the constant function 0, in which case f is the constant function

$$f(x) = f(a) \quad \text{for all } x \in [a,b],$$

and we can choose c to be any point in (a,b). But if one of the extreme values of g is not 0, then the condition (7.6) implies that the extremum is achieved at a point c other than a and b. In other words, $c \in (a,b)$. By Theorem 7.5 we now conclude that $f'(c) = g'(c) = 0$. □

There is a simple geometric interpretation of Rolle's theorem: The curve which represents a differentiable function f, and which intersects a horizontal line (parallel to the x-axis) at two points, must have a horizontal tangent at some point between them. Of course there may be more than one such point, as shown in Figure 7.5.

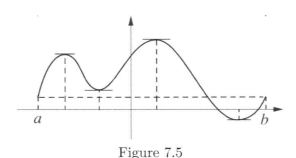

Figure 7.5

Note that the theorem does not require the existence of $f'(a)$ or $f'(b)$, only the continuity of f at these points, for the function

$$f(x) = \begin{cases} x, & 0 \le x < 1 \\ 0, & x = 1 \end{cases}$$

satisfies all the conditions of Rolle's theorem except continuity at $x = 1$, but $f'(x) = 1$ for all $x \in (0,1)$.

Of the three conditions on f in the statement of Rolle's theorem, the third seems to be the least essential. If we drop the requirement that $f(a) = f(b)$, we arrive at a more general result, also known as the *mean value theorem* of differential calculus.

Theorem 7.7 (Mean Value Theorem)
If the function f is continuous on $[a, b]$ and differentiable on (a, b), then there is a point c in (a, b) where

$$f(b) - f(a) = (b-a)f'(c).$$

Proof. Define the function $g : [a, b] \to \mathbb{R}$ by

$$g(x) = f(x) - \left[f(a) + \frac{f(b) - f(a)}{b - a}(x - a) \right]. \qquad (7.7)$$

It is worth noting in equation (7.7) that the function in the square brackets is represented by the straight line joining the points $(a, f(a))$ and $(b, f(b))$, so $g(x)$ is the difference, at x, between the y-coordinates on this line and on the curve which represents f (see Figure 7.6).

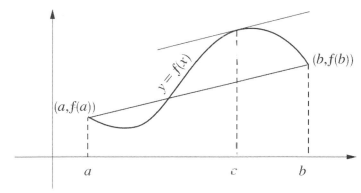

Figure 7.6

Since $g(a) = g(b) = 0$, the function g satisfies the conditions of Rolle's theorem, so there is a point $c \in (a, b)$ at which

$$0 = g'(c) = f'(c) - \frac{f(b) - f(a)}{b - a},$$

hence $f(b) - f(a) = (b-a)f'(c)$. □

The mean value theorem has a corresponding geometric interpretation: The curve of a function which satisfies the conditions of the theorem has a tangent at some point in (a,b) which is parallel to the straight line that joins $(a, f(a))$ and $(b, f(b))$.

The next two theorems indicate how the mean value theorem can be used to determine the behaviour of a differentiable function.

Theorem 7.8
Let $f : [a,b] \to \mathbb{R}$ be a continuous function which is differentiable on (a,b).
(i) If $f'(x) = 0$ for all $x \in (a,b)$, f is constant on $[a,b]$.
(ii) If, for all $x \in (a,b)$, $f'(x) \neq 0$, f is injective on $[a,b]$.

Proof
(i) Take any two points $x_1, x_2 \in [a,b]$. We shall prove that $f(x_1) = f(x_2)$. Suppose $x_1 < x_2$, and apply the mean value theorem to the interval $[x_1, x_2]$ to conclude that that there is a point $c \in (x_1, x_2)$ such that
$$f(x_2) - f(x_1) = (x_2 - x_1)f'(c).$$
Since $c \in (x_1, x_2)$, $f'(c) = 0$ and therefore $f(x_1) = f(x_2)$.

(ii) If f is not injective, there would be two points $x_1, x_2 \in [a,b]$ such that $x_1 < x_2$ and $f(x_1) = f(x_2)$. Applying the mean value theorem to $[x_1, x_2]$ we would then conclude that there is a $c \in (x_1, x_2) \subseteq (a,b)$ where
$$f'(c) = \frac{f(x_2) - f(x_1)}{x_2 - x_1} = 0.$$
This contradicts the hypothesis that $f'(x) \neq 0$ for all $x \in (a,b)$. □

Corollary 7.8 If the functions f and g are continuous on $[a,b]$, differentiable on (a,b), and
$$f'(x) = g'(x) \text{ for all } x \in (a,b),$$
then there is a constant c such that
$$f(x) = g(x) + c \text{ for all } x \in [a,b].$$

Proof. Apply Theorem 7.8(i) to the function $f - g$.

Theorem 7.9
Let $f : [a, b] \to \mathbb{R}$ be a continuous function which is differentiable on (a, b).
(i) If $f'(x) \geq 0$ for all $x \in (a, b)$, f is increasing on $[a, b]$.
(ii) If $f'(x) \leq 0$ for all $x \in (a, b)$, f is decreasing on $[a, b]$.
(iii) If $f'(x) > 0$ for all $x \in (a, b)$, f is strictly increasing on $[a, b]$.
(iv) If $f'(x) < 0$ for all $x \in (a, b)$, f is strictly decreasing on $[a, b]$.

Proof. Suppose $x_1, x_2 \in [a, b]$ and $x_1 < x_2$. From the mean value Theorem, applied to the interval $[x_1, x_2]$, there is a $c \in (x_1, x_2) \subseteq (a, b)$ such that
$$f(x_2) - f(x_1) = (x_2 - x_1) f'(c).$$
This implies that $f(x_2) - f(x_1)$ and $f'(c)$ have the same sign, and that $f(x_1) = f(x_2)$ if and only if $f'(c) = 0$. □

Remarks 7.3
1. If we drop the condition that f be continuous on $[a, b]$ in Theorems 7.8 and 7.9 and Corollary 7.8, then the conclusions of the theorems remain valid on (a, b) instead of $[a, b]$. To see that, observe that, for any $x_1, x_2 \in (a, b)$, f is differentiable (and hence continuous) on $[x_1, x_2]$, so the mean value theorem applies to f on $[x_1, x_2]$.

2. If f is differentiable and increasing on (a, b), then, for any $c \in (a, b)$,
$$\frac{f(x) - f(c)}{x - c} \geq 0 \quad \text{for all } x \in (a, b) \setminus \{c\},$$
hence
$$f'(c) = \lim_{x \to c} \frac{f(x) - f(c)}{x - c} \geq 0.$$
This means we can restate parts (i) and (ii) of Theorem 7.9 in stronger terms: a differentiable function f is increasing (decreasing) on $[a, b]$ if and only if $f'(x) \geq 0$ ($f'(x) \leq 0$) for all $x \in (a, b)$. But this does not apply to parts (iii) and (iv), since a function may be strictly increasing on $[a, b]$ without its derivative being positive on (a, b). Consider, for example, the function $f(x) = x^3$, which is strictly increasing on $\mathbb{R}$ though $f'(0) = 0$.

Example 7.11 Using the mean value theorem, we shall now prove that
(i) $-x \leq \sin x \leq x$ for all $x \geq 0$,
(ii) $\sin x \geq \dfrac{2}{\pi} x$ for all $x \in [0, \pi/2]$. \hfill (7.1)

Inequality (i) was already established in Example 5.10 using a geometric argument. It clearly holds when $x = 0$. If $x > 0$, we apply Theorem 7.7 to the sine function over the interval $[0, x]$ to conclude that
$$\sin x - \sin 0 = (x - 0) \cos c,$$
where $c \in (0, x)$. Since $|\cos c| \leq 1$, this implies $|\sin x| \leq x$. Since $\sin(-x) = -\sin x$, we conclude that $|\sin x| \leq |x|$ for all $x \in \mathbb{R}$.

To prove (ii), consider the function $f(x) = \sin x - 2x/\pi$, which is 0 at $x = 0$ and $x = \pi/2$. Its derivative $f'(x) = \cos x - 2/\pi$ is 0 at $c = \cos^{-1}(2/\pi) \in (0, \pi/2)$. Since $\cos x$ is a decreasing function on $[0, \pi/2]$, $f'(x)$ is positive on $(0, c)$ and negative on $(c, \pi/2)$. By Theorem 7.9 the function $f(x)$ is therefore increasing on $[0, c]$ and decreasing on $[c, \pi/2]$. Consequently, we have
$$f(x) \geq f(0) = 0 \text{ for all } x \in [0, c],$$
$$f(x) \geq f(\pi/2) = 0 \text{ for all } x \in [c, \pi/2],$$
which implies $f(x) \geq 0$ for all $x \in [0, \pi/2]$.

Example 7.12 For any rational number r, let
$$f(x) = (1 + x)^r \text{ for all } x > -1.$$
We shall now prove that, if $r > 1$, then
$$f(x) > 1 + rx \text{ for all } x > -1, x \neq 0.$$
This inequality is known as *Bernoulli's inequality*.

Let $g : (-1, \infty) \to \mathbb{R}$ be defined by
$$g(x) = f(x) - (1 + rx).$$
Using the rules of differentiation, we have
$$g'(x) = r(1 + x)^{r-1} - r$$
$$= r[(1 + x)^{r-1} - 1].$$
If $x > 0$ then $(1 + x)^{r-1} > 1$ and therefore $g'(x) > 0$, and if $-1 < x < 0$ then $(1 + x)^{r-1} < 1$ and $g'(x) < 0$. By Theorem 7.9, g is strictly decreasing on $(-1, 0]$ and strictly increasing on $[0, \infty)$, which means
$$g(x) > g(0) \text{ for all } x \in (-1, \infty) \setminus \{0\}.$$

But, since $g(0) = 0$,

$$g(x) > 0 \text{ for all } x \in (-1, \infty) \backslash \{0\},$$

and hence $f(x) > 1 + rx$ on $(-1, \infty)$, $x \neq 0$.

Besides Theorems 7.8 and 7.9, the mean value theorem has many deep consequences which will occupy us for the remainder of this chapter. In the following theorem, the sign of the derivative of a function is used to classify its critical points.

Theorem 7.10
Let c be a critical point and a point of continuity for the function $f : D \to \mathbb{R}$.
(i) If there is an open interval $I \subseteq D$, which contains c, such that

$$f'(x) < 0 \text{ for all } x \in I, x < c$$
$$f'(x) > 0 \text{ for all } x \in I, x > c,$$

then $f(c)$ is a local minimum for f.
(ii) If there is an open interval $I \subseteq D$, which contains c, such that

$$f'(x) > 0 \text{ for all } x \in I, x < c$$
$$f'(x) < 0 \text{ for all } x \in I, x > c,$$

then $f(c)$ is a local maximum for f.
(iii) If there is an open interval $I \subseteq D$, which contains c, such that $f'(x)$ has the same sign on $I \backslash \{c\}$, then $f(c)$ is not a local extremum for f.

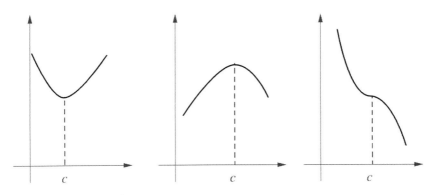

Figure 7.7

Proof. Figure 7.7 illustrates the three cases of the theorem. To prove case (i), note that, by Theorem 7.9, f is decreasing on the interval $I \cap (-\infty, c]$ and increasing on the interval $I \cap [c, \infty)$, hence

$$f(x) \geq f(c) \text{ for all } x \in I, x < c,$$
$$f(x) \geq f(c) \text{ for all } x \in I, x > c.$$

Thus $f(x) \geq f(c)$ for all $x \in I$, which means $f(c)$ is a local minimum. The other two cases can be proved by similar arguments. □

Theorem 7.4 gives the conditions under which the inverse function may be differentiated, and the form of its derivative. The following more general result shows the influence of the derivative of a function on the local behaviour of the inverse function. It is the one-dimensional version of a theorem that is central to the analysis of functions of several variables.

Theorem 7.11 (Inverse Function Theorem)
Let f be a differentiable function whose derivative is continuous on the open interval I. If $f'(b) \neq 0$ for some $b \in I$, then there is an open interval $J \subseteq I$ which contains b such that $f|_J$ is injective, and the function $(f|_J)^{-1}$ is differentiable on $f(J)$, where its derivative is

$$((f|_J)^{-1})'(f(x)) = \frac{1}{f'(x)} \text{ for all } x \in J.$$

Proof. Since f' is continuous at b, by Theorem 5.5, there are positive numbers r and δ such that

$$|f'(x)| > r \text{ for all } x \in (b-\delta, b+\delta),$$

which means $f'(x) \neq 0$ for all x in the interval $J = (b-\delta, b+\delta)$. By Theorem 7.8, $f|_J$ is injective and, by Theorem 7.4, $(f|_J)^{-1}$ is differentiable at every $f(x) \in f(J)$ with derivative

$$((f|_J)^{-1})'(f(x)) = \frac{1}{f'(x)}. \quad \square$$

We end this section with a rather subtle question: Can any function be a derivative (of another function)? We have already seen in Example

7.7 that the function

$$g(x) = \begin{cases} x^2 \sin \dfrac{1}{x}, & x \neq 0 \\ 0, & x = 0 \end{cases}$$

is differentiable on $\mathbb{R}$, and that its derivative is given by

$$g'(x) = \begin{cases} 2x \sin \dfrac{1}{x} - \cos \dfrac{1}{x}, & x \neq 0 \\ 0, & x = 0. \end{cases}$$

The function g' is clearly not continuous at $x = 0$, as it has no limit at that point. g' is bounded in a neighborhood of 0, so its discontinuity there is an *oscillatory*, not a *jump*, discontinuity (see Section 6.1), and this is no coincidence; for g' has to satisfy the mean value property, as the following theorem stipulates.

Theorem 7.12 (Darboux)
Let the function $f : [a, b] \to \mathbb{R}$ be differentiable. If λ is a number between $f'(a)$ and $f'(b)$, then there is a point $c \in (a, b)$ such that $f'(c) = \lambda$.

Proof. By using $-f$ instead of f if necessary, we may assume that

$$f'(a) < \lambda < f'(b). \tag{7.9}$$

The function defined by

$$g(x) = f(x) - \lambda x \text{ for all } x \in [a, b]$$

is differentiable on $[a, b]$, and

$$g'(x) = f'(x) - \lambda \text{ for all } x \in [a, b].$$

If there is no $c \in (a, b)$ such that $f'(c) = \lambda$, then, for every $x \in (a, b)$, $g'(x) \neq 0$. By Theorem 7.8, this means that g is injective on $[a, b]$. Since g is also continuous, it is strictly monotonic (by Theorem 6.7). If g is strictly increasing, then

$$\frac{g(x) - g(a)}{x - a} > 0 \text{ for all } x \in (a, b),$$

hence

$$g'(a) = \lim_{x \to a} \frac{g(x) - g(a)}{x - a} \geq 0.$$

This contradicts the assumption (7.9) that $g'(a) = f'(a) - \lambda < 0$. If g, on the other hand, is strictly decreasing, then

$$\frac{g(x) - g(b)}{x - b} < 0 \text{ for all } x \in (a, b),$$

and we obtain $g'(b) \leq 0$ in the limit as $x \to b$, thereby contradicting the assumption that $g'(b) = f'(b) - \lambda > 0$. Thus there is a $c \in (a, b)$ where $g'(c) = f'(c) - \lambda = 0$. $\square$

According to Darboux's theorem, not every function can be a derivative of another function over an interval. For example, the function

$$f(x) = \begin{cases} 1, & x > 0 \\ 0, & x \leq 0 \end{cases}$$

is not the derivative of any function in a neighborhood of 0, as it fails to satisfy the intermediate value property on that neighborhood.

EXERCISES 7.2

1. Define a function $f : [0, 1] \to \mathbb{R}$ which is continuous on $[0, 1]$, differentiable on $(0, 1)$, and satisfies $f(0) = f(1) = 0$, but is not differentiable at $x = 0$ and $x = 1$. Find a point $c \in (0, 1)$ where $f'(c) = 0$.

2. Give an example to show that, if there is a point in $(0, 1)$ where f is not differentiable, the conclusion of the mean value theorem may not hold.

3. For each of the following functions, determine whether f satisfies the conditions of the mean value theorem. If it does, find the point c, and if not, explain why.

   (a) $f(x) = x^3$, $x \in [-1, 1]$
   (b) $f(x) = |x|$, $x \in [-1, 1]$
   (c) $f(x) = x|x|$, $x \in [-1, 1]$
   (d) $f(x) = |x|^3$, $x \in [-1, 2]$
   (e) $f(x) = \dfrac{1}{1+x}$, $x \in [0, 2]$

(f) $f(x) = \begin{cases} 1/x, & x \in (0,1] \\ 0, & x = 0 \end{cases}$

(g) $f(x) = \begin{cases} x^2, & x \in [-1,1] \\ 1, & x \in (1,2]. \end{cases}$

4. Prove that $|\cos x - \cos y| \leq |x - y|$ for all $x, y \in \mathbb{R}$.

5. If $f''(x) = 0$ for all $x \in (a,b)$, prove that there are two constants c_1 and c_2 such that $f(x) = c_1 x + c_2$ for all $x \in (a,b)$. What can you say about the form of f if $f'''(x) = 0$ for all $x \in (a,b)$?

6. Show that the function

$$f(x) = \begin{cases} x^2 \sin \dfrac{1}{x} - \dfrac{1}{2}x, & x \neq 0 \\ 0, & x = 0 \end{cases}$$

satisfies $f'(0) > 0$, but is not increasing on any neighborhood of $x = 0$. How is this consistent with Theorem 7.9?

7. Prove the following:

(a) $x < \tan x$ for all $x \in (0, \pi/2)$

(b) If $f(x) = \dfrac{x}{\sin x}$, f is an increasing function on $(0, \pi/2)$

(c) $\sqrt{1+x} \leq 1 + \dfrac{1}{2}x$ for all $x > 0$

(d) $\lim_{x \to \infty}(\sqrt{x+k} - \sqrt{x}) = 0$, where k is any positive constant.

8. If f and g are differentiable on $\mathbb{R}$, and $f'(x) < g'(x)$ for all $x \in \mathbb{R}$, prove that for every $a \in \mathbb{R}$ we have

$$f(x) - f(a) \leq g(x) - g(a) \text{ for all } x \in [a, \infty).$$

9. Suppose f is differentiable on $\mathbb{R}$ and $1 \leq f'(x) \leq 2$ for all $x \in \mathbb{R}$. If $f(0) = 0$ prove that

$$|x| \leq |f(x)| \leq 2|x| \text{ for all } x \in \mathbb{R}.$$

10. Suppose $f : (a,b) \to \mathbb{R}$ is differentiable and f' is bounded on (a,b).

(a) Prove that f satisfies a Lipschitz condition on (a, b), i.e., there is a real constant K such that
$$|f(x) - f(y)| \le K |x - y| \quad \text{for all } x, y \in (a, b).$$
Conclude that f is uniformly continuous.

(b) Give an example of a function $f : (0, 1) \to \mathbb{R}$ which is differentiable and uniformly continuous, but which does not satisfy a Lipschitz condition on $(0, 1)$.

(c) Is the function defined on $\mathbb{R}$ by
$$f(x) = \begin{cases} x^2 \sin \dfrac{1}{x}, & x \ne 0 \\ 0, & x = 0 \end{cases}$$
uniformly continuous on $\mathbb{R}$?

11. Let f be continuous on $[a, b]$ and differentiable on (a, b). If $\lim_{x \to a} f'(x)$ exists, prove that f is differentiable at a and $f'(a) = \lim_{x \to a} f'(x)$.

12. Determine the local extreme values for each of the following functions, and the intervals where each is increasing or decreasing:

(a) $f(x) = 3x - 4x^2$, $x \in \mathbb{R}$
(b) $g(x) = \dfrac{x}{x^2 + 1}$, $x \in \mathbb{R}$
(c) $h(x) = x + \dfrac{1}{x}$, $x \ne 0$.

13. Find the points where each function assumes its extreme values on the given interval:

(a) $f(x) = |x^2 - 4|$ on $[-3, 4]$
(b) $g(x) = 1 + x^{2/3}$ on $[-2, 1]$
(c) $h(x) = x |x^2 - 1|$ on $[-2, 2]$.

14. Suppose the function f is continuous on $[0, \infty)$ and differentiable on $(0, \infty)$. If f' is increasing on $(0, \infty)$ and $f(0) = 0$, prove that the function $g(x) = f(x)/x$ is increasing on $(0, \infty)$.

Hint: Differentiate g and use the mean value theorem applied to f.

15. If $0 < a < b$ and $n \in \{2, 3, 4, \cdots\}$, prove that
$$b^{1/n} - a^{1/n} < (b-a)^{1/n}.$$

Hint: Show that the function $f(x) = x^{1/n} - (x-1)^{1/n}$ is strictly decreasing on $[1, \infty)$, and apply the mean value theorem over a suitable interval.

16. Let $f : \mathbb{R} \to \mathbb{R}$ be differentiable, and $f(0) = 0$, $f(1) = f(2) = -1$. Prove that there are three points a, b, and c in $(0, 2)$ such that $f'(a) = -1/2$, $f'(b) = -3/4$, and $f'(c) = -1/11$.

Hint: Apply Darboux's theorem to f'.

17. Suppose c is a critical point of the function f. If f' is differentiable in a neighborhood of c, prove the following:

   (a) If $f''(c) > 0$ then $f(c)$ is a local minimum value of f.
   (b) If $f''(c) < 0$ then $f(c)$ is a local maximum value of f.

This is known as the *second derivative test* for classifying critical points. Theorem 7.10 provides a *first derivative test* for such a classification.

18. Provide examples to show that, if $f''(c) = 0$, then $f(c)$ may be a local minimum, maximum, or neither.

7.3 L'Hôpital's Rule

When two functions f and g have limits at the point c,
$$\lim_{x \to c} f(x) = \ell, \quad \lim_{x \to c} g(x) = m,$$
we can write
$$\lim_{x \to c} \frac{f(x)}{g(x)} = \frac{\ell}{m} \qquad (7.10)$$
provided $m \neq 0$. But if $m = 0$, then there are two possibilities:
  (i) $\ell \neq 0$, in which case the limit (7.10) does not exist in $\mathbb{R}$, and there is nothing more to be said.

(ii) $\ell = 0$, in which case the limit (7.10) may exist, such as

$$\lim_{x \to 0} \frac{\sin x}{x},$$

or it may not exist, as in

$$\lim_{x \to 0} \frac{\sin x}{|x|}.$$

In case (ii), that is, when

$$\lim_{x \to c} f(x) = \lim_{x \to c} g(x) = 0,$$

we shall develop a method for deciding when the limit of f/g exists at c, and how it can be evaluated.

We begin with a generalized version of the mean value theorem:

Theorem 7.13 (Cauchy's Mean Value Theorem)
If the functions f and g are continuous on $[a,b]$ and differentiable on (a,b), then there is a point $c \in (a,b)$ such that

$$[f(b) - f(a)]g'(c) = [g(b) - g(a)]f'(c).$$

Proof. Let h be the function defined on $[a,b]$ by

$$h(x) = [f(b) - f(a)]g(x) - [g(b) - g(a)]f(x),$$

which is clearly continuous on $[a,b]$ and differentiable on (a,b). Since $h(a) = h(b)$, there is a $c \in (a,b)$ where

$$h'(c) = [f(b) - f(a)]g'(c) - [g(b) - g(a)]f'(c) = 0$$

by Rolle's theorem. □

Note that the mean value theorem is a special case of this theorem in which $g(x) = x$. Now we turn to the main result of this section.

Theorem 7.14 (L'Hôpital's Rule)
Let the functions f and g be continuous on a real interval I and differentiable on $I \setminus \{c\}$, where c is a point in I. If
(i) $g'(x) \neq 0$ for all $x \in I \setminus \{c\}$,
(ii) $f(c) = g(c) = 0$,
(iii) $\lim_{x \to c} \dfrac{f'(x)}{g'(x)}$ exists in $\bar{\mathbb{R}} = \mathbb{R} \cup \{-\infty, \infty\}$,

then
$$\lim_{x \to c} \frac{f(x)}{g(x)} = \lim_{x \to c} \frac{f'(x)}{g'(x)}.$$

Proof. Let (x_n) be any sequence in I such that $x_n \neq c$ and $x_n \to c$. According to Theorem 5.1, it suffices to prove that
$$\lim_{n \to \infty} \frac{f(x_n)}{g(x_n)} = \lim_{x \to c} \frac{f'(x)}{g'(x)}.$$

Applying Cauchy's mean value theorem, we can find a sequence (c_n) in $I \setminus \{c\}$ such that

1. c_n lies between x_n and c,
2. $[f(x_n) - f(c)]g'(c_n) = [g(x_n) - g(c)]f'(c_n)$ for all $n \in \mathbb{N}$.

Note that condition 1 implies $g'(c_n) \neq 0$ for all $n \in \mathbb{N}$; and since $f(c) = g(c) = 0$, condition 2 becomes
$$f(x_n)g'(c_n) = g(x_n)f'(c_n). \qquad (7.11)$$

The fact that $g'(x) \neq 0$ on $I \setminus \{c\}$ means that g is injective on the interval whose end-points are x_n and c, hence
$$g(x_n) \neq g(c) = 0.$$

This allows us to divide equation (7.11) by $g(x_n)g'(c_n)$ to obtain
$$\frac{f(x_n)}{g(x_n)} = \frac{f'(c_n)}{g'(c_n)}.$$

As the sequence $(f'(c_n)/g'(c_n))$ converges (in the extended sense) to
$$\lim_{x \to c} \frac{f'(x)}{g'(x)},$$
the sequence $(f(x_n)/g(x_n))$ also converges, and to the same limit. □

Since this proof remains valid if c is an end-point of I, L'Hôpital's rule may also be used to evaluate the one-sided limits of the function f/g when the conditions of the theorem are satisfied.

Example 7.13 In Chapter 5 we evaluated the limit at $x = 0$ of the function
$$h(x) = \frac{\sin x}{x}, \quad x \in \mathbb{R}\setminus\{0\}$$
using a geometric argument. Now we observe that the functions
$$f(x) = \sin x, \quad g(x) = x$$
satisfy the conditions of Theorem 6.14 on any neighborhood of 0, therefore
$$\lim_{x \to 0} \frac{\sin x}{x} = \lim_{x \to 0} \frac{\cos x}{1} = 1.$$

The function $\dfrac{\sin x}{|x|}$ also satisfies the conditions of Theorem 6.14 on the intervals $[0, 1]$ and $[-1, 0]$ at $c = 0$, hence

$$\lim_{x \to 0^+} \frac{\sin x}{|x|} = \lim_{x \to 0^+} \frac{\sin x}{x} = \lim_{x \to 0^+} \frac{\cos x}{1} = 1,$$

$$\lim_{x \to 0^-} \frac{\sin x}{|x|} = \lim_{x \to 0^-} \frac{\sin x}{-x} = \lim_{x \to 0^-} \frac{\cos x}{-1} = -1.$$

Thus the limit of $\sin x / |x|$ does not exist at 0.

The next result extends L'Hôpital's rule to the case when $c = \infty$.

Theorem 7.15
Let the functions f and g be differentiable on $[a, \infty)$ and suppose
$$\lim_{x \to \infty} f(x) = \lim_{x \to \infty} g(x) = 0,$$
$$g'(x) \neq 0 \quad \text{for all } x > a.$$

If $\lim\limits_{x \to \infty} \dfrac{f'(x)}{g'(x)}$ exists in $\bar{\mathbb{R}}$, then

$$\lim_{x \to \infty} \frac{f(x)}{g(x)} = \lim_{x \to \infty} \frac{f'(x)}{g'(x)}.$$

Proof. Clearly, we can take $a > 0$. Define the functions F and G on $(0, 1/a]$ by
$$F(x) = f\left(\frac{1}{x}\right), \quad G(x) = g\left(\frac{1}{x}\right).$$

This means
$$f(x) = F\left(\frac{1}{x}\right), \quad g(x) = G\left(\frac{1}{x}\right), \quad x \in [a, \infty),$$
and so the limit we have to evaluate is
$$\lim_{x \to \infty} \frac{F(1/x)}{G(1/x)}.$$
But since $x \to \infty$ if and only if $t = 1/x \to 0^+$, this is the same as the limit
$$\lim_{t \to 0^+} \frac{F(t)}{G(t)}.$$
By the chain rule,
$$F'(t) = f'\left(\frac{1}{t}\right) \cdot \left(-\frac{1}{t^2}\right),$$
$$G'(t) = G'\left(\frac{1}{t}\right) \cdot \left(-\frac{1}{t^2}\right).$$
Hence
$$\lim_{t \to 0^+} \frac{F'(t)}{G'(t)} = \lim_{t \to 0^+} \frac{f'(1/t)}{g'(1/t)} = \lim_{x \to \infty} \frac{f'(x)}{g'(x)}.$$
To complete the proof, we need to show that
$$\lim_{t \to 0^+} \frac{F(t)}{G(t)} = \lim_{t \to 0^+} \frac{F'(t)}{G'(t)}. \tag{7.12}$$
The functions F and G are differentiable on $(0, 1/a)$, and they may be extended as continuous functions to $[0, 1/a)$ by the definitions
$$F(0) = \lim_{t \to 0^+} F(t) = \lim_{x \to \infty} f(x) = 0,$$
$$G(0) = \lim_{t \to 0^+} G(t) = \lim_{x \to \infty} g(x) = 0.$$
This allows us to apply Theorem 7.14 and obtain (7.12). □

Limits of the type $\lim_{x \to c} f(x)/g(x)$, where
$$\lim_{x \to c} f(x) = \lim_{x \to c} g(x) = 0$$
are denoted by the *indeterminate form* 0/0. There are other indeterminate forms, such as ∞/∞, 1^∞, $0 \cdot \infty$, $\infty - \infty$, 0^0, and ∞^0. The next

242   Elements of Real Analysis

theorem gives L'Hôpital's rule for the form ∞/∞, and the others will be discussed through some representative examples.

Theorem 7.16
Let the functions f and g be differentiable on (a,b) and $g'(x) \neq 0$ for any $x \in (a,b)$. If

$$\lim_{x \to a^+} g(x) = \lim_{x \to a^+} f(x) = \infty$$

and $\lim_{x \to a^+} \dfrac{f'(x)}{g'(x)}$ *exists in* $\bar{\mathbb{R}}$, *then*

$$\lim_{x \to a^+} \frac{f(x)}{g(x)} = \lim_{x \to a^+} \frac{f'(x)}{g'(x)}.$$

Proof. We shall prove the theorem in the case when

$$\lim_{x \to a^+} \frac{f'(x)}{g'(x)} = \ell$$

with $\ell \in \mathbb{R}$, and leave the cases when $\ell = \pm\infty$ as an exercise. Given $\varepsilon > 0$, there is a $\delta > 0$ such that

$$\left| \frac{f'(x)}{g'(x)} - \ell \right| < \varepsilon \quad \text{for all } x \in (a, a+\delta). \tag{7.13}$$

Choose any point c in $(a, a+\delta)$. Since $f(x) \to \infty$ as $x \to a^+$, there is a point $d \in (a, c)$ (see Figure 7.8) such that

$$x \in (a, d) \Rightarrow f(x) > f(c).$$

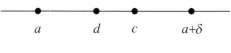

Figure 7.8

Define the function h on (a, d) by

$$h(x) = \frac{1 - f(c)/f(x)}{1 - g(c)/g(x)},$$

and note that $1 - g(c)/g(x) \neq 0$ because g is injective on (a, b), and $h(x) \neq 0$ because $f(x) > f(c)$ on (a, d). Since $f(c)/f(x) \to 0$ and $g(c)/g(x) \to 0$ as $x \to a^+$, it follows that $\lim_{x \to a^+} h(x) = 1$, so there is a $t \in (a, d)$ such that

$$x \in (a, t) \Rightarrow |h(x) - 1| < \varepsilon' = \min\{\varepsilon, 1/2\}$$
$$\Rightarrow 1 - \varepsilon' < h(x) < 1 + \varepsilon'$$
$$\Rightarrow \frac{1}{|h(x)|} < \frac{1}{1 - \varepsilon'} \leq 2. \tag{7.14}$$

Let $x \in (a, t)$. We shall complete the proof by showing that

$$\left| \frac{f(x)}{g(x)} - \ell \right| < k\varepsilon$$

for some constant k. Observe that

$$\frac{f(x)}{g(x)} = \frac{f(x)}{g(x)} \frac{h(x)}{h(x)} = \frac{f(x) - f(c)}{g(x) - g(c)} \frac{1}{h(x)}.$$

If we apply Cauchy's mean value theorem to f and g on $[x, c]$, and use the above equation, we obtain, for some $y \in (x, c)$,

$$\frac{f'(y)}{g'(y)} = \frac{f(x) - f(c)}{g(x) - g(c)} = \frac{f(x)}{g(x)} h(x) \quad \text{for all } x \in (a, d).$$

In view of (7.13) and (7.14), and with $a < x < t < d < c < a + \delta$, we have

$$\left| \frac{f(x)}{g(x)} - \ell \right| = \left| \frac{f'(y)}{g'(y)} \frac{1}{h(x)} - \ell \right|$$
$$= \left| \frac{f'(y)}{g'(y)} - \ell h(x) \right| \frac{1}{|h(x)|}$$
$$\leq 2 \left(\left| \frac{f'(y)}{g'(y)} - \ell \right| + |\ell - \ell h(x)| \right)$$
$$\leq 2(\varepsilon + |\ell|\varepsilon)$$
$$\leq 2(1 + |\ell|)\varepsilon. \quad \square$$

This theorem does not preclude the possibility that the interval (a, b) extends to infinity at either end, or both ends, which allows us to deal with the limits

$$\lim_{x \to \infty} \frac{f(x)}{g(x)}, \quad \lim_{x \to -\infty} \frac{f(x)}{g(x)}$$

when the limits of f and g at $\pm\infty$ are ∞. But in the proof we have to use the appropriate form for the neighborhoods of ∞ and $-\infty$.

In the following examples, we assume the validity of the rules for differentiating the exponential function e^x and the logarithmic function $\log x$,

$$\frac{d}{dx} e^x = e^x, \quad x \in \mathbb{R},$$

$$\frac{d}{dx} \log x = \frac{1}{x}, \quad x > 0,$$

which will be proved in Chapter 9.

Example 7.14 To determine

$$\lim_{x \to \infty} \frac{\log x}{x},$$

where $\log x$ and x both tend to ∞, we note that Theorem 7.16 applies to the interval $(0, \infty)$. Thus

$$\lim_{x \to \infty} \frac{\log x}{x} = \lim_{x \to \infty} \frac{1/x}{1} = 0.$$

Note that L'Hôpital's rule does not apply to the limit at 0^+ (since $\log x \to -\infty$ while $x \to 0$), and it gives the false result

$$\lim_{x \to 0^+} \frac{\log x}{x} = \lim_{x \to 0^+} \frac{1/x}{1} = \infty.$$

Example 7.15 The limit $\lim_{x \to 0^+} x \log x$ is of the form $0 \cdot (-\infty)$. We can write

$$x \log x = \frac{\log x}{1/x},$$

which satisfies the conditions of Theorem 7.16 on $(0, \infty)$, since $\log x \to -\infty$ and $1/x \to \infty$. Note that the negative sign of $\lim_{x \to 0^+} \log x$ does not violate the conditions of the theorem, since we can apply the rule to $-x \log x$ and change the sign of the result. Thus

$$\lim_{x \to 0^+} \frac{\log x}{1/x} = \lim_{x \to 0^+} \frac{1/x}{-1/x^2} = 0.$$

It may seem, on the other hand, that writing

$$x \log x = \frac{x}{1/\log x}$$

allows us to apply Theorem 7.14 for the indeterminate form 0/0. But the limit of
$$\frac{f'(x)}{g'(x)} = -x(\log x)^2$$
as $x \to 0^+$ is even more difficult to evaluate than the original limit.

Example 7.16 To evaluate the limit of x^x as $x \to 0^+$, we use the properties of the exponential function
$$x^x = (e^{\log x})^x = e^{x \log x} \quad \text{for all } x > 0.$$
Assuming the continuity of $e^x = \exp x$ on $(0, \infty)$,
$$\begin{aligned}
\lim_{x \to 0^+} x^x &= \lim_{x \to 0^+} \exp(x \log x) \\
&= \exp\left(\lim_{x \to 0^+} x \log x\right) \\
&= \exp(0) \\
&= 1.
\end{aligned}$$

Example 7.17 The limit of the function $(1 + p/x)^x$ at ∞ leads to the indeterminate form 1^∞, and at 0 to the form ∞^0.
(i) In the first case we have
$$\left(1 + \frac{p}{x}\right)^x = \exp\left[x \log\left(1 + \frac{p}{x}\right)\right].$$
The limit
$$\lim_{x \to \infty} x \log\left(1 + \frac{p}{x}\right) = \lim_{x \to \infty} \frac{\log(1 + p/x)}{1/x}$$
is of the form 0/0 to which Theorem 7.14 applies, so we may use L'Hôpital's rule to obtain
$$\begin{aligned}
\lim_{x \to \infty} x \log\left(1 + \frac{p}{x}\right) &= \lim_{x \to \infty} \frac{-px^{-2}/(1 + p/x)}{-x^{-2}} \\
&= \lim_{x \to \infty} \frac{px}{x + p} \\
&= p.
\end{aligned}$$
Hence,
$$\lim_{x \to \infty} \left(1 + \frac{p}{x}\right)^x = \exp p = e^p,$$
as would be expected, in view of example 3.12.

(ii) In the second case,
$$\lim_{x \to 0^+} x \log\left(1 + \frac{p}{x}\right) = \lim_{x \to 0^+} \frac{px}{x+p} = 0.$$

Therefore
$$\lim_{x \to 0^+} \left(1 + \frac{1}{x}\right)^x = \exp 0 = 1.$$

EXERCISES 7.3

1. If $\lim_{x \to c} f(x)/g(x)$ exists in $\mathbb{R}$ and $\lim_{x \to c} g(x) = 0$, prove that $\lim_{x \to c} f(x) = 0$.

2. Define two functions, f and g, which are differentiable on $(0, \infty)$ such that $g(x) \neq 0$ for any $x \in (0, \infty)$, $\lim_{x \to 0^+} f(x) = \lim_{x \to 0^+} g(x) = 0$, but $\lim_{x \to 0^+} f(x)/g(x)$ does not exist in $\mathbb{R}$.

3. Define two functions, f and g, in a neighborhood of 0 such that $\lim_{x \to 0} f(x)/g(x) = 0$ but $\lim_{x \to 0} f'(x)/g'(x)$ does not exist.

4. Evaluate the following limits:

   (a) $\lim_{x \to 0} \dfrac{\cos x - 1}{x^2}$

   (b) $\lim_{x \to 0} \dfrac{\sqrt{1+x} - \sqrt{1-x}}{x}$

   (c) $\lim_{x \to 0} \dfrac{x^3}{\sin x - x}$

   (d) $\lim_{x \to 0} \dfrac{1 - \cos x}{\sin^2 x}.$

5. Evaluate the following limits:

   (a) $\lim_{x \to 0^+} \dfrac{\log(1+x)}{\sin x}$

   (b) $\lim_{x \to 0^+} \dfrac{\tan x}{x}$

   (c) $\lim_{x \to \infty} \dfrac{\log x}{\sqrt{x}}$

(d) $\lim\limits_{x \to \infty} \dfrac{x^4}{e^x}$.

6. Evaluate the following limits:

   (a) $\lim\limits_{x \to \infty} x^{1/x}$

   (b) $\lim\limits_{x \to 0^+} \left(1 + \dfrac{2}{x}\right)^x$

   (c) $\lim\limits_{x \to 0^+} \left(\dfrac{1}{x} - \dfrac{1}{\arctan x}\right)$

   (d) $\lim\limits_{x \to (\pi/2)^-} (\sec x - \tan x)$.

7.4 Taylor's Theorem

Suppose f is a real polynomial on $\mathbb{R}$ of degree n,
$$f(x) = a_0 + a_1 x + a_2 x^2 + \cdots + a_n x^n.$$

By differentiating f successively and setting $x = 0$ in each derivative, we can evaluate the coefficients

$$a_0 = f(0), \quad a_1 = \dfrac{f'(0)}{1!}, \quad a_2 = \dfrac{f''(0)}{2!}, \quad \cdots, \quad a_n = \dfrac{f^{(n)}(0)}{n!}.$$

More generally, if
$$f(x) = a_0 + a_1(x - x_0) + a_2(x - x_0)^2 + \cdots + a_n(x - x_0)^n,$$

where x_0 is fixed, then we have

$$a_0 = f(x_0), \quad a_k = \dfrac{f^{(k)}(x_0)}{k!}, \quad k = 1, 2, \cdots, n,$$

in which case we can write

$$f(x) = f(x_0) + \dfrac{f'(x_0)}{1!}(x - x_0) + \dfrac{f''(x_0)}{2!}(x - x_0)^2 + \cdots$$
$$+ \dfrac{f^{(n)}(x - x_0)}{n!}(x - x_0)^n. \qquad (7.15)$$

Now if f is not a polynomial, but a function which is differentiable n times in a neighborhood of x_0, can the right-hand side of equation

(7.15) be considered an approximation of $f(x)$, and in what sense? The answer to this question is provided by *Taylor's theorem*. This theorem, in a certain sense, is another generalization of the mean value theorem with deep significance in analytic function theory.

Theorem 7.17 (Taylor's Theorem)
Let the function f and its derivatives $f', f'', \cdots, f^{(n)}$ be continuous on $[a, b]$, and suppose $f^{(n)}$ is differentiable on (a, b). If x_0 is a point in $[a, b]$ then, for every $x \in [a, b]\setminus\{x_0\}$, there is a point c between x_0 and x such that

$$f(x) = f(x_0) + \frac{f'(x_0)}{1!}(x - x_0) + \frac{f''(x_0)}{2!}(x - x_0)^2 + \cdots$$
$$+ \frac{f^{(n)}(x_0)}{n!}(x - x_0)^n + \frac{f^{(n+1)}(c)}{(n + 1)!}(x - x_0)^{n+1}. \qquad (7.16)$$

Remarks 7.4

1. We retrieve the mean value theorem by setting $n = 0$ in (7.16).
2. The number c depends on x and n.

Proof. Define the function g on the closed interval I, whose end-points are x_0 and x, by

$$g(t) = f(x) - f(t) - f'(t)(x - t) - \frac{f''(t)}{2!}(x - t)^2 - \cdots$$
$$- \frac{f^{(n)}(t)}{n!}(x - t)^n,$$

so that we need only prove the equality

$$g(x_0) = \frac{f^{(n+1)}(c)}{(n + 1)!}(x - x_0)^{n+1}.$$

for some $c \in I$. Differentiating g,

$$g'(t) = -f'(t) + f'(t) - f''(t)(x - t) + f''(t)(x - t) - \cdots$$
$$- \frac{f^{(n)}(t)}{(n - 1)!}(x - t)^{n-1} + \frac{f^{(n)}(t)}{(n - 1)!}(x - t) - \frac{f^{(n+1)}(t)}{n!}(x - t)^n$$
$$= -\frac{f^{(n+1)}(t)}{n!}(x - t)^n, \qquad (7.17)$$

and setting

$$h(t) = g(t) - g(x_0)\left(\frac{x - t}{x - x_0}\right)^{n+1}, \quad t \in I,$$

we see that

$$h(x_0) = g(x_0) - g(x_0) = 0,$$
$$h(x) = g(x) - 0 = 0.$$

The function h, therefore, satisfies the conditions of Rolle's theorem on I, so there is a point c between x_0 and x such that

$$0 = h'(c) = g'(c) + (n+1)\frac{(x-c)^n}{(x-x_0)^{n+1}} g(x_0),$$

which, using (7.17), implies

$$g(x_0) = -g'(c)\frac{(x-x_0)^{n+1}}{(n+1)(x-c)^n}$$
$$= \frac{f^{(n+1)}(c)}{n!}(x-c)^n \cdot \frac{(x-x_0)^{n+1}}{(n+1)(x-c)^n}$$
$$= \frac{f^{(n+1)}(c)}{(n+1)!}(x-x_0)^{n+1}. \ \square$$

Taylor's theorem allows us to express the function f as

$$f(x) = p_n(x) + R_n(x),$$

where $p_n(x)$ is the polynomial

$$p_n(x) = f(x_0) + \sum_{k=1}^{n} \frac{f^{(k)}(x_0)}{k!}(x-x_0)^k,$$

and $R_n(x)$, the difference between $f(x)$ and $p_n(x)$, is called *Taylor's remainder* and is given by

$$R_n(x) = \frac{f^{(n+1)}(c)}{(n+1)!}(x-x_0)^{n+1} \tag{7.18}$$

in the so-called *Lagrange form*. There is another representation of R_n, known as the *Cauchy form*, given by

$$R_n(x) = \frac{f^{(n+1)}(d)}{n!}(x-x_0)(x-d)^n, \tag{7.19}$$

where d lies between x_0 and x. To prove the validity of (7.19), we have to show that there is a point d in I, the interval bounded by x_0 and x, such that

$$f(x) = f(x_0) + f'(x_0)(x-x_0) + \cdots + \frac{f^{(n)}(x_0)}{n!}(x-x_0)^n$$
$$+ \frac{f^{(n+1)}(d)}{n!}(x-x_0)(x-d)^n.$$

If we apply the mean value theorem to the function g (defined in the proof of Theorem 7.17) on the interval I, we can find a point $d \in I$ where

$$\frac{g(x) - g(x_0)}{x - x_0} = g'(d)$$
$$\frac{0 - g(x_0)}{x - x_0} = -\frac{f^{(n+1)}(d)}{n!}(x-d)^n,$$

by (7.17). Hence

$$g(x_0) = \frac{f^{(n+1)}(d)(x-x_0)(x-d)^n}{n!}.$$

Since $g(x_0) = f(x) - p_n(x)$, Equation (7.19) follows.

If the remainder term $R_n(x)$ in Taylor's theorem tends to 0 as n increases, the theorem provides an effective means for approximating functions by polynomials, $R_n(x)$ being the *error* in the approximation.

Example 7.18 To approximate the function $f(x) = \sqrt{x+1}$ on $(-1, 1)$ by a polynomial of degree 3, we take $n = 3$, $x_0 = 0$, and calculate the first four derivatives of the function at 0 :

$$f'(x) = \frac{1}{2}(x+1)^{-1/2}, \quad f'(0) = \frac{1}{2},$$
$$f''(x) = -\frac{1}{4}(x+1)^{-3/2}, \quad f''(0) = -\frac{1}{4},$$
$$f'''(x) = \frac{3}{8}(x+1)^{-5/2}, \quad f'''(0) = \frac{3}{8},$$
$$f^{(4)}(x) = -\frac{15}{16}(x+1)^{-7/2}, \quad f^{(4)}(0) = -\frac{15}{16}.$$

Substituting into Taylor's formula (7.16), we see that

$$\sqrt{x+1} = 1 + \frac{1}{2}x - \frac{1}{4}\frac{1}{2!}x^2 + \frac{3}{8}\frac{1}{3!}x^3 + R_3(x),$$

where
$$R_3(x) = -\frac{15}{16}\frac{(1+c)^{-7/2}}{4!}x^4$$
and $0 < c < x$. On the interval $[0, 1/2]$, for example, the error in the approximation
$$\sqrt{x+1} \simeq 1 + \frac{1}{2}x - \frac{1}{8}x^2 + \frac{1}{16}x^3$$
is
$$|R_3(x)| < \frac{15}{16}\frac{1}{4!}\frac{1}{2^4} < 0.0025.$$

Example 7.19 To approximate the number e, take
$$f(x) = e^x, \quad x_0 = 0.$$
Taylor's formula yields
$$e^x = 1 + \frac{x}{1!} + \frac{x^2}{2!} + \cdots + \frac{x^n}{n!} + \frac{x^{n+1}}{(n+1)!}e^c, \quad 0 < c < x.$$
Choose $x = 1$, so that
$$e = 1 + 1 + \frac{1}{2} + \cdots + \frac{1}{n!} + \frac{1}{(n+1)!}e^c, \quad 0 < c < 1.$$
Suppose we wish to approximate e to a margin of error not exceeding 10^{-6}. Then we select n so that
$$\frac{1}{(n+1)!}e^c \leq 10^{-6}, \quad 0 < c < 1. \tag{7.20}$$
Since $e^c < e < 3$, (7.20) will be satisfied if $(n+1)! \geq 3 \times 10^6$, that is, if $n \geq 9$. For $n = 9$, we obtain
$$e \simeq 1 + 1 + \frac{1}{2} + \cdots + \frac{1}{9!} = 2.718282$$
to within an error of 10^{-6}.

Example 7.20 Let
$$f(x) = \frac{1}{1-x}, \quad x \in (-1, 1), \quad x_0 = 0.$$
Therefore
$$f'(0) = 1, \ f''(0) = 2, \cdots, \ f^{(k)}(0) = k!$$

and hence
$$\frac{1}{1-x} = 1 + x + x^2 + \cdots + x^n + R_n(x).$$
Using Lagrange's form of the remainder, we have
$$R_n(x) = \frac{(n+1)!}{(n+1)!(1-c)^{n+2}} x^{n+1} = \frac{x^{n+1}}{(1-c)^{n+2}}, \quad (7.21)$$
where c lies between 0 and x. With $-1 < x < 1$ it is not clear from (7.21) that $R_n(x) \to 0$ as $n \to \infty$. Cauchy's form gives
$$\begin{aligned}R_n(x) &= \frac{f^{(n+1)}(d)}{n!} x(x-d)^n \\ &= \frac{(n+1)!}{n!(1-d)^{n+2}} x(x-d)^n \\ &= (n+1)x^{n+1} \frac{(1-\theta)^n}{(1-\theta x)^{n+2}},\end{aligned}$$
where $0 < \theta < 1$ and $d = \theta x$. Since $-1 < x < 1$, it follows that
$$1 - \theta x > 1 - \theta > 0,$$
and that
$$|1 - \theta x| \geq 1 - |\theta x| > 1 - |x|.$$
Therefore
$$|R_n(x)| \leq \frac{(n+1)|x|^{n+1}}{(1-\theta x)^2} < \frac{(n+1)|x|^{n+1}}{(1-|x|)^2}.$$
With $|x| < 1$, and in view of Exercise 3.2.6, we have $n|x|^n \to 0$, hence $|R_n(x)| \to 0$ as $n \to \infty$ for every $x \in (-1, 1)$.

Exercise 7.4.6 provides an example of a function for which $R_n(x) \not\to 0$ at any $x \neq 0$, though the function is differentiable to any order. This subject will be followed up in Chapter 9, where the distinction between such functions and those for which $R_n(x) \to 0$ is clarified.

In Taylor's theorem it was assumed that $f^{(n)}$ was differentiable on the interval (a, b) containing c. The following result, often referred to as *Young's theorem*, requires that $f^{(n)}$ be differentiable only at x_0.

Theorem 7.18
Let the function f and all its derivatives up to order n be continuous on $[a, b]$, and suppose $f^{(n)}$ is differentiable at the point $x_0 \in [a, b]$. If $x \in [a, b]$ then

$$f(x) = f(x_0) + \frac{f'(x_0)}{1!}(x - x_0) + \frac{f''(x_0)}{2!}(x - x_0)^2 + \cdots$$
$$+ \frac{f^{(n)}(x_0)}{n!}(x - x_0)^n + \frac{f^{(n+1)}(x_0)}{(n+1)!}(x - x_0)^{n+1} + E_n(x), \quad (7.22)$$

where $E_n(x)/(x - x_0)^{n+1} \to 0$ as $x \to x_0$.

Proof. Let p be the polynomial

$$p(x) = f(x_0) + \frac{f'(x_0)}{1!}(x - x_0) + \frac{f''(x_0)}{2!}(x - x_0)^2 + \cdots + \frac{f^{(n)}(x_0)}{n!}(x - x_0)^n,$$

and define

$$P(x) = f(x) - p(x), \quad Q(x) = (x - x_0)^{n+1}.$$

To prove the theorem, we have to show that

$$\lim_{x \to x_0} \frac{P(x)}{Q(x)} = \frac{f^{(n+1)}(x_0)}{(n+1)!}.$$

Observe that

$$p(x_0) = f(x_0),$$
$$p^{(k)}(x_0) = f^{(k)}(x_0), \quad k = 1, 2, \cdots, n,$$

which implies

$$P(x_0) = P'(x_0) = \cdots = P^{(n)}(x_0) = 0.$$

Also

$$Q^{(k)}(x) = \frac{(n+1)!}{(n+1-k)!}(x - x_0)^{n+1-k}$$
$$\Rightarrow Q(x_0) = Q'(x_0) = \cdots = Q^{(n)}(x_0) = 0.$$

To evaluate the limit of P/Q at x_0, we apply L'Hôpital's rule n times:

$$\lim_{x \to x_0} \frac{P(x)}{Q(x)} = \lim_{x \to x_0} \frac{P'(x)}{Q'(x)}$$
$$= \lim_{x \to x_0} \frac{P''(x)}{Q''(x)}$$
$$= \cdots$$
$$= \lim_{x \to x_0} \frac{P^{(n)}(x)}{Q^{(n)}(x)}$$
$$= \lim_{x \to x_0} \frac{f^{(n)}(x) - p^{(n)}(x)}{(n+1)!(x - x_0)}$$
$$= \lim_{x \to x_0} \frac{f^{(n)}(x) - f^{(n)}(x_0)}{(n+1)!(x - x_0)}$$
$$= \frac{f^{(n+1)}(x_0)}{(n+1)!}. \quad \Box$$

Theorem 7.18 may be used to classify the critical points of a function where it is differentiable a sufficient number of times, and provides a clear improvement on the so-called second derivative test given in Exercise 7.2.17.

Corollary 7.18 *Suppose $f'(c) = f''(c) = \cdots = f^{(m-1)}(c) = 0$, and $f^{(m)}(c) \neq 0$. Then*
(i) For $f(c)$ to be an extremum of the function f, m has to be an even number.
(ii) When m is even, $f(c)$ is a local maximum if $f^{(m)}(c) < 0$, and a local minimum if $f^{(m)}(c) > 0$.

Proof. Setting $x_0 = c$ in (7.22), we have

$$f(x) = f(c) + \frac{f^{(m)}(c)}{m!}(x - c)^m + E,$$

where $E/(x - c)^m \to 0$ as $x \to c$. Hence

$$\lim_{x \to c} \frac{f(x) - f(c)}{(x - c)^m} = \frac{f^{(m)}(c)}{m!}.$$

(i) Let m be an odd positive integer, and suppose $f^{(m)}(c) > 0$ (otherwise replace f by $-f$). It then follows that

$$\lim_{x \to c} \frac{f(x) - f(c)}{(x - c)^m} > 0,$$

so there is a $\delta > 0$ such that
$$\frac{f(x) - f(c)}{(x-c)^m} > 0 \quad \text{for all } x \in (c-\delta, c+\delta). \tag{7.23}$$

m being odd, we therefore have
$$f(x) - f(c) < 0 \quad \text{for all } x \in (c-\delta, c),$$
$$f(x) - f(c) > 0 \quad \text{for all } x \in (c, c+\delta),$$

which implies $f(c)$ is neither a local minimum nor a local maximum.

(ii) Now let m be even. If $f^{(m)}(c) > 0$, (7.23) implies $f(x) - f(c) \geq 0$ on $(c-\delta, c+\delta)$, hence $f(c)$ is a local minimum; but if $f^{(m)}(c) < 0$, then $f(x) - f(c) \leq 0$ on $(c-\delta, c+\delta)$, and $f(c)$ is a local maximum. □

EXERCISES 7.4

1. Use Taylor's theorem with $n = 2$ to obtain a suitable approximation of each of the following roots:

    (a) $\sqrt[3]{1.2}$
    (b) $\sqrt{0.9}$.

2. Prove that Taylor's remainder for $f(x) = \sin x$ in the representation (7.16) tends to 0 as $n \to \infty$ for any $x_0, x \in \mathbb{R}$.

3. Prove that, for $x > 0$,
$$\left| \log(1+x) - \left[x - \frac{x^2}{2} + \frac{x^3}{3} - \cdots + (-1)^{n-1}\frac{x^n}{n} \right] \right| < \frac{x^{n+1}}{n+1}.$$
Use this inequality to approximate $\log 1.3$ to within 0.001.

4. Approximate the function $\cos x$ on $[-1, 1]$ by a polynomial of degree 6. Give an estimate of the error in the approximation.

5. Use Corollary 7.18 to decide whether $f(0)$ is an extremum value of f in each case:

    (a) $f(x) = \cos x - 1$
    (b) $f(x) = \sin x - x + \frac{1}{6}x^3$

(c) $f(x) = x \sin x$.

6. Let
$$f(x) = \begin{cases} e^{-1/x^2}, & x \neq 0 \\ 0, & x = 0. \end{cases}$$
Prove that f is differentiable on $\mathbb{R}$ any number of times, and that $f^{(k)}(0) = 0$ for every $k \in \mathbb{N}$. Show that $R_n(x) \not\to 0$ as $n \to \infty$ for any $x \in \mathbb{R}\setminus\{0\}$.

8

The Riemann Integral

In this chapter we introduce the *Riemann integral* and deduce its salient properties. In all probability the reader is already familiar with the integral and its applications from earlier calculus courses. Despite the apparent technicality of the various definitions of the integral that we offer in Sections 8.1 and 8.2, the concept arose from our need to handle a simple and intuitive idea: how do we define and calculate the areas of shapes in the plane?

Starting with a square whose side length is 1 as the unit area, intuition suggests that the area of a shape should be the number of non-overlapping unit squares needed to cover that shape exactly. With this in mind, we quickly obtain laws for the areas of simple shapes like rectangles, triangles, and more general polygonal shapes. Significant difficulties arise, however, when we want to deal with shapes bounded by curves rather than straight edges. The first attempt in recorded history at working out such areas is credited to *Archimedes* (287-212 BC), if we overlook some evidence that the ancient Egyptians had quite accurately calculated the area of the circle. Archimedes determined the area under the parabola $y = x^2$ over the interval $[0, 1]$ by approximating it using polygonal shapes. This is the very idea on which the *Riemann Integral*, named after *B.Riemann*, is founded.

Section 8.4 deals with the relationship between integration and differentiation. The all important "Fundamental Theorem of Calculus", derived independently by *Newton* and *Leibnitz*, reveals in striking simplicity how one operation is essentially the inverse of the other. This is rather surprising, since the derivative of a function measures its rate of change, or the slope of its curve, while its integral measures the area under its curve, and there is no apparent connection between the two concepts.

8.1 Riemann Integrability

Unless otherwise stated, $[a, b]$ is a closed and bounded interval in $\mathbb{R}$ and f is a bounded real-valued function on $[a, b]$.

Definition 8.1 A finite ordered set of points $P = \{x_0, x_1, x_2, \cdots, x_n\}$ is called a *partition* of $[a, b]$ if

$$a = x_0 < x_1 < x_2 < ... < x_n = b.$$

The interval $[x_i, x_{i+1}]$ is called the *i-th subinterval* of P. These subintervals are clearly non-overlapping, except at the end-points, and they cover $[a, b]$, which justifies the term "partition". There are obviously many ways to partition $[a, b]$, and we shall denote the class of all partitions of $[a, b]$ by $\mathcal{P}(a, b)$.

Let $P = \{x_0, x_1, x_2, \cdots, x_n\}$ be a partition of $[a, b]$. For each $0 \leq i \leq n - 1$, let

$$M_i = \sup\{f(x) : x \in [x_i, x_{i+1}]\},$$
$$m_i = \inf\{f(x) : x \in [x_i, x_{i+1}]\}.$$

Since f is bounded, both M_i and m_i are real numbers for all $i = 0, 1, \cdots, n - 1$.

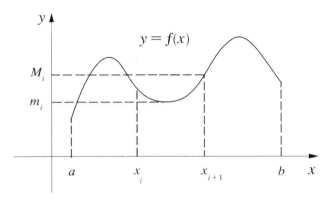

Figure 8.1

Definition 8.2 The *upper sum* $U(f, P)$ and the *lower sum* $L(f, P)$ of the function f, corresponding to the partition P, are defined as

$$U(f, P) = \sum_{i=0}^{n-1} M_i (x_{i+1} - x_i)$$
$$L(f, P) = \sum_{i=0}^{n-1} m_i (x_{i+1} - x_i).$$

If $f \geq 0$, as shown in Figure 8.2, then $U(f, P)$ is the area of the union of the rectangles D_i, D_i being the rectangle whose base is the subinterval $[x_i, x_{i+1}]$ and whose height is M_i. Likewise $L(f, P)$ is the area of the union of the rectangles D'_i, where D'_i has the same base as D_i but has height m_i (Figure 8.3).

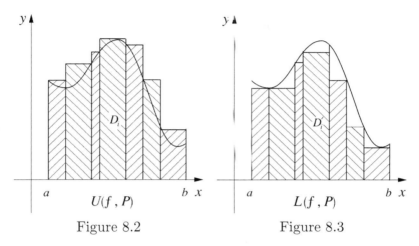

Figure 8.2   Figure 8.3

Since $M_i \geq m_i$ and $x_{i+1} - x_i > 0$ for all i, we see that

$$U(f, P) \geq L(f, P) \text{ for all } P \in \mathcal{P}(a, b). \tag{8.1}$$

Definition 8.3 We say a partition Q is *finer* than the partition P if, as sets, $P \subseteq Q$.

To obtain a partition Q finer than P we need only add points from (a, b) to P and rearrange the resulting set in ascending order. Observe that each subinterval $[x_i, x_{i+1}]$ of P is then a union of subintervals of Q whose number depends on how many points were added within $[x_i, x_{i+1}]$. In Figure (8.4) we refine P by adding four points u, v, w and z to obtain the partition Q, with the result that

$$[x_0, x_1] = [x_0, x_1],$$
$$[x_1, x_2] = [x_1, u] \cup [u, x_2],$$
$$[x_2, x_3] = [x_2, v] \cup [v, w] \cup [w, x_3].$$

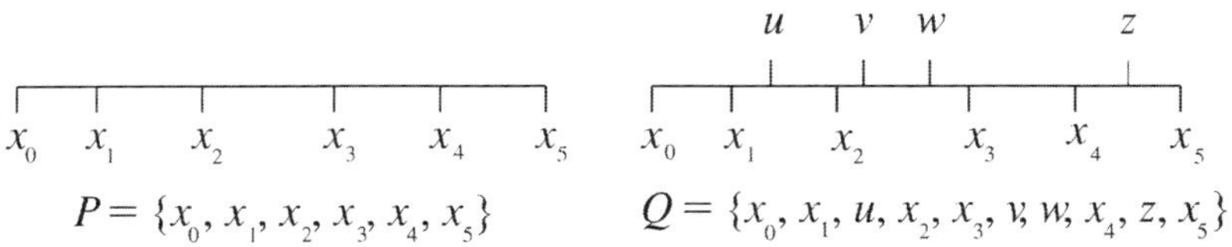

$P = \{x_0, x_1, x_2, x_3, x_4, x_5\}$ $Q = \{x_0, x_1, u, x_2, x_3, v, w, x_4, z, x_5\}$

Figure 8.4

At the risk of a slight abuse of notation, we shall extend the use of some symbols, normally reserved for operations and relations on sets, to partitions. Thus, for two partitions P and Q we shall write $P \cup Q$ for the partition resulting from the union of the sets of points in P and Q. We have already written $P \subseteq Q$ to indicate that Q refines P.

Lemma 8.1 *If the partition Q is finer than the partition P, then*

$$U(f, Q) \leq U(f, P), \qquad (8.2)$$
$$L(f, Q) \geq L(f, P). \qquad (8.3)$$

In other words, refinement reduces upper sums and increases lower sums.

Proof. Let $P = \{x_0, x_1, x_2, \cdots, x_n\}$, and suppose first that Q has just one additional point $u \in (x_k, x_{k+1})$, so that

$$Q = \{x_0, x_1, \cdots, x_k, u, x_{k+1}, \cdots, x_n\}.$$

Let

$$M_i = \sup\{f(x) : x \in [x_i, x_{i+1}]\}, \ i = 0, 1, ..., n-1,$$
$$M' = \sup\{f(x) : x \in [x_k, u]\},$$
$$M'' = \sup\{f(x) : x \in [u, x_{k+1}]\}.$$

From the definition of the upper sum, we have

$$U(f, Q) = \sum_{i=0}^{k-1} M_i(x_{i+1} - x_i) + M'(u - x_k) + M''(x_{k+1} - u)$$
$$+ \sum_{i=k+1}^{n-1} M_i(x_{i+1} - x_i).$$

Since both $[x_k, u]$ and $[u, x_{k+1}]$ are subsets of $[x_k, x_{k+1}]$, we must have

$$M' \leq M_k, \quad M'' \leq M_k.$$

Hence

$$U(f,Q) \le \sum_{i=0}^{k-1} M_i(x_{i+1} - x_i) + M_k(u - x_k) + M_k(x_{k+1} - u)$$
$$+ \sum_{i=k+1}^{n-1} M_i(x_{i+1} - x_i)$$
$$= \sum_{i=0}^{k-1} M_i(x_{i+1} - x_i) + M_k(x_{k-1} - x_k) + \sum_{i=k+1}^{n-1} M_i(x_{i+1} - x_i)$$
$$= U(f,P), \tag{8.4}$$

which proves (8.2) in this case.

Suppose now that Q has r more points than P. Adding these points one at a time, we obtain a sequence of nested partitions $P \subseteq Q_1 \subseteq Q_2 \subseteq \cdots Q_{r-1} \subseteq Q$, in which each partition exceeds its predecessor by exactly one point. Using (8.4), we see that

$$U(f,Q) \le U(f,Q_{r-1}) \le \cdots \le U(f,Q_2) \le U(f,Q_1) \le U(f,P)$$

and thereby (8.2) is proved. The proof of (8.3) is similar; and we leave the obvious modifications to the reader. □

Lemma 8.2 *If P and Q are any two partitions of $[a,b]$, then*

$$U(f,P) \ge L(f,Q). \tag{8.5}$$

Proof. Since $P \cup Q$ refines both P and Q, we have

$$U(f,P) \ge U(f, P \cup Q)$$
$$L(f, P \cup Q) \ge L(f,Q).$$

But (8.1) asserts that $U(f, P \cup Q) \ge L(f, P \cup Q)$. Therefore

$$U(f,P) \ge L(f,Q). \ \square$$

Lemma 8.2 indicates that the set

$$E = \{U(f,P) : P \in \mathcal{P}(a,b)\}$$

is bounded below by $L(f, P_0)$ for any $P_0 \in \mathcal{P}(a,b)$, while

$$F = \{L(f,P) : P \in \mathcal{P}(a,b)\}$$

is bounded above by $U(f, P_0)$. The completeness axiom ensures the existence of $\inf E$ and $\sup F$ in $\mathbb{R}$.

Definition 8.4 The *upper integral* $U(f)$ and the *lower integral* $L(f)$ of f over $[a, b]$ are defined as

$$U(f) = \inf E = \inf \{U(f, P) : P \in \mathcal{P}(a, b)\}$$
$$L(f) = \sup F = \sup \{L(f, P) : P \in \mathcal{P}(a, b)\}.$$

Since, for any $P \in \mathcal{P}(a, b)$, $L(f, P)$ is a lower bound for E, we must have

$$U(f) \geq L(f, P) \quad \text{for all } P \in \mathcal{P}(a, b).$$

Thus $U(f)$ is an upper bound for F and we conclude that

$$U(f) \geq L(f). \tag{8.6}$$

Definition 8.5 Let $f : [a, b] \to \mathbb{R}$ be bounded. We say that f is *Riemann integrable* over $[a, b]$ if $U(f) = L(f)$, in which case the common value is called the *Riemann integral* of f over $[a, b]$, and is denoted by $I(f)$ or $\int_a^b f$. The function f in $\int_a^b f$ is called the *integrand* and a and b are the *limits of integration*. The class of functions that are Riemann integrable over $[a, b]$ will be denoted by $\mathcal{R}(a, b)$.

Remarks 8.1

1. It is also customary to denote the integral of f over $[a, b]$ by $\int_a^b f(x)dx$, and we shall presently see some justification for this nomenclature. It should be clear, however, that the value of the integral is not affected by the choice of the symbol for the variable x. Thus

$$\int_a^b f = \int_a^b f(x)dx = \int_a^b f(t)dt = \int_a^b f(u)du = \cdots. \tag{8.7}$$

The justification for this redundancy will become apparent later on when we make the connection between integration and differentiation.

2. Figures 8.2 and 8.3 suggest that, no matter how we define the area A under the graph of f, we must have

$$L(f, P) \leq A \leq U(f, P).$$

Consequently, we must have

$$L(f) \leq A \leq U(f),$$

and so, if $f \in \mathcal{R}(a,b)$, then A has to equal $I(f)$. In other words, the integral $I(f)$ is a measure of the area under the curve of $f(x)$ from $x = a$ to $x = b$.

Example 8.1 Suppose f is the constant function
$$f(x) = c \text{ for all } x \in [a,b].$$
If $P = \{x_0, x_1, x_2, ..., x_n\}$ is any partition of $[a,b]$, then $M_i = m_i = c$, and hence
$$U(f,P) = L(f,P) = \sum_{i=0}^{n-1} c(x_{i+1} - x_i) = c(b-a).$$
Therefore
$$U(f) = L(f) = c(b-a)$$
and we conclude that $f \in \mathcal{R}(a,b)$ and that $\int_a^b c\, dx = c(b-a)$.

Example 8.2 Let $f : [a,b] \to \mathbb{R}$ be defined by
$$f(x) = \begin{cases} 1, & x \in \mathbb{Q} \cap [a,b] \\ 0, & x \notin \mathbb{Q} \cap [a,b] \end{cases}$$
This is often referred to as the *Dirichlet function*. If
$$P = \{x_0, x_1, x_2, \cdots, x_n\} \in \mathcal{P}(a,b),$$
then, $\mathbb{Q}$ and $\mathbb{Q}^c$ being dense in $\mathbb{R}$, we have
$$m_i = 0, \ M_i = 1, \ i = 0, 1, 2, \ldots, n-1.$$
Consequently
$$U(f,P) = \sum_{i=0}^{n-1} 1 \cdot (x_{i+1} - x_i) = b - a$$
$$L(f,P) = \sum_{i=0}^{n-1} 0 \cdot (x_{i+1} - x_i) = 0,$$
which implies
$$U(f) = b - a, \ L(f) = 0$$
and we conclude that f is not Riemann integrable over $[a,b]$.

Calculating upper and lower sums is usually more straightforward than evaluating $U(f)$ and $L(f)$. The following result is therefore useful in that it enables us to determine the integrability of a function without having to evaluate its upper and lower integrals.

Theorem 8.1 (Riemann's Criterion)
The following statements are equivalent:
(i) $f \in \mathcal{R}(a,b)$.
(ii) For all $\varepsilon > 0$, there is a partition $P \in \mathcal{P}(a,b)$ such that $U(f,P) - L(f,P) < \varepsilon$.

Proof

Suppose $f \in \mathcal{R}(a,b)$, and let $\varepsilon > 0$ be given. The definitions of $U(f)$ and $L(f)$ imply the existence of $P_1, P_2 \in \mathcal{P}(a,b)$ such that

$$U(f, P_1) < U(f) + \frac{\varepsilon}{2}, \quad L(f, P_2) > L(f) - \frac{\varepsilon}{2}.$$

Let $P = P_1 \cup P_2$. Using Lemma 8.1, we have

$$U(f, P_1) \geq U(f, P), \quad L(f, P_2) \leq L(f, P).$$

Since $U(f) = L(f)$, we must have

$$\begin{aligned}
U(f,P) &\leq U(f, P_1) \\
&< U(f) + \frac{\varepsilon}{2} \\
&= L(f) + \frac{\varepsilon}{2} \\
&< L(f, P_2) + \varepsilon \\
&\leq L(f, P) + \varepsilon.
\end{aligned}$$

Hence

$$U(f, P) - L(f, P) < \varepsilon.$$

Now assume that (ii) is valid. Given any $\varepsilon > 0$, choose $P \in \mathcal{P}(a,b)$ such that

$$U(f, P) < L(f, P) + \varepsilon.$$

Since $U(f) \leq U(f, P)$ and $L(f, P) \leq L(f)$, we must have

$$U(f) \leq U(f, P) < L(f, P) + \varepsilon \leq L(f) + \varepsilon.$$

But $\varepsilon > 0$ is arbitrary, therefore $U(f) \leq L(f)$. Since we always have $U(f) \geq L(f)$, it follows that $U(f) = L(f)$ and hence $f \in \mathcal{R}(a,b)$. □

Corollary 8.1 $f \in \mathcal{R}(a,b)$ if, and only if, there exists a sequence (P_n) in $\mathcal{P}(a,b)$ such that

$$U(f, P_n) - L(f, P_n) \to 0,$$

in which case
$$\int_a^b f = \lim_{n\to\infty} U(f, P_n) = \lim_{n\to\infty} L(f, P_n).$$

Proof. Suppose $f \in \mathcal{R}(a, b)$. Given $\varepsilon = 1/n$, we deduce from the Riemann criterion the existence of a partition $P_n \in \mathcal{P}(a, b)$ such that
$$U(f, P_n) - L(f, P_n) < \frac{1}{n}.$$
Clearly $\lim_{n\to\infty} U(f, P_n) - L(f, P_n) = 0$.

On the other hand, if (P_n) is a sequence in $\mathcal{P}(a, b)$ such that
$$\lim_{n\to\infty} U(f, P_n) - L(f, P_n) = 0,$$
then, for all $\varepsilon > 0$, there is an $N \in \mathbb{N}$ such that
$$U(f, P_n) - L(f, P_n) < \varepsilon \text{ for all } n \geq N.$$
Thus the Riemann criterion is satisfied and therefore $f \in \mathcal{R}(a, b)$. Moreover, since
$$L(f, P_n) \leq L(f) = I(f) = U(f) \leq U(f, P_n),$$
we have
$$0 \leq I(f) - L(f, P_n) \leq U(f, P_n) - L(f, P_n).$$
Hence,
$$\lim_{n\to\infty} L(f, P_n) = I(f)$$
and, furthermore,
$$U(f, P_n) = L(f, P_n) + [U(f, P_n) - L(f, P_n)] \to I(f). \quad \square$$

Example 8.3 Let $f : [a, b] \to \mathbb{R}$ be defined by $f(x) = x$. We shall prove that $f \in \mathcal{R}(a, b)$ and calculate $\int_a^b f$. If $P = \{x_0, x_1, x_2, \cdots, x_n\}$ is any partition of $[a, b]$, then, as figure 8.5 indicates,
$$M_i = x_{i+1}, \quad m_i = x_i, \quad i = 0, 1, \cdots, n-1.$$

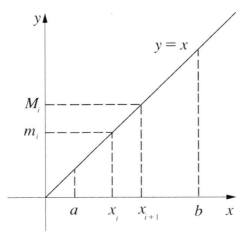

Figure 8.5

Thus
$$U(f,P) - L(f,P) = \sum_{i=0}^{n-1} (x_{i+1} - x_i)^2.$$

Choosing P_n to be the uniform partition into n (equal) parts, so that $x_i = a + i(b-a)/n$, we get

$$U(f,P_n) - L(f,P_n) = \sum_{i=0}^{n-1} \frac{(b-a)^2}{n^2}$$
$$= \frac{(b-a)^2}{n} \to 0,$$

which means that $f \in \mathcal{R}(a,b)$. Furthermore,

$$L(f,P_n) = \sum_{i=0}^{n-1} \left[a + \frac{i(b-a)}{n}\right] \frac{b-a}{n}$$
$$= a(b-a) + \frac{(b-a)^2}{n^2} \sum_{i=0}^{n-1} i$$
$$= a(b-a) + \frac{(b-a)^2}{n^2} \frac{n(n-1)}{2}$$
$$\to a(b-a) + \frac{(b-a)^2}{2} = \frac{1}{2}(b^2 - a^2).$$

Hence
$$\int_a^b x\,dx = \frac{1}{2}(b^2 - a^2).$$

Example 8.4 Suppose $f : [0,1] \to \mathbb{R}$ is defined by

$$f(x) = \begin{cases} 0, & x \in [0,1] \setminus \mathbb{Q} \\ 1/q, & x = p/q \in [0,1] \cap \mathbb{Q}, \end{cases}$$

where p and q have no common factors, symbolically expressed by writing $(p,q) = 1$. We shall show that $f \in \mathcal{R}(0,1)$.

The density of $\mathbb{Q}^c$ in $[0,1]$ implies that $L(f,P) = 0$ for all $P \in \mathcal{P}(0,1)$. Suppose $\varepsilon > 0$ is given. We shall show how to choose a partition P satisfying $U(f,P) < \varepsilon$. Riemann's criterion will then ensure the integrability of f.

Let A be the set of all rationals $p/q \in [0,1]$ satisfying $(p,q) = 1$ and $q \leq 2/\varepsilon$. There are finitely many integers q not exceeding $2/\varepsilon$, and for each such q there are finitely many integers p such that $p/q \in [0,1]$ with $(p,q) = 1$. Thus A is a finite set, call it $\{r_1, r_2, \cdots, r_k\}$. Choose $n \in \mathbb{N}$ such that $n > 4k/\varepsilon$ and let $P_n = \{x_0, x_1, x_2, \cdots, x_n\}$ be the uniform partition of $[0,1]$ into n parts. If we set

$$J_1 = \{i : [x_i, x_{i+1}] \cap A \neq \varnothing\},$$
$$J_2 = \{i : [x_i, x_{i+1}] \cap A = \varnothing\},$$

then

$$U(f, P_n) = \sum_{i \in J_1} M_i (x_{i+1} - x_i) + \sum_{i \in J_2} M_i (x_{i+1} - x_i).$$

Since $M_i \leq 1$ and the number of elements of J_1 cannot exceed $2k$ (for every point in A cannot intersect more than two subintervals of P_n), we obtain

$$\sum_{i \in J_1} M_i (x_{i+1} - x_i) \leq \sum_{i \in J_1} \frac{1}{n} \leq \frac{2k}{n} \leq \frac{\varepsilon}{2}.$$

On the other hand, if $i \in J_2$, then for each $x \in [x_i, x_{i+1}]$ either $x \notin \mathbb{Q}$, and hence $f(x) = 0$, or $x = p/q$ in which case $f(x) = 1/q < \varepsilon/2$ because $x \notin A$. In either case we see that $0 \leq f(x) \leq \varepsilon/2$. Consequently

$$\sum_{i \in J_2} M_i (x_{i+1} - x_i) \leq \sum_{i=0}^{n-1} \frac{\varepsilon}{2} (x_{i+1} - x_i) = \frac{\varepsilon}{2},$$

and

$$U(f, P_n) < \frac{\varepsilon}{2} + \frac{\varepsilon}{2} = \varepsilon$$

as promised.

Since $L(f, P) = 0$ for all partitions P, $L(f) = 0$, and therefore $\int_0^1 f = 0$.

Example 8.2 shows that not every bounded function is integrable. A search for suitable conditions for integrability is thus both meaningful and desirable. Using Riemann's criterion, we shall soon prove the integrability of two wide classes of functions bounded on $[a, b]$. We begin with a definition.

Definition 8.6 The *norm* of a partition $P = \{x_0, x_1, x_2, \cdots, x_n\}$ is defined to be the positive number

$$\|P\| = \max\{x_{i+1} - x_i : i = 0, 1, \cdots, n-1\}.$$

That is, $\|P\|$ is the length of the longest subinterval of P.

Theorem 8.2
If f is monotonic on $[a, b]$, then $f \in \mathcal{R}(a, b)$.

Proof. Let us assume first that f is increasing. If $f(a) = f(b)$, f would be constant and therefore integrable (Example 8.1), so suppose $f(a) \neq f(b)$.

Given $\varepsilon > 0$, choose a partition $P = \{x_0, x_1, x_2, \cdots, x_n\}$ such that

$$\|P\| < \frac{\varepsilon}{f(b) - f(a)}.$$

Since f is increasing,

$$M_i = f(x_{i+1}), \quad m_i = f(x_i) \quad \text{for all } i = 0, 1, \cdots, n-1.$$

Consequently,

$$U(f, P) - L(f, P) = \sum_{i=0}^{n-1} [f(x_{i+1}) - f(x_i)][x_{i+1} - x_i]$$
$$\leq \|P\| \sum_{i=0}^{n-1} [f(x_{i+1}) - f(x_i)]$$
$$\leq \|P\| [f(b) - f(a)] < \varepsilon.$$

Riemann's criterion now implies that $f \in \mathcal{R}(a, b)$.

The case where f is decreasing is handled by an obvious modification of the above argument. We leave the details to the reader. $\square$

Theorem 8.3
If f is continuous on $[a,b]$, then $f \in \mathcal{R}(a,b)$.

Proof. Let $\varepsilon > 0$ be given. Since the interval $[a,b]$ is compact, f is uniformly continuous on $[a,b]$, and hence there is a $\delta > 0$ such that

$$x, t \in [a,b], |x - t| < \delta \Rightarrow |f(x) - f(t)| < \frac{\varepsilon}{b-a}. \tag{8.8}$$

Choose a partition $P = \{x_0, x_1, x_2, \cdots, x_n\}$ that satisfies $\|P\| < \delta$. Since f is continuous, it attains its maximum and minimum on $[x_i, x_{i+1}]$ and therefore there exist points $u_i, v_i \in [x_i, x_{i+1}]$ such that

$$M_i = f(v_i), \ m_i = f(u_i) \ \text{for all} \ i = 0, 1, \cdots, n-1.$$

Since $|u_i - v_i| < \delta$, we see from (8.8) that $M_i - m_i < \dfrac{\varepsilon}{b-a}$, and therefore

$$U(f,P) - L(f,P) = \sum_{i=0}^{n-1} (M_i - m_i)(x_{i+1} - x_i)$$

$$< \frac{\varepsilon}{b-a} \sum_{i=0}^{n-1} (x_{i+1} - x_i)$$

$$= \frac{\varepsilon}{b-a}(b - a) = \varepsilon.$$

Consequently, f is integrable by Riemann's criterion. □

This theorem, like its predecessor, gives only a sufficient condition for Riemann integrability, for monotonic functions need not be continuous, just as continuous functions need not be monotonic. It can be strengthened to assert the integrability of any bounded function whose set of discontinuities in $[a,b]$ is finite. Here we prove integrability for the case when f has one point of discontinuity $c \in (a,b)$, but the argument can be extended to the more general case.

Let $\varepsilon > 0$ be given. Since f is bounded, there is a number K such that $|f(x)| \leq K$ for all $x \in [a,b]$. Choose $\varepsilon' > 0$ such that $a < c - \varepsilon'$, $b > c + \varepsilon'$ and $\varepsilon' < \varepsilon/(4K + 2)$. Since f is continuous, and therefore integrable (by Theorem 8.3), on $[a, c - \varepsilon']$, there is a partition $P_1 = \{x_0, x_1, x_2, ..., x_{k-1}\}$ of $[a, c - \varepsilon']$ such that

$$U(f, P_1) - L(f, P_1) = \sum_{i=0}^{k-2}(M_i - m_i)(x_{i+1} - x_i) < \varepsilon'.$$

Using Theorem 8.3 on $[c+\varepsilon', b]$, we also have a partition $P_2 = \{x_{k+1}, x_{k+2}, ..., x_n\}$ of $[c+\varepsilon', b]$ such that

$$U(f, P_2) - L(f, P_2) = \sum_{i=k+1}^{n-1} (M_i - m_i)(x_{i+1} - x_i) < \varepsilon'.$$

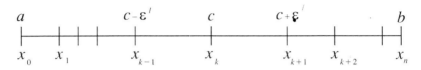

Figure 8.6

It is clear that $P = \{x_0, x_1, \ldots, x_{k-1}, x_k, x_{k+1}\ldots, x_n\}$, where $x_k = c$, is a partition of $[a, b]$. Since

$$U(f, P) - L(f, P) \le \sum_{i=0}^{k-2} (M_i - m_i)(x_{i+1} - x_i) + 4K\varepsilon'$$
$$+ \sum_{i=k+1}^{n-1} (M_i - m_i)(x_{i+1} - x_i)$$
$$< 2\varepsilon' + 4K\varepsilon' < \varepsilon,$$

we conclude that $f \in \mathcal{R}(a, b)$.

But the condition that the set of discontinuities be finite is not necessary for integrability, for Example 8.4 exhibits an integrable function on $[0, 1]$ which is discontinuous at each point in the infinite set $\mathbb{Q} \cap [0, 1]$ (see Example 5.3). We shall take up the relationship between continuity and Riemann integrability in Chapter 11, where we give a necessary and sufficient condition for a function to be Riemann integrable.

EXERCISES 8.1

In Exercises 1 to 6, determine whether f is integrable over $[0, 1]$ and evaluate $\int_a^b f(x)\,dx$ whenever it exists.

1. $f(x) = \begin{cases} 1, & x \in [0, 1/2) \\ 0, & x \in [1/2, 1] \end{cases}.$

2. $f(x) = \begin{cases} 2x, & x \in [0, 1/2) \\ x - 2, & x \in [1/2, 1] \end{cases}.$

3. $f(x) = \begin{cases} x, & x \in \mathbb{Q} \\ 0, & x \notin \mathbb{Q}. \end{cases}$

4. $f(x) = \begin{cases} 1/x, & x \neq 0 \\ 0, & x = 0. \end{cases}$

5. $f(x) = x^2$.

   Use the formula $\sum_{i=1}^{m} i^2 = \frac{1}{6} m(m+1)(2m+1)$.

6. $f(x) = x^3$.

   Use $\sum_{i=1}^{m} i^3 = [\frac{1}{2} m(m+1)]^2$.

7. Suppose that $f \in \mathcal{R}(a,b)$ and let g be defined by $g(x) = f(x-c)$ for all $x \in [a+c, b+c]$, where c is a real constant. Prove that $g \in \mathcal{R}(a+c, b+c)$ and that
$$\int_{a+c}^{b+c} g(x)\,dx = \int_a^b f(x)\,dx.$$

8. Suppose that $f \in \mathcal{R}(a,b)$ and let g be defined by $g(x) = f(x/c)$ for all $x \in [ac, bc]$, where $c > 0$. Prove that $g \in \mathcal{R}(ac, bc)$ and that $\int_{ac}^{bc} g(x)\,dx = c \int_a^b f(x)\,dx$.

9. Suppose the two functions f and g are equal everywhere on $[a,b]$ except on the finite subset $\{c_1, c_2, \ldots, c_n\}$. If $f \in \mathcal{R}(a,b)$, prove that $g \in \mathcal{R}(a,b)$ and that $\int_a^b f = \int_a^b g$.

10. Let f be continuous and non-negative on $[a,b]$. If $\int_a^b f(x)\,dx = 0$, prove that $f(x) = 0$ for all $x \in [a,b]$.

11. Let f be continuous on $[a,b]$. If $\int_a^b fg = 0$ for all $g \in \mathcal{R}(a,b)$, prove that $f(x) = 0$ for all $x \in [a,b]$.

12. Suppose f is bounded on $[a,b]$. If $f \in \mathcal{R}(c,b)$ for every $c \in (a,b)$, prove that $f \in \mathcal{R}(a,b)$ and that
$$\int_a^b f(x)\,dx = \lim_{c \to a^+} \int_c^b f(x)\,dx.$$

8.2 Darboux's Theorem and Riemann Sums

The attentive reader may have noticed how we obtained the desired results in the last two theorems, and some of the preceding examples, by choosing partitions of small enough norms. This is not altogether accidental as the following theorem indicates.

Theorem 8.4 (Darboux)
Suppose the function $f : [a,b] \to \mathbb{R}$ is bounded. Given $\varepsilon > 0$, there is a $\delta > 0$ such that, if P is any partition of $[a,b]$ satisfying $\|P\| < \delta$, then

$$U(f,P) - U(f) < \varepsilon, \quad L(f) - L(f,P) < \varepsilon.$$

In the language of limits, we may (loosely) express this result as

$$\lim_{\|P\| \to 0} U(f,P) = U(f), \quad \lim_{\|P\| \to 0} L(f,P) = L(f).$$

Proof. Let $\varepsilon > 0$ be given. From the definition of $U(f)$, there is a partition $P_0 = \{u_0, u_1, \ldots u_r\}$ such that

$$U(f, P_0) < U(f) + \varepsilon/2. \tag{8.9}$$

Since f is bounded, there is a number K such that $|f(x)| \leq K$ for all $x \in [a,b]$. Take $\delta_1 = \varepsilon/8rK$, and suppose $P = \{x_0, x_1, x_2, \ldots, x_n\}$ is any partition satisfying $\|P\| < \delta_1$. Set $P_1 = P_0 \cup P$ and let us calculate $U(f,P) - U(f,P_1)$ by calculating the contribution of each subinterval $[x_i, x_{i+1}]$ to this difference.

If the subinterval does not contain any member of P_0, then its contribution to $U(f,P)$ equals its contribution to $U(f,P_1)$ and thereby contributes nothing to the difference. Suppose the subinterval contains an element of P_0. Since its contribution to $U(f,P)$ is at most $K(x_{i+1} - x_i)$, while its contribution to $U(f,P_1)$ is at least $-K(x_{i+1} - x_i)$, we see that its contribution to the difference does not exceed $2K(x_{i+1} - x_i)$. Hence

$$U(f,P) - U(f,P_1) \leq \sum_{i \in J} 2K(x_{i+1} - x_i),$$

where $J = \{i : [x_i, x_{i+1}] \cap P_0 \neq \varnothing\}$.

Since each member of P_0 can belong to at most two subintervals of P, J cannot have more than $2r$ members. Consequently

$$U(f,P) - U(f,P_1) \leq \sum_{i \in J} 2K\|P\| \leq 4Kr\|P\| < \varepsilon/2.$$

With $U(f, P_1) \leq U(f, P_0)$, we obtain

$$U(f, P) - U(f, P_0) < \varepsilon/2$$

and deduce from (8.9) that

$$U(f, P) - U(f) = [U(f, P) - U(f, P_0)] + [U(f, P_0) - U(f)] < \varepsilon.$$

In a similar fashion we can find a $\delta_2 > 0$ such that

$$\|P\| < \delta_2 \Rightarrow L(f) - L(f, P) < \varepsilon.$$

Now the number $\delta = \min\{\delta_1, \delta_2\}$ meets the requirements. $\square$

Definition 8.7 Let $P = \{x_0, x_1, x_2, ..., x_n\}$ be a partition of $[a, b]$. We say that $\boldsymbol{\alpha} = (\alpha_0, \alpha_1, \ldots, \alpha_{n-1})$ is a *mark* on P if $\alpha_i \in [x_i, x_{i+1}]$ for all $i \in \{0, 1, \ldots, n-1\}$. We then call the sum

$$S(f, P, \boldsymbol{\alpha}) = \sum_{i=0}^{n-1} f(\alpha_i)(x_{i+1} - x_i)$$

the *Riemann sum* of f on P with mark $\boldsymbol{\alpha}$.

Since $m_i \leq f(\alpha_i) \leq M_i$ for every i, we always have

$$L(f, P) \leq S(f, P, \boldsymbol{\alpha}) \leq U(f, P).$$

for any partition P of $[a, b]$ and any mark $\boldsymbol{\alpha}$ on P.

The following theorem captures an equivalent definition of the Riemann integral often used in calculus courses.

Theorem 8.5
The following statements are equivalent:
(i) $f \in \mathcal{R}(a, b)$ with integral over $[a, b]$ equal to A.
(ii) For any $\varepsilon > 0$, there is a $\delta > 0$ such that, if P is any partition satisfying $\|P\| < \delta$ and $\boldsymbol{\alpha}$ is any mark on P, then

$$|S(f, P, \boldsymbol{\alpha}) - A| < \varepsilon,$$

that is,

$$\lim_{\|P\| \to 0} S(f, P, \boldsymbol{\alpha}) = A.$$

Proof
Assume (i) is true. Then
$$A = U(f) = L(f)$$
and, if $\varepsilon > 0$ is given, we know from Darboux's theorem that there is a $\delta > 0$ such that, if P is any partition satisfying $\|P\| < \delta$, then
$$A - L(f, P) < \varepsilon, \quad U(f, P) - A < \varepsilon.$$
Since, for every mark $\boldsymbol{\alpha}$ on P, we have
$$L(f, P) \leq S(f, P, \boldsymbol{\alpha}) \leq U(f, P),$$
it follows that
$$|S(f, P, \boldsymbol{\alpha}) - A| < \varepsilon,$$
as required in (ii).

Now assume the validity of (ii). Let $\varepsilon > 0$ be given and suppose that $\delta > 0$ is chosen so that
$$\|P\| < \delta \Rightarrow |S(f, P, \boldsymbol{\alpha}) - A| < \varepsilon/2$$
for every mark $\boldsymbol{\alpha}$ on P. Take any partition $P = \{x_0, x_1, x_2, ..., x_n\}$ satisfying $\|P\| < \delta$. From the definitions of m_i and M_i, we are assured of the existence of α_i and β_i in $[x_i, x_{i+1}]$ such that
$$m_i > f(\alpha_i) - \frac{\varepsilon}{2(b-a)}, \quad M_i < f(\beta_i) + \frac{\varepsilon}{2(b-a)}.$$
This implies that $\boldsymbol{\alpha} = (\alpha_0, \alpha_1, \ldots, \alpha_n)$ and $\boldsymbol{\beta} = (\beta_0, \beta_1, \ldots, \beta_n)$ are marks on P and that
$$L(f, P) > S(f, P, \boldsymbol{\alpha}) - \varepsilon/2, \quad U(f, P) < S(f, P, \boldsymbol{\beta}) + \varepsilon/2.$$
Since $\|P\| < \delta$, we get
$$L(f, P) > A - \varepsilon, \quad U(f, P) < A + \varepsilon,$$
and since $L(f, P) \leq L(f)$ and $U(f, P) \geq U(f)$, we obtain
$$L(f) > A - \varepsilon, \quad U(f) < A + \varepsilon.$$
Letting $\varepsilon \to 0$, and recalling that $L(f) \leq U(f)$, we conclude that
$$U(f) = L(f) = A. \quad \square$$

Corollary 8.5 *Suppose $f \in \mathcal{R}(a,b)$. If (P_n) is a sequence of partitions such that $\|P_n\| \to 0$, then, for any choice of marks $\boldsymbol{\alpha}_n$ on P_n, we have*

$$\int_a^b f = \lim_{n \to \infty} S(f, P_n, \boldsymbol{\alpha}_n).$$

We leave the proof as an exercise. This corollary shows that, once we know that f is integrable, we need not work with general partitions to evaluate its integral. A uniform partition into n equal parts, whose norm $(b-a)/n$ tends to 0, can always be used.

Example 8.5 Using Riemann sums, evaluate $\int_0^1 (x - x^2)\, dx$.

Solution. Since the function $f(x) = x - x^2$ is continuous, it is integrable. Let $P_n = \{x_0, x_1, x_2, ..., x_n\}$ be the uniform partition of $[0,1]$, in which case

$$x_i = \frac{i}{n}, \quad i = 0, 1, \ldots, n.$$

Choosing the mark $\boldsymbol{\alpha}_n = (x_0, x_1, ..., x_{n-1})$, we obtain

$$\begin{aligned}
S(f, P_n, \boldsymbol{\alpha}_n) &= \sum_{i=0}^{n-1} \left(\frac{i}{n} - \frac{i^2}{n^2} \right) \frac{1}{n} \\
&= \frac{1}{n^2} \sum_{i=0}^{n-1} i - \frac{1}{n^3} \sum_{i=0}^{n-1} i^2 \\
&= \frac{1}{n^2} \frac{n(n-1)}{2} - \frac{1}{n^3} \frac{(n-1)n(2n-1)}{6},
\end{aligned}$$

and since $\|P_n\| = 1/n \to 0$, we have

$$\begin{aligned}
\int_0^1 (x - x^2)\, dx &= \lim_{n \to \infty} S(f, P_n, \boldsymbol{\alpha}_n) \\
&= \frac{1}{2} - \frac{1}{3} = \frac{1}{6}.
\end{aligned}$$

EXERCISES 8.2

1. If $f \in \mathcal{R}(0,1)$, prove that

$$\lim_{n \to \infty} \frac{1}{n} \sum_{k=1}^n f(k/n) = \int_0^1 f(x)\, dx.$$

2. Use Exercise 8.2.1 to write each of the following limits as an integral:

   (a) $\lim_{n\to\infty} \sum_{k=1}^{n} \dfrac{1}{n+k}$.

   (b) $\lim_{n\to\infty} \sum_{k=1}^{n} \dfrac{k}{n^2+k^2}$.

3. Evaluate the following limits:

   (a) $\lim_{n\to\infty} \sum_{k=1}^{n} \dfrac{1}{n} \cos(k\pi/n)$.

   (b) $\lim_{n\to\infty} \sum_{k=1}^{n} \dfrac{n}{n^2+k^2}$.

   (c) $\lim_{n\to\infty} \sum_{k=1}^{2^n} \dfrac{k^3}{2^{4n}}$.

4. Prove Corollary 8.5.

5. If f is continuous and increasing on $[a,b]$, prove that $\int_a^b f = bf(b) - af(a) - \int_{f(a)}^{f(b)} f^{-1}$. Evaluate $\int_0^1 \sqrt{x}\,dx$.

   Hint: Let $P = \{x_0, x_1, \cdots, x_n\}$ be a partition of $[a,b]$, and set $Q = \{f(x_0), f(x_1), \cdots, f(x_n)\}$; then prove that $L(f^{-1}, Q) + U(f, P) = bf(b) - af(a)$.

6. Prove that the following statements are equivalent:

   (a) $f \in \mathcal{R}(a,b)$ and $\int_a^b f = A$.

   (b) For any $\varepsilon > 0$, there is a partition $P_\varepsilon \in \mathcal{P}(a,b)$ such that, if $P_\varepsilon \subseteq P$, then $|S(f, P, \alpha) - A| < \varepsilon$.

8.3 Properties of the Integral

We present in this section the basic properties of the Riemann integral. Since we have at hand more than one criterion for integrability, most of the following theorems can have alternative proofs. The reader is invited to explore these, if only to test her/his grasp of these definitions.

Theorem 8.6 (Linearity Property)

If $f_1, f_2 \in \mathcal{R}(a, b)$ and $k_1, k_2 \in \mathbb{R}$, then $k_1 f_1 + k_2 f_2 \in \mathcal{R}(a, b)$ and

$$\int_a^b k_1 f_1 + k_2 f_2 = k_1 \int_a^b f_1 + k_2 \int_a^b f_2.$$

Proof. The result is immediate if $k_1 = k_2 = 0$, so we assume that $|k_1| + |k_2| > 0$. Let $\varepsilon > 0$ be given. Using Theorem 8.5, we can choose $\delta_i > 0$ $(i = 1, 2)$ such that, if $P \in \mathcal{P}(a, b)$ satisfies $\|P\| < \delta_i$ and $\boldsymbol{\alpha}$ is a mark on P, then

$$\left| S(f_i, P, \boldsymbol{\alpha}) - \int_a^b f_i \right| < \frac{\varepsilon}{|k_1| + |k_2|}, \quad i = 1, 2.$$

Put $\delta = \min\{\delta_1, \delta_2\}$ and let $P \in \mathcal{P}(a, b)$ satisfy $\|P\| < \delta$. Since

$$S(k_1 f_1 + k_2 f_2, P, \boldsymbol{\alpha}) = k_1 S(f_1, P, \boldsymbol{\alpha}) + k_2 S(f_2, P, \boldsymbol{\alpha}),$$

we clearly have

$$\left| S(k_1 f_1 + k_2 f_2, P, \boldsymbol{\alpha}) - \left(k_1 \int_a^b f_1 + k_2 \int_a^b f_2 \right) \right|$$

$$\leq |k_1| \left| S(f_1, P, \boldsymbol{\alpha}) - \int_a^b f_1 \right| + |k_2| \left| S(f_2, P, \boldsymbol{\alpha}) - \int_a^b f_2 \right| < \varepsilon.$$

Theorem 8.5 now yields the desired result. □

Theorem 8.7 (Positivity Property)

If $f \in \mathcal{R}(a, b)$ is non-negative, then

$$\int_a^b f \geq 0.$$

Proof. Let $P = \{x_0, x_1, \ldots, x_n\}$ be any partition of $[a, b]$. Since $f \geq 0$ on $[a, b]$, $m_i \geq 0$ for all $i \in \{0, 1, \ldots, n-1\}$ and therefore $L(f, P) \geq 0$. Now

$$\int_a^b f = L(f) = \sup\{L(f, P) : P \in \mathcal{P}(a, b)\} \geq 0. \square$$

Corollary 8.7. (Monotonicity Property)

If $f, g \in \mathcal{R}(a, b)$ and $f(x) \leq g(x)$ for all $x \in [a, b]$, then

$$\int_a^b f \leq \int_a^b g.$$

Proof. By the linearity property the function $h = g - f$ is Riemann integrable and
$$\int_a^b h = \int_a^b g - \int_a^b f.$$
Since h is non-negative, the positivity property implies that $\int_a^b h \geq 0$, and the required result follows. □

Theorem 8.8
Let $c \in (a, b)$. Then $f \in \mathcal{R}(a, b)$ if and only if $f \in \mathcal{R}(a, c) \cap \mathcal{R}(c, b)$, in which case
$$\int_a^b f = \int_a^c f + \int_c^b f. \tag{8.10}$$

Proof
Suppose first that $f \in \mathcal{R}(a, b)$. If $\varepsilon > 0$ is given, we can choose a partition $P = \{x_0, x_1, ..., x_n\}$ of $[a, b]$ such that
$$U(f, P) - L(f, P) < \varepsilon.$$
By refining P if needed, we may assume that c is a member of P, say $c = x_k$. It is obvious that $Q = \{x_0, x_1, ..., x_k\}$ is a partition of $[a, c]$ and that $Q' = \{x_k, x_{k+1}, \cdots, x_n\}$ is a partition $[c, b]$. Since $M_i - m_i \geq 0$ for all $i = 0, 1, \cdots, n-1$,
$$U(f, Q) - L(f, Q) = \sum_{i=0}^{k-1} (M_i - m_i)(x_{i+1} - x_i)$$
$$\leq \sum_{i=0}^{n-1} (M_i - m_i)(x_{i+1} - x_i)$$
$$= U(f, P) - L(f, P) < \varepsilon.$$
This proves that $f \in \mathcal{R}(a, c)$. Similarly, using the partition $Q' = \{x_k, x_{k+1}, \ldots, x_n\}$ of $[c, b]$, we can show that $f \in \mathcal{R}(c, b)$.

On the other hand, if $f \in \mathcal{R}(a, c) \cap \mathcal{R}(c, b)$, then, given $\varepsilon > 0$, we can choose a partition $Q = \{x_0, x_1, ..., x_k\}$ of $[a, c]$ and a partition $Q' = \{x_k, x_{k+1}, \ldots, x_n\}$ of $[c, b]$ such that
$$U(f, Q) - L(f, Q) < \varepsilon/2, \ U(f, Q') - L(f, Q') < \varepsilon/2.$$
The partition $P = \{x_0, x_1, ..., x_n\}$ satisfies
$$U(f, P) - L(f, P) = U(f, Q) - L(f, Q) + U(f, Q') - L(f, Q') < \varepsilon,$$

which proves $f \in \mathcal{R}(a,b)$. Furthermore, if $\|Q\| \to 0$ and $\|Q'\| \to 0$, then $\|P\| \to 0$. Since

$$U(f,P) = U(f,Q) - U(f,Q'),$$

equation (8.10) follows from Darboux's theorem. □

Remark 8.2 If $f(x) \geq 0$ on $[a,b]$ and $f(x_0) > 0$ for some point x_0 in $[a,b]$, the monotonicity property does not justify the conclusion that $\int_a^b f > 0$. If, however, we have $f(x) \geq k$ for some $k > 0$ on some subinterval $[c,d]$ of $[a,b]$, then Theorems 8.7 and 8.8 give

$$\int_a^b f = \int_a^c f + \int_c^d f + \int_d^b f \geq \int_c^d k = k(d-c) > 0.$$

A particular instance where this occurs is when f is non-negative, continuous, and $f(x_0) > 0$ for some $x_0 \in [a,b]$. It then follows that there is a positive number k and a neighborhood $U = (x_0 - \delta, x_0 + \delta)$ of x_0 such that $f(x) \geq k$ for all $x \in U \cap [a,b]$ (see Exercise 6.1.4). This clearly implies that $\int_a^b f > 0$.

Thus we have proved

Corollary 8.8 *If the function f is non-negative and continuous on $[a,b]$, then $\int_a^b f = 0$ if, and only if, $f(x) = 0$ for all $x \in [a,b]$.*

By convention, we write

$$\int_b^a f = -\int_a^b f \tag{8.11}$$

This makes (8.10) valid even if $c \notin [a,b]$ provided that f is integrable on the largest interval constructible from the points a, b and c. For instance if $c < a$ and $f \in \mathcal{R}(c,b)$, then

$$\int_c^b f = \int_c^a f + \int_a^b f$$

$$\int_a^b f = \int_c^b f - \int_c^a f$$

$$= \int_a^c f + \int_c^b f.$$

Observe that (8.11) implies

$$\int_a^a f = 0,$$

and this allows us to consider $\int_a^b f$ as an integral over any one of the intervals (a,b), $[a,b)$, $(a,b]$, or $[a,b]$. In other words, $\mathcal{R}(a,b)$ denotes the set of integrable functions over any one of these intervals.

The next theorem is known as the *mean value theorem of integral calculus*.

Theorem 8.9 (Mean Value Theorem)
If f is continuous on $[a,b]$, then there exists a point $c \in (a,b)$ such that
$$\int_a^b f(x)\,dx = f(c)(b-a).$$

Proof. The case where f is constant on $[a,b]$ can be dismissed, for we can then choose c to be any point in (a,b). Since f is continuous on the compact interval $[a,b]$, it attains its maximum value M and its minimum value m on $[a,b]$; i.e., there are points $x_1, x_2 \in [a,b]$ such that
$$m = f(x_1) \leq f(x) \leq f(x_2) = M \quad \text{for all } x \in [a,b].$$
Since f is not constant, there are points $t_1, t_2 \in [a,b]$ such that
$$f(t_1) > m, \quad f(t_2) < M.$$
It follows from Remark 8.2 applied to the functions $f(x) - m$ and $M - f(x)$ that
$$\int_a^b m\,dx < \int_a^b f(x)\,dx < \int_a^b M\,dx.$$
Using the result of Example 8.1,
$$m(b-a) < \int_a^b f(x)\,dx < M(b-a)$$
$$m < \frac{1}{b-a}\int_a^b f(x)\,dx < M$$
$$f(x_1) < \frac{1}{b-a}\int_a^b f(x)\,dx < f(x_2).$$
By the intermediate value property for continuous functions (Theorem 6.6), there is a point c between x_1 and x_2 where
$$f(c) = \frac{1}{b-a}\int_a^b f(x)\,dx. \quad \square$$

Theorem 8.10
Let $f : [a,b] \to [c,d]$ be Riemann integrable. If $\varphi : [c,d] \to \mathbb{R}$ is continuous, then $\varphi \circ f$ is Riemann integrable on $[a,b]$.

Proof. Let $\varepsilon > 0$ be given. We shall prove the existence of a partition $P \in \mathcal{P}(a,b)$ such that
$$U(\varphi \circ f) - L(\varphi \circ f) < \varepsilon.$$

Since φ is continuous on the compact interval $[c,d]$, it is bounded and uniformly continuous. Consequently there is a real constant K such that
$$|\varphi(t)| \leq K \quad \text{for all } t \in [c,d], \tag{8.12}$$
and if we set $\varepsilon' = \dfrac{\varepsilon}{2K + (b-a)}$, we know that there is a $\delta > 0$ such that
$$s, t \in [c,d], \ |t - s| < \delta \Rightarrow |\varphi(t) - \varphi(s)| < \varepsilon'. \tag{8.13}$$

On the other hand, since $f \in \mathcal{R}(a,b)$, there is a partition $P = \{x_0, x_1, ..., x_n\}$ such that
$$U(f, P) - L(f, P) < \varepsilon'\delta. \tag{8.14}$$

Let
$$M_i = \sup\{f(x) : x \in [x_i, x_{i+1}]\}$$
$$M_i^* = \sup\{\varphi(f(x)) : x \in [x_i, x_{i+1}]\}$$
$$m_i = \inf\{f(x) : x \in [x_i, x_{i+1}]\}$$
$$m_i^* = \inf\{\varphi(f(x)) : x \in [x_i, x_{i+1}]\}.$$

We then have
$$U(\varphi \circ f) - L(\varphi \circ f) = \sum_{i=0}^{n-1}(M_i^* - m_i^*)(x_{i+1} - x_i)$$
$$= \sum_{i \in J_1}(M_i^* - m_i^*)(x_{i+1} - x_i)$$
$$+ \sum_{i \in J_2}(M_i^* - m_i^*)(x_{i+1} - x_i), \tag{8.15}$$

where
$$J_1 = \{i \in \{0, 1, \ldots, n-1\} : M_i - m_i < \delta\}$$
$$J_2 = \{i \in \{0, 1, \ldots, n-1\} : M_i - m_i \geq \delta\}.$$

If $i \in J_1$, then
$$|f(x) - f(y)| < \delta \text{ for all } x, y \in [x_i, x_{i+1}],$$
and therefore, using (8.13),
$$|\varphi(f(x)) - \varphi(f(y))| < \varepsilon' \text{ for all } x, y \in [x_i, x_{i+1}],$$
which implies that $M_i^* - m_i^* \leq \varepsilon'$. Therefore
$$\sum_{i \in J_1} (M_i^* - m_i^*)(x_{i+1} - x_i) \leq \sum_{i \in J_1} \varepsilon'(x_{i+1} - x_i) \leq \varepsilon'(b - a).$$
Now from (8.14) we see that
$$\varepsilon'\delta > \sum_{i=0}^{n-1} (M_i - m_i)(x_{i+1} - x_i)$$
$$\geq \sum_{i \in J_2} (M_i - m_i)(x_{i+1} - x_i)$$
$$\geq \delta \sum_{i \in J_2} (x_{i+1} - x_i),$$
hence
$$\sum_{i \in J_2} (x_{i+1} - x_i) < \varepsilon'.$$
Since $M_i^* - m_i^* \leq 2K$, we must have
$$\sum_{i \in J_2} (M_i^* - m_i^*)(x_{i+1} - x_i) < 2K\varepsilon'.$$
Using these inequalities in (8.15) yields
$$U(\varphi \circ f) - L(\varphi \circ f) < \varepsilon'(b-a) + 2K\varepsilon' = \varepsilon. \quad \square$$

Corollary 8.10.1. *If $f \in \mathcal{R}(a,b)$, then $|f| \in \mathcal{R}(a,b)$ and*
$$\left|\int_a^b f\right| \leq \int_a^b |f|.$$

Proof. Choose K such that $|f(x)| \leq K$ on $[a,b]$. Define $\varphi : [-K, K] \to \mathbb{R}$ by
$$\varphi(x) = |x|.$$

Since φ is continuous, it follows from Theorem 8.10 that $\varphi \circ f \in \mathcal{R}(a,b)$. But $\varphi \circ f$ is none other than $|f|$, hence $|f| \in \mathcal{R}(a,b)$. To obtain the inequality, observe that

$$f(x) \le |f(x)| \text{ and } -f(x) \le |f(x)| \text{ for all } x \in [a,b].$$

The monotonicity and linearity properties then give

$$\int_a^b f \le \int_a^b |f| \text{ and } -\int_a^b f \le \int_a^b |f|,$$

from which the desired result follows. □

Corollary 8.10.2 *If* $f \in \mathcal{R}(a,b)$, *then* $f^n \in \mathcal{R}(a,b)$ *for all* $n \in \mathbb{N}$.

Proof. Use Theorem 8.10 with $\varphi(x) = x^n$ on $[-K, K]$, where K is chosen to satisfy $|f(x)| \le K$ for all $x \in [a,b]$. □

Corollary 8.10.3 *If* $f, g \in \mathcal{R}(a,b)$, *then* $fg \in \mathcal{R}(a,b)$.

Proof. Using the identity

$$fg = \frac{1}{2}\left[(f+g)^2 - f^2 - g^2\right],$$

the conclusion follows from the linearity property and Corollary 8.10.2. □

The following example shows that we cannot relax the condition of continuity of φ to mere integrability.

Example 8.6 Let $f : [0,1] \to [0,1]$ be defined by

$$f(x) = \begin{cases} 0, & x \notin \mathbb{Q} \\ 1/q, & x = p/q \in \mathbb{Q}, \end{cases}$$

where p/q is in simplest form, and let $\varphi : [0,1] \to \mathbb{R}$ be given by

$$\varphi(x) = \begin{cases} 0, & x = 0 \\ 1, & x \neq 0. \end{cases}$$

f was shown to be integrable in Example 8.4, while φ is integrable because it is continuous except at the single point 0. However

$$(\varphi \circ f)(x) = \begin{cases} 0, & x \notin \mathbb{Q} \\ 1, & x \in \mathbb{Q}, \end{cases}$$

which was shown to be non-integrable in Example 8.2.

Exercises 8.3

1. If f is bounded on $[a, b]$ and $k > 0$, prove that
$$L(kf) = kL(f), \quad U(kf) = kU(f).$$
What if $k \leq 0$? Deduce that if $f \in \mathcal{R}(a, b)$, then for any real number k, $kf \in \mathcal{R}(a, b)$ and $\int_a^b kf = k \int_a^b f$.

2. Let f and g be bounded on $[a, b]$. Prove that
$$L(f + g) \geq L(f) + L(g), \quad U(f + g) \leq U(f) + U(g).$$
Prove that if f and g are integrable, then so is $f + g$, in which case $\int_a^b (f + g) = \int_a^b f + \int_a^b g$.

3. Let f and g be bounded on $[a, b]$. If $f \leq g$, prove that
$$L(f) \leq L(g), \quad U(f) \leq U(g).$$

4. If $f, g \in \mathcal{R}(a, b)$, prove the *Cauchy-Schwarz inequality*
$$\int_a^b fg \leq \left[\int_a^b f^2\right]^{1/2} \left[\int_a^b g^2\right]^{1/2}.$$
If $b - a = 1$, show that $\int_a^b f \leq \left[\int_a^b f^2\right]^{1/2}$.

5. For any two functions f and g, which are continuous on $[a, b]$, set
$$\rho(f, g) = \int_a^b |f - g|.$$
Prove the following:

(a) $\rho(f, g) \geq 0$
(b) $f = g \Leftrightarrow \rho(f, g) = 0$
(c) $\rho(f, g) = \rho(g, f)$
(d) $\rho(f, h) \leq \rho(f, g) + \rho(g, h)$, where h is also continuous on $[a, b]$.

Any function ρ satisfying these conditions is called a *metric*.

Let $f_n : [0,1] \to \mathbb{R}$ be defined as follows (see Figure 8.7):

$$f_n(x) = \begin{cases} 0, & x \leq 1/2 \\ n(x-1/2), & 1/2 < x \leq 1/2 + 1/n \\ 1, & x > 1/2 + 1/n. \end{cases}$$

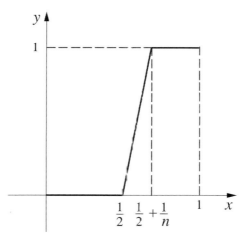

Figure 8.7

Prove that $\rho(f_n, f_m) = \int_a^b |f_n - f_m| \to 0$ as $n, m \to \infty$.

6. Let f be continuous on $[0,1]$. If $g_n(x) = f(x^n)$, prove that $\int_0^1 g_n \to f(0)$ as $n \to \infty$.

7. If $f : [a,b] \to [0, \infty)$ is continuous and $M = \sup\{f(x) : x \in [a,b]\}$, prove that
$$\lim_{n \to \infty} \left\{ \int_a^b [f(x)]^n \, dx \right\}^{1/n} = M.$$

8. Suppose that f is continuous on $[0, \infty)$ and $\lim_{x \to \infty} f(x) = a$. Show that
$$\lim_{x \to \infty} \frac{1}{x} \int_0^x f(t) \, dt = a.$$
Hint: Choose N large enough and use the mean value theorem on $[N, x]$.

9. Let f be continuous on $[a,b]$ and suppose that $g \in \mathcal{R}(a,b)$ is non-negative. Prove the existence of a point $c \in [a,b]$ such that

$$\int_a^b f(x)g(x)\,dx = f(c)\int_a^b g(x)\,dx.$$

This result is called the *second mean value theorem*. Give an example to show that the condition $g \geq 0$ cannot be dropped.

10. Give an example of a function $f \notin \mathcal{R}(a,b)$ such that $|f| \in \mathcal{R}(a,b)$.

8.4 The Fundamental Theorem of Calculus

In this section the integral is considered as a function of the upper limit, a function that proves to have smoother properties than the integrand. This leads to the *fundamental theorem of calculus* and the connection between integration and differentiation.

Theorem 8.11
Suppose that $f \in \mathcal{R}(a,b)$, and let $F : [a,b] \to \mathbb{R}$ be defined by

$$F(x) = \int_a^x f(t)\,dt.$$

Then
(i) F is continuous and satisfies a Lipschitz condition on $[a,b]$.
(ii) If f is continuous at $c \in [a,b]$, F is differentiable at c and

$$F'(c) = f(c).$$

Proof

(i) Since f is bounded, there is a number K such that

$$|f(x)| \leq K \quad \text{for all } x \in [a,b].$$

If $x, y \in [a,b]$ and $x < y$, then, from the properties of the integral, we have

$$|F(y) - F(x)| = \left| \int_a^y f(t)\, dt - \int_a^x f(t)\, dt \right|$$

$$= \left| \int_x^y f(t)\, dt \right|$$

$$\leq \int_x^y |f(t)|\, dt$$

$$\leq \int_x^y K\, dt = K|y - x|,$$

which means F is Lipschitz continuous on $[a, b]$.

(ii) Suppose f is continuous at c. Then, given any $\varepsilon > 0$, there is a $\delta > 0$ such that

$$x \in [a, b],\ |x - c| < \delta \Rightarrow |f(x) - f(c)| < \varepsilon.$$

If t lies between c and x, then $|t - c| < \delta$ and therefore $|f(t) - f(c)| < \varepsilon$. Noting that

$$f(c) = \frac{1}{x - c} \int_c^x f(c)\, dt,\ x \neq c,$$

we can write

$$\left| \frac{F(x) - F(c)}{x - c} - f(c) \right| = \left| \frac{1}{x - c} \int_c^x [f(t) - f(c)]\, dt \right|$$

$$\leq \frac{1}{|x - c|} \left| \int_c^x |f(t) - f(c)|\, dt \right|$$

$$\leq \frac{1}{|x - c|} \left| \int_c^x \varepsilon\, dt \right| = \varepsilon,$$

which confirms that $\lim_{x \to c} \frac{F(x) - F(c)}{x - c} = f(c)$, as required. □

Corollary 8.11 *If f is continuous on $[a, b]$ and*

$$F(x) = \int_a^x f(t)\, dt,\ x \in [a, b], \tag{8.16}$$

then F is differentiable on $[a, b]$ and $F' = f$.

A function whose derivative on an interval I is f is called an *anti-derivative*, or *primitive function*, of f on I. So this corollary asserts the existence of an anti-derivative F for any continuous function f :

$[a, b] \to \mathbb{R}$, given by (8.16). But the following example shows that this is not true for every integrable function. Let $f : [0, 1] \to \mathbb{R}$ be given by

$$f(x) = \begin{cases} 0, & x = 1 \\ 1, & x \neq 1. \end{cases}$$

f is integrable, as its set of discontinuities is a single point, but since f does not satisfy the intermediate value property on $[0, 1]$, it cannot, by Theorem 7.12, be the derivative of any function on $[a, b]$.

Suppose f is continuous on $[a, b]$ and let F be defined as in (8.16). If G is any anti-derivative of f, then

$$f(x) = G'(x) = F'(x), \ x \in [a, b],$$

and we conclude from Corollary 7.8 the existence of a real constant C such that

$$G(x) = F(x) + C, \ x \in [a, b].$$

Putting $x = a$ we obtain $C = G(a)$, and putting $x = b$ we arrive at $F(b) = G(b) - G(a)$, or

$$\int_a^b G'(t)\, dt = G(b) - G(a).$$

The continuity of f is not really required for the validity of this result. The following celebrated theorem asserts its validity for any integrable function.

Theorem 8.12 (Fundamental Theorem of Calculus)
If F is differentiable on $[a, b]$ and $F' \in \mathcal{R}(a, b)$, then

$$\int_a^b F'(x)\, dx = F(b) - F(a). \tag{8.17}$$

Proof. If $P = \{x_0, x_1, ..., x_n\}$ is any partition of $[a, b]$, then

$$F(b) - F(a) = \sum_{i=0}^{n-1} [F(x_{i+1}) - F(x_i)].$$

Applying the mean value theorem to F on $[x_i, x_{i+1}]$, we can choose $\alpha_i \in (x_i, x_{i+1})$ such that

$$F(x_{i+1}) - F(x_i) = F'(\alpha_i)(x_{i+1} - x_i).$$

Consequently

$$F(b) - F(a) = \sum_{i=0}^{n-1} F'(\alpha_i)(x_{i+1} - x_i) = S(F', P, \boldsymbol{\alpha}),$$

where $\boldsymbol{\alpha} = (\alpha_0, \alpha_1, \ldots, \alpha_{n-1})$ is a mark on P.

Now choose a sequence (P_k) of partitions such that $\|P_k\| \to 0$, and a corresponding sequence of marks $(\boldsymbol{\alpha}_k)$ determined by the mean value theorem as above. We then have

$$F(b) - F(a) = S(F', P_k, \boldsymbol{\alpha}_k), \quad k \in \mathbb{N}.$$

Since $F' \in \mathcal{R}(a,b)$, the right hand side converges to $\int_a^b F'(x)\,dx$ as $k \to \infty$. $\square$

Remark 8.3 It is common to describe integration as the process inverse to differentiation. This may suggest that derivatives are automatically integrable, so that the requirement of integrability of F' in the theorem is redundant, which is not true. We exhibit here a function on $[0,1]$ whose derivative is not Riemann integrable. Let $F : [0,1] \to \mathbb{R}$ be given by

$$F(x) = \begin{cases} 0, & x = 0 \\ x^2 \sin(1/x^2), & 0 < x \leq 1. \end{cases}$$

A simple calculation will confirm that F is differentiable on $[0,1]$ but that F' is not integrable (F' is in fact unbounded).

We now employ Theorems 8.11 and 8.12 to prove the validity of the two most widely used methods for evaluating integrals.

Theorem 8.13 (First Substitution Rule)
Suppose φ is differentiable on $[a,b]$ and its derivative φ' is continuous. If f is continuous on the range of φ, then

$$\int_a^b f(\varphi(t))\,\varphi'(t)\,dt = \int_{\varphi(a)}^{\varphi(b)} f(x)\,dx. \tag{8.18}$$

Proof. Let J be the interval whose end-points are $\varphi(a)$ and $\varphi(b)$. Since φ is continuous, its range I is an interval, and since both $\varphi(a)$ and $\varphi(b)$ lie in I, $I \supseteq J$. f is therefore continuous on J and the integral on the right-hand side of (8.18) exists. The continuity of f, φ, and φ' ensures the existence of the integral on the left-hand side of (8.18). All that remains is to prove the equality.

Let us define $F : I \to \mathbb{R}$ and $G : [a, b] \to \mathbb{R}$ as follows:

$$F(y) = \int_{\varphi(a)}^{y} f(x)\, dx, \quad G = F \circ \varphi.$$

From the chain rule and Theorem 8.11, we have

$$G'(t) = F'(\varphi(t))\, \varphi'(t) = f(\varphi(t))\, \varphi'(t), \quad t \in [a, b].$$

Noting that $G(a) = 0$, and using Theorem 8.12, we see that

$$G(b) = \int_{a}^{b} f(\varphi(t))\, \varphi'(t)\, dt,$$

which proves (8.18). □

Example 8.7 To evaluate the integral $\int_{1}^{2} t\,(t^2 + 1)^{7/2}\, dt$, let

$$\varphi(t) = t^2 + 1, \quad f(x) = \frac{1}{2} x^{7/2}.$$

Then

$$\int_{1}^{2} t\,(t^2 + 1)^{7/2}\, dt = \int_{1}^{2} f(\varphi(t))\, \varphi'(t)\, dt,$$

and from Theorem 8.13 we obtain

$$\int_{1}^{2} t\,(t^2 + 1)^{7/2}\, dt = \int_{2}^{5} f(x)\, dx.$$

Since $x \mapsto x^{9/2}/9$ is an anti-derivative of f,

$$\int_{2}^{5} f(x)\, dx = \left. \frac{x^{9/2}}{9} \right|_{2}^{5} = \frac{5^{9/2}}{9} - \frac{2^{9/2}}{9},$$

where, following a standard convention, we write $F(x)|_{a}^{b}$ for $F(b) - F(a)$. Hence

$$\int_{1}^{2} t\,(t^2 + 1)^{7/2}\, dt = \frac{5^{9/2}}{9} - \frac{2^{9/2}}{9}.$$

Remark 8.4 To calculate an integral of the form $\int_{a}^{b} f(\varphi(t))\, \varphi'(t)\, dt$ all we have to do is set $x = \varphi(t)$, formally write $dx = \varphi'(t)\, dt$, and replace the limits of integration on t by the corresponding limits on x.

This leads to $\int_{\varphi(a)}^{\varphi(b)} f(x)\,dx$, which theorem 8.13 asserts is the correct value of the integral. For instance, in Example 8.7, we have

$$x = \varphi(t) = t^2 + 1,$$
$$dx = 2t\,dt, \quad t\,dt = dx/2.$$

Furthermore, $x = 2$ when $t = 1$ and $x = 5$ when $t = 2$. The given integral can then be expressed in terms of the variable x as

$$\int_1^2 t\left(t^2+1\right)^{7/2} dt = \int_2^5 \frac{x^{7/2}}{2}\,dx = \frac{1}{2}\int_2^5 x^{7/2}\,dx.$$

Theorem 8.14 (Second Substitution Rule)
Let the function $\varphi : [a,b] \to \mathbb{R}$ have a continuous derivative that does not vanish anywhere in (a,b). If f is continuous on the range of φ, and ψ is the inverse of φ, then

$$\int_a^b f(\varphi(t))\,dt = \int_{\varphi(a)}^{\varphi(b)} f(x)\,\psi'(x)\,dx. \qquad (8.19)$$

Proof. Once again, we note that the range of φ is an interval I that contains the interval J with end points $\varphi(a)$ and $\varphi(b)$. Since $\varphi'(t) \neq 0$ for all $t \in (a,b)$, the inverse function theorem (Theorem 7.11) ensures the existence of $\psi = \varphi^{-1}$ and the continuity of ψ' on I. To prove the equality (8.19), we use (8.18) with $f \circ \varphi$ in place of f, ψ in place of φ, $\varphi(a)$ in place of a and $\varphi(b)$ in place of b. We then obtain

$$\int_{\varphi(a)}^{\varphi(b)} (f \circ \varphi)(\psi(t))\,\psi'(t)\,dt = \int_{\psi(\varphi(a))}^{\psi(\varphi(b))} (f \circ \varphi)(x)\,dx.$$

This gives

$$\int_{\varphi(a)}^{\varphi(b)} f(t)\,\psi'(t)\,dt = \int_a^b f(\varphi(x))\,dx,$$

which coincides with (8.19) after interchanging the dummy indices t and x. $\square$

To evaluate an integral of the form $\int_a^b f(\varphi(t))\,dt$ using this theorem, we set $x = \varphi(t)$ so that $t = \varphi^{-1}(x) = \psi(x)$ and $dt = \psi'(x)\,dx$, and replace the limits on t by the corresponding limits on x.

Example 8.8 We can use Theorem 8.14 to evaluate
$$\int_1^4 \frac{1}{(1+\sqrt{t})^3} dt$$
by first setting $x = \varphi(t) = 1 + \sqrt{t}$, so that
$$t = \psi(x) = (x-1)^2$$
$$dt = 2(x-1)\, dx.$$
Hence
$$\int_1^4 \frac{1}{(1+\sqrt{t})^3} dt = \int_2^3 \frac{2(x-1)}{x^3} dx$$
$$= 2\int_2^3 (x^{-2} - x^{-3})\, dx.$$
Since $-x^{-1} + \frac{1}{2}x^{-2}$ is a primitive of $x^{-2} - x^{-3}$, we obtain
$$2\int_2^3 (x^{-2} - x^{-3})dx = 2\left[\left(-\frac{1}{3} + \frac{1}{18}\right) - \left(-\frac{1}{2} + \frac{1}{8}\right)\right] = 7/36.$$

Theorem 8.15 (Integration by Parts)
Let $f, g : [a,b] \to \mathbb{R}$ be differentiable on $[a,b]$. If $f', g' \in \mathcal{R}(a,b)$, then
$$\int_a^b f(x)g'(x)dx = f(b)g(b) - f(a)g(a) - \int_a^b f'(x)g(x)dx. \quad (8.20)$$

Proof. Corollary 8.10.3 ensures the existence of both integrals in (8.20). To prove the equality, put $h = fg$. From the product rule, we have
$$h' = f'g + fg'$$
and we deduce from the fundamental theorem of calculus that
$$f(b)g(b) - f(a)g(a) = h(b) - h(a)$$
$$= \int_a^b [f'(x)g(x) + f(x)g'(x)]dx$$
$$= \int_a^b f'(x)g(x)dx + \int_a^b f(x)g'(x)dx. \quad \square$$

EXERCISES 8.4

1. Use Theorem 8.9 for integrals to obtain an alternative proof of Corollary 8.11.

2. If f is continuous on $[a,b]$ and H is given by $H(x) = \int_x^b f(t)\,dt$, calculate $H'(x)$.

3. Suppose f is continuous on $[a,b]$ and $g, h : [c,d] \to [a,b]$ are differentiable. If $F : [c,d] \to \mathbb{R}$ is defined by

$$F(x) = \int_{h(x)}^{g(x)} f(t)\,dt,$$

prove that

$$F'(x) = f(g(x))g'(x) - f(h(x))h'(x).$$

4. Calculate $F'(x)$ if

   (a) $F(x) = \int_0^{g(x)} \cos t\,dt$, $g(x) = \int_{-}^{x^2} \sin t^2\,dt$,

   (b) $F(x) = \int_{2x}^{x^2} \sin\sqrt{1+t^2}\,dt$.

5. If f is continuous on $(0,1)$, prove that

$$\int_0^x f(u)(x-u)\,du = \int_0^x \left[\int_0^u f(t)\,dt\right] du, \quad x \in (0,1).$$

Hint: Differentiate both sides.

6. The function $\log : (0,\infty) \to \mathbb{R}$ is often defined by

$$\log x = \int_1^x \frac{1}{t}\,dt.$$

Use this definition to verify the following:

   (a) $\dfrac{d}{dx}\log x = \dfrac{1}{x}$, $x > 0$.
   (b) $\log(xy) = \log x + \log y$, $x, y > 0$.
   (c) $\log x^p = p\log x$, $x > 0$, $p \in \mathbb{Q}$.

Use (b) and Exercise 6.2.10 to verify that this definition of $\log x$ coincides with the one given in Section 2.4.

7. Let f be continuous on $[a,b]$. If
$$\int_a^x f(t)\,dt = \int_x^b f(t)\,dt \quad \text{for all } x \in [a,b],$$
prove that
$$f(x) = 0 \quad \text{for all } x \in [a,b].$$
Deduce an alternative proof of Exercise 8.1.10.

8. Let f be continuous on $[0,\infty)$ and suppose that it does not vanish on $(0,\infty)$. If
$$[f(x)]^2 = \int_0^x f(t)\,dt \quad \text{for all } x > 0,$$
prove that
$$f(x) = \frac{1}{2}x \quad \text{for all } x \in [0,\infty).$$

9. Evaluate the following integrals:
   (a) $\int_1^4 \dfrac{\sqrt{1+\sqrt{x}}}{\sqrt{x}}\,dx.$
   (b) $\int_0^4 x\sqrt{3x+4}\,dx.$
   (c) $\int_0^{\pi/4} x^2 \sin x\,dx.$

10. Prove the following form of Taylor's theorem:
    If f and its derivatives $f', f'', \ldots, f^{(n)}, f^{(n+1)}$ are continuous on $[a,b]$, then
    $$f(b) = f(a) + \sum_{k=0}^{n} \frac{f^{(k)}(a)}{k!}(b-a)^k + R_n,$$
    where
    $$R_n = \frac{1}{n!} \int_a^b (b-t)^n f^{(n+1)}(t)\,dt.$$
    Hint: Integrate by parts a suitable number of times.

8.5 Improper Integrals

Throughout the preceding sections we had consistently assumed that the integrand f and the interval of integration $[a, b]$ were bounded. This places a considerable limitation on applications. For instance, we may wish to calculate the area under the graph of the unbounded function

$$f(x) = \begin{cases} 1/x, & x \neq 0 \\ 0, & x = 0 \end{cases}$$

over $[0, 1]$, or the area under the graph of

$$g(x) = 1/x^2$$

over the unbounded interval $[1, \infty)$. Our aim in this section is to extend the definition of the integral to cover such cases.

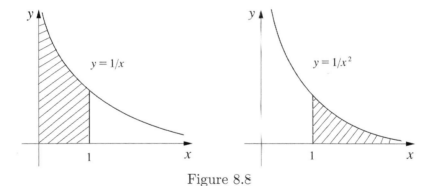

Figure 8.8

8.5.1 Unbounded Integrand

Let f be a real valued function on $[a, b]$ that is not necessarily bounded in the neighborhood of a. We say that f has an *improper integral* over $[a, b]$ if $f \in \mathcal{R}(c, b)$ for all $c \in (a, b)$ and $\lim_{c \to a^+} \int_c^b f$ exists. When it exists, this limit represents the *value* of the improper integral, and will be denoted by $\int_{a+}^b f$ or $\int_{a+}^b f(x) dx$.

Observe that if f is bounded on $[a, b]$, then f has an improper integral on $[a, b]$ if and only if $f \in \mathcal{R}(a, b)$ (see Exercise 8.1.12), in which case

$$\int_{a+}^b f = \int_a^b f. \tag{8.21}$$

Thus no confusion would arise if, as is customarily done, we drop the $+$ sign in the notation of the improper integral and simply write $\int_a^b f$.

Example 8.9 The function

$$f(x) = \begin{cases} 1/\sqrt{x}, & x \neq 0 \\ k, & x = 0, \end{cases}$$

where k is a real constant, is unbounded in the neighborhood of 0. Since f is continuous on $[c, 1]$ for any $c \in (0, 1)$, $f \in \mathcal{R}(c, 1)$, and a simple calculation gives

$$\int_c^1 f = 2\left(1 - \sqrt{c}\right).$$

Since $\sqrt{c} \to 0$ as $c \to 0^+$, f has an improper integral over $[0, 1]$ whose value is 2.

Note that the value of the improper integral (8.21) does not depend on the definition of f at a, the point where its limit becomes unbounded, and in many cases (as in the next example) f is not even defined at a. This is consistent with the observation made earlier that the value of the integral over $[a, b]$ is the same as its value over $(a, b]$.

Example 8.10 The function $f(x) = 1/x^3$, $x \in (0, 1]$, is unbounded in the neighborhood of 0 but is integrable on $[c, 1]$ for every $0 < c < 1$. Since

$$\lim_{c \to 0^+} \int_c^1 \frac{1}{x^3} dx = \lim_{c \to 0^+} \frac{1}{2}\left(\frac{1}{c^2} - 1\right) = +\infty,$$

the improper integral $\int_0^1 1/x^3 dx$ does not exist.

The case where the function is unbounded in the neighborhood of b is similarly treated. We define the resulting improper integral $\int_a^b f$ to be

$$\int_a^{b^-} f = \lim_{c \to b^-} \int_a^c f$$

whenever the limit exists.

Suppose now that f is unbounded in the neighborhood of a point $c \in (a, b)$. We say that f has an improper integral on $[a, b]$ if both improper integrals $\int_a^{c^-} f$ and $\int_{c^+}^b f$ exist. We then define the improper integral to be $\int_a^{c^-} f + \int_{c^+}^b f$ and denote it by $\int_a^b f$. Observe that

$$\int_a^b f = \int_a^{c^-} f + \int_{c^+}^b f = \lim_{\varepsilon_1 \to 0^+} \int_a^{c-\varepsilon_1} f + \lim_{\varepsilon_2 \to 0^+} \int_{c+\varepsilon_2}^b f,$$

where ε_1 and ε_2 are independent. It is obvious that the existence of these limits implies the existence of

$$\lim_{\varepsilon \to 0^+} \left\{ \int_a^{c-\varepsilon} f - \int_{c+\varepsilon}^b f \right\}. \tag{8.22}$$

The limit (8.22), when it exists, is called *The Cauchy principal value* of the integral and is commonly denoted by $\operatorname{PV} \int_a^b f$. The existence of this value does not, however, ensure the existence of the improper integral. Here is an example.

Example 8.11 Let $f : [-1, 1] \to \mathbb{R}$ be defined by

$$f(x) = \begin{cases} 1/x^3, & x \neq 0 \\ k, & x = 0. \end{cases}$$

A simple calculation yields

$$\int_{-1}^{-\varepsilon} f = \frac{1}{2}\left(1 - \frac{1}{\varepsilon^2}\right), \quad \int_{\varepsilon}^{1} f = \frac{1}{2}\left(\frac{1}{\varepsilon^2} - 1\right),$$

so that

$$\operatorname{PV} \int_{-1}^{1} f = 0.$$

However, we have already seen in Example 8.10 that $\int_{0+}^{1} f$ does not exist.

8.5.2 *Unbounded Interval*

If $f : [a, \infty) \to \mathbb{R}$ is Riemann integrable on $[a, c]$ for every $c > a$, and if

$$\lim_{c \to \infty} \int_a^c f(x)\, dx \tag{8.23}$$

exists, then we say that f has an improper integral on $[a, \infty)$, which is denoted by $\int_a^\infty f$ or $\int_a^\infty f(x)\, dx$, whose value is the limit (8.23). Thus

$$\int_a^\infty f(x)\, dx = \lim_{c \to \infty} \int_a^c f(x)\, dx.$$

In a similar fashion the improper integral $\int_{-\infty}^a f(x)\, dx$ is defined as $\lim_{c \to -\infty} \int_c^a f(x)\, dx$ whenever this limit exists.

To integrate over the unbounded interval $(-\infty, \infty)$, we pick any real number a and consider the improper integrals $\int_{-\infty}^{a} f(x)\,dx$ and $\int_{a}^{\infty} f(x)\,dx$. If both exist, we say that the improper integral $\int_{-\infty}^{\infty} f(x)\,dx$ exists and define its value by

$$\int_{-\infty}^{\infty} f(x)\,dx = \int_{-\infty}^{a} f(x)\,dx + \int_{a}^{\infty} f(x)\,dx.$$

Note that, in view of Theorem 8.8, the existence of $\int_{-\infty}^{\infty} f(x)\,dx$ requires the integrability of f over every bounded interval. Hence, when this improper integral exists, its value does not depend on a. We can also write

$$\int_{-\infty}^{\infty} f(x)\,dx = \lim_{c \to -\infty} \int_{c}^{a} f(x)\,dx + \lim_{d \to \infty} \int_{a}^{d} f(x)\,dx,$$

where c and d are independent. The limit

$$\lim_{c \to \infty} \int_{-c}^{c} f(x)\,dx = \lim_{c \to \infty} \left\{ \int_{-c}^{a} f(x)\,dx + \int_{a}^{c} f(x)\,dx \right\}$$

is called the *Cauchy principal value* of the integral $\int_{-\infty}^{\infty} f$, and is denoted by $\mathrm{PV} \int_{-\infty}^{\infty} f$. Here again the existence of $\mathrm{PV} \int_{-\infty}^{\infty} f$ does not guarantee that of the improper integral $\int_{-\infty}^{\infty} f$.

Example 8.12 Consider $f(x) = 1/x^2$ on $[1, \infty)$. We have

$$\lim_{c \to \infty} \int_{1}^{c} 1/x^2 \, dx = \lim_{c \to \infty} \left(1 - \frac{1}{c}\right) = 1.$$

Therefore the improper integral $\int_{1}^{\infty} (1/x^2)\,dx$ exists and has value 1.

Example 8.13. Let $f(x) = x$ on $\mathbb{R}$. Since

$$\lim_{c \to \infty} \int_{0}^{c} x\,dx = \lim_{c \to \infty} \frac{c^2}{2} = \infty,$$

the improper integral $\int_{0}^{\infty} x\,dx$ does not exist, and therefore neither does $\int_{-\infty}^{\infty} x\,dx$. However

$$\lim_{c \to \infty} \int_{-c}^{c} x\,dx = \lim_{c \to \infty} \left(\frac{c^2}{2} - \frac{c^2}{2}\right) = 0,$$

so that $\mathrm{PV} \int_{-\infty}^{\infty} x\,dx = 0$.

Having been defined by a limiting process, it is customary to call an improper integral *convergent* if it exists and *divergent* otherwise.

To conclude this chapter, we return to the tests of convergence for series, as given in Chapter 4, and present yet another test which utilizes the Riemann integral.

Theorem 8.16 (Integral Test)
Suppose $f : [1, \infty) \to [0, \infty)$ is decreasing and Riemann integrable on $[1, b]$ for all $b > 1$. Then
(i) The series $\sum_{n=1}^{\infty} f(n)$ is convergent if and only if the improper integral $\int_1^{\infty} f(t)\, dt$ is convergent.
(ii) In case the series is convergent, we have

$$\int_{n+1}^{\infty} f(t)\, dt \leq \sum_{k=n+1}^{\infty} f(k) \leq \int_n^{\infty} f(t)\, dt, \ n \in \mathbb{N}.$$

Proof
(i) Since f is decreasing and integrable over $[k-1, k]$, we have

$$f(k) \leq \int_{k-1}^{k} f(t)\, dt \leq f(k-1). \tag{8.24}$$

Summing up from $k = 2$ to $k = n$, we obtain

$$\sum_{k=2}^{n} f(k) \leq \int_1^n f(t)\, dt \leq \sum_{k=1}^{n-1} f(k).$$

The convergence of the series ensures that the increasing sequence of integrals $\int_1^n f(t)\, dt$ is bounded above by $\sum_{k=1}^{\infty} f(k)$ and is therefore convergent. On the other hand, if the improper integral $\int_1^{\infty} f(t)\, dt$ exists, then it becomes an upper bound for the partial sums of the series, and we conclude that the series is convergent.

(ii) By summing over k from $n + 1$ to m in (8.24) we see that

$$\sum_{k=n+1}^{m} f(k) \leq \int_n^m f(t)\, dt \leq \sum_{k=n}^{m-1} f(k)$$

$$\Rightarrow \int_{n+1}^{m+1} f(t)\, dt \leq \sum_{k=n+1}^{m} f(k) \leq \int_n^m f(t)\, dt.$$

300   Elements of Real Analysis

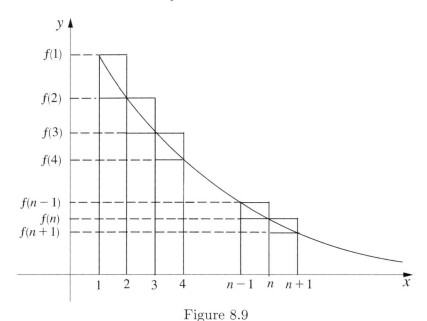

Figure 8.9

If the series converges then so does the improper integral, and the desired result is obtained by letting $m \to \infty$. $\square$

To illustrate the effectiveness of this test, we reconsider in the next example the series $\sum 1/n^p$, whose convergence properties have already been derived in Examples 4.6, 4.7 and 4.8.

Example 8.14 The function $f(t) = 1/t^p$ satisfies the conditions of Theorem 8.14. Since the integral

$$\int_1^n t^{-p} dt = \begin{cases} \dfrac{1}{1-p} \left(n^{1-p} - 1\right), & p \neq 1 \\ \log n, & p = 1 \end{cases}$$

converges to $1/(p-1)$ if $p > 1$ and to ∞ if $p \leq 1$, we conclude that $\sum 1/n^p$ is convergent for all $p > 1$ and divergent for all $p \leq 1$.

EXERCISES 8.5

1. Evaluate the following improper integrals:

    (a) $\displaystyle\int_0^1 \log x\, dx$

(b) $\int_1^2 \dfrac{1}{x \log x} dx$

(c) $\int_{-\infty}^{\infty} \dfrac{1}{1+x^2} dx$

(d) $\int_0^{\infty} \dfrac{dx}{\sqrt{x}\,(x+1)}$

(e) $\int_{-2}^{2} \dfrac{dx}{\sqrt{4-x^2}}$.

2. Let $\alpha < 0$. Prove that the function $f(x) = x^{\alpha}$ has a convergent improper integral on $[0,1]$ if $\alpha > -1$, and on $[1, \infty)$ if $\alpha < -1$.

3. Let f be defined on $(0, \infty)$ by

$$f(x) = \begin{cases} 1/\sqrt{x}, & x \in (0,1] \\ 1/x^2, & x > 1. \end{cases}$$

Define and evaluate $\int_0^{\infty} f(x)\, dx$.

4. Suppose $f \in \mathcal{R}(a, c)$ for all $c > a$. Prove the equivalence of the following statements:

   (a) f has a convergent improper integral on $[a, \infty)$.

   (b) For all $\varepsilon > 0$, there is a number N such that

   $$b, c > N \Rightarrow \left| \int_b^c f(x)\, dx \right| < \varepsilon.$$

5. Suppose $f, g \in \mathcal{R}(a, c)$ for all $c > a$, and that

   $$|f(x)| \le g(x) \text{ for all } x \in [a, \infty).$$

   If $\int_a^{\infty} g(x)\, dx$ is convergent, prove that $\int_a^{\infty} f(x)\, dx$ is also convergent. Use this to prove the existence of the integral $\int_0^{\infty} \dfrac{1}{1+x^4}\, dx$.

6. If $f : [0, \infty) \to [0, \infty)$ is monotonic, and the integral $\int_a^{\infty} f(x)\, dx$ is convergent, prove that

   $$\lim_{x \to \infty} x f(x) = 0.$$

7. Use the integral test to determine the convergence or divergence of the following series:

(a) $\displaystyle\sum_{n=2}^{\infty} \frac{1}{n \log n}$

(b) $\displaystyle\sum_{n=2}^{\infty} \frac{\log n}{n^2}$

(c) $\displaystyle\sum_{n=2}^{\infty} \frac{1}{n (\log n)^2}.$

8. Prove that the improper integral
$$\int_0^\infty \frac{\sin x}{x} dx$$
exists, but that
$$\int_0^\infty \frac{|\sin x|}{x} dx$$
does not.

9

Sequences and Series of Functions

Having discussed sequences and series of *numbers* in Chapters 3 and 4, we now consider sequences and series of *functions*. For the purposes of this study, these have two modes of convergence, *pointwise convergence* and *uniform convergence*. Our interest in the second type stems from the fact that it imparts to the limit function some of the smoothness properties of the members of the sequence. This naturally leads to series of functions, as a special class of sequences of functions, and their corresponding convergence properties. Of particular significance in this context are the power series, which are taken up towards the end of the chapter and used to define some well known analytic functions.

9.1 Sequences of Functions

Let $D \subseteq \mathbb{R}$ and suppose that, for each $n \in \mathbb{N}$, we have a function $f_n : D \to \mathbb{R}$. We then say that we have a *sequence*

$$(f_n : n \in \mathbb{N})$$

of real-valued functions on D. Evaluating f_n, for each $n \in \mathbb{N}$, at any fixed point $x \in D$ generates a sequence $(f_n(x) : n \in \mathbb{N})$ of real *numbers* which may or may not converge. If the sequence $(f_n(x))$ converges for every $x \in D$, the sequence of functions (f_n) is said to *converge pointwise* on D. The function $f : D \to \mathbb{R}$ defined by

$$f(x) = \lim_{n \to \infty} f_n(x) \qquad (9.1)$$

is called the (pointwise) *limit* of (f_n), and we express this symbolically by writing

$$\lim f_n = f \quad \text{or} \quad f_n \to f \quad \text{on } D.$$

Recalling Definition 3.2, we therefore have

Definition 9.1 The sequence of functions (f_n) is said to *converge pointwise* to f on D if, given $\varepsilon > 0$, there is, for each $x \in D$, a natural number $N = N(\varepsilon, x)$ (which depends on ε and x) such that

$$n \geq N \Rightarrow |f_n(x) - f(x)| < \varepsilon.$$

Example 9.1 Let $f_n : [0,1] \to \mathbb{R}$ be defined by

$$f_n(x) = x^n.$$

For any fixed $x \in [0,1)$, we get

$$\lim_{n \to \infty} f_n(x) = 0.$$

Since $f_n(1) = 1$ for all n, $\lim f_n(1) = 1$, and we see that $\lim f_n = f$ is given by (see Figure 9.1)

$$f(x) = \begin{cases} 0, & 0 \leq x < 1 \\ 1, & x = 1. \end{cases} \tag{9.2}$$

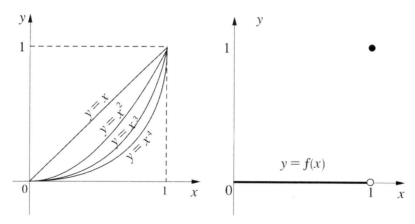

Figure 9.1

Observe, in this example, that f_n is continuous on $[0,1]$ for each n, but that $\lim f_n$ is not, since $\lim_{x \to 1} f(x) \neq f(1)$. Thus continuity is not preserved by pointwise convergence in this case. Equivalently, we can write

$$\lim_{x \to 1} \lim_{n \to \infty} f_n(x) \neq \lim_{n \to \infty} \lim_{x \to 1} f_n(x),$$

and we see that interchanging the order of the limiting processes can produce different results under pointwise convergence.

Example 9.2 Let $f_n : [0, 1] \to \mathbb{R}$ be defined by

$$f_n(x) = \begin{cases} 1, & x = p/q \in \mathbb{Q}, q \leq n \\ 0, & \text{otherwise} \end{cases} \tag{9.3}$$

where the rational number p/q is assumed to be in simplest form. Thus, for all $n \geq q$, we have $f_n(p/q) = 1 \to 1$ as $n \to \infty$. If, on the other hand, $x \notin \mathbb{Q}$, then $f_n(x) = 0$ for all n, and $\lim_{n\to\infty} f_n(x) = 0$. Therefore the sequence (f_n) converges to the function

$$f(x) = \begin{cases} 1, & x \in \mathbb{Q} \cap [0, 1] \\ 0, & x \in \mathbb{Q}^c \cap [0, 1]. \end{cases} \tag{9.4}$$

Here we note that $f_n \in \mathcal{R}(0, 1)$ for all n, since f_n is continuous except at a finite number of points (where it equals 1), but $f \notin \mathcal{R}(0, 1)$ as we saw in Example 8.2. So we conclude that, under pointwise convergence, the limit of a sequence of Riemann integrable functions may not be Riemann integrable, even though, as is the case in this example, the sequence $(f_n(x))$ actually *increases* to $f(x)$ for every $x \in [0, 1]$.

Example 9.3 Let

$$f_n(x) = nx(1 - x^2)^n, \ x \in [0, 1].$$

For any $x \in (0, 1]$ it is easy to show that $f_n(x) \to 0$ (see Exercise 3.2.6). Since $f_n(0) = 0$ for all n, we see that

$$\lim f_n(x) = 0 \ \text{ for all } x \in [0, 1].$$

Here f_n is integrable and

$$\int_0^1 f_n(x)dx = n \int_0^1 x(1 - x^2)^n dx$$

$$= \left[\frac{-n}{2(n+1)} (1 - x^2)^{n+1} \right]_0^1$$

$$= \frac{n}{2(n+1)} \to \frac{1}{2}.$$

Since $\int_0^1 f(x)dx = 0$, we conclude that

$$\lim_{n\to\infty} \int_0^1 f_n(x)dx \neq \int_0^1 \lim_{n\to\infty} f_n(x)dx.$$

Example 9.4 Let $f_n : \mathbb{R} \to \mathbb{R}$ be given by

$$f_n(x) = \begin{cases} -1, & x < -1/n \\ \sin(n\pi x/2), & -1/n \leq x \leq 1/n \\ 1, & x > 1/n. \end{cases} \quad (9.5)$$

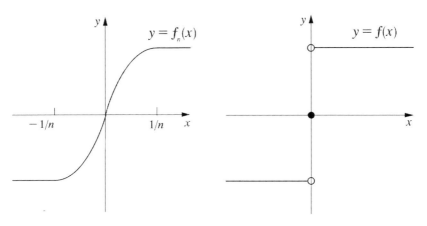

Figure 9.2

If $x > 0$ then, by the theorem of Archimedes, there is a positive integer N such that $x > 1/N$. It then follows that $x > 1/n$ for all $n \geq N$, and therefore $f_n(x) = 1$ for all $n \geq N$. Consequently $f_n(x) \to 1$.

Similarly, if $x < 0$, then $f_n(x) \to -1$. Since $f_n(0) = 0$ for all n, we see that

$$f(x) = \lim_{n \to \infty} f_n(x) = \begin{cases} -1, & x < 0 \\ 0, & x = 0 \\ 1, & x > 0. \end{cases} = \operatorname{sgn} x. \quad (9.6)$$

It is straightforward to verify that f_n is differentiable on $\mathbb{R}$ for all $n \in \mathbb{N}$, yet f is not even continuous.

Example 9.5 Suppose

$$f_n(x) = \frac{\sin nx}{n}, \quad x \in \mathbb{R}, \ n \in \mathbb{N}.$$

Clearly $f_n \to 0$ on $\mathbb{R}$. Here the limit function is differentiable, but its derivative is far from equal to the limit of f_n'. In fact $f_n'(x) = \cos nx$ does not even converge on $\mathbb{R}$ (except at integral multiples of 2π).

The above examples highlight the need for a mode of convergence other than pointwise convergence if the limit function is to inherit any of the smoothness properties of the sequence.

Going back to Example 9.1, we see in Figure 9.1 that the deviation of the graph of f_n from that of f becomes more pronounced as x approaches 1 from the left, so that $|f_n - f|$ cannot be made arbitrarily small *over the whole interval* $[0, 1]$ by taking n large enough.

Definition 9.2 A sequence (f_n) of real functions on $D \subseteq \mathbb{R}$ converges *uniformly* to a function f on D, symbolically expressed by

$$f_n \stackrel{u}{\to} f,$$

if, given $\varepsilon > 0$, there is a positive integer N (which depends on ε) such that
$$n \geq N \Rightarrow |f_n(x) - f(x)| < \varepsilon \text{ for all } x \in D.$$

According to this definition, for any $\varepsilon > 0$, we can always "force" the graph of f_n into a band of width less than 2ε centred about the graph of f, and extending over the *whole domain* D, by taking n large enough, as shown in Figure 9.3.

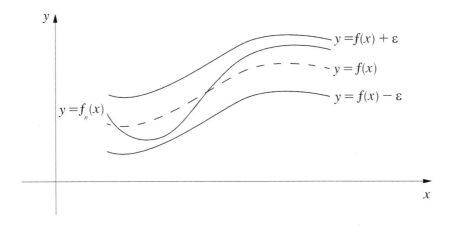

Figure 9.3

It is as if this band is a *neighborhood* of f in some sense. Recalling the definition of convergence for a numerical sequence in terms of neighborhoods, we immediately see that the corresponding "natural" mode of convergence for a sequence of functions is uniform, not pointwise, convergence. The basic difference between these two modes is, of course, that the integer N in Definition 9.2 does not depend on x, as is generally the case in Definition 9.1.

Remarks 9.1

1. It is worth noting that, if $f_n \to f$ uniformly, then $f_n \to f$ pointwise. This implies that, if a sequence of functions converges uniformly, it converges to the pointwise limit of the sequence.

2. As with numerical sequences, the inequality $|f_n(x) - f(x)| < \varepsilon$ in Definitions 9.1 and 9.2 may be replaced by $|f_n(x) - f(x)| \leq \varepsilon$ or $|f_n(x) - f(x)| < C\varepsilon$, where C is a positive number which does not depend on ε.

Since the statement $|f_n(x) - f(x)| \leq \varepsilon$, for all $x \in D$, is equivalent to the requirement that

$$\sup_{x \in D} |f_n(x) - f(x)| \leq \varepsilon,$$

we see that $f_n \xrightarrow{u} f$ if and only if, for every $\varepsilon > 0$, there is a positive integer N such that

$$n \geq N \Rightarrow \sup_{x \in D} |f_n(x) - f(x)| \leq \varepsilon,$$

which, in turn, is equivalent to the statement

$$\sup_{x \in D} |f_n(x) - f(x)| \to 0 \quad as \quad n \to \infty.$$

This observation is useful enough to be recorded as a theorem.

Theorem 9.1

The sequence $f_n : D \to \mathbb{R}$ converges to f uniformly on D if, and only if,

$$\sup_{x \in D} |f_n(x) - f(x)| \to 0 \quad as \quad n \to \infty.$$

Let us now apply this test to our previous examples:

1. The sequence of functions $f_n(x) = x^n$, $x \in [0, 1]$, converges pointwise to the function defined in (9.2). Here we have

   $$\sup_{x \in [0,1]} |f_n(x) - f(x)| = \sup_{x \in [0,1)} x^n = 1 \not\to 0,$$

   hence f_n does not converge uniformly to f on $[0, 1]$, or, for that matter, on $[0, 1)$.

2. In Example 9.2 the sequence f_n defined on $[0, 1]$ by (9.3) converges pointwise to the function f expressed by (9.4). Since

$$\sup_{x \in [0,1]} |f_n(x) - f(x)| \geq \left| f_n\left(\frac{1}{n+1}\right) - f\left(\frac{1}{n+1}\right) \right| = |0 - 1| = 1$$

does not tend to 0 as $n \to \infty$, the sequence (f_n) does not converge uniformly to f on $[0, 1]$.

3. The sequence $f_n(x) = nx(1 - x^2)^r$, $x \in [0, 1]$, converges pointwise to 0, as we found in Example 9.3. To evaluate

$$\sup_{x \in [0,1]} |f_n(x) - f(x)| = \sup_{x \in [0,1]} nx(1 - x^2)^n,$$

we use differential calculus to determine the maximum value of the polynomial $nx(1 - x^2)^n$ over the interval $[0, 1]$. The point where this polynomial attains its maximum value is $1/\sqrt{2n+1}$, therefore

$$\sup_{x \in [0,1]} |f_n(x) - f(x)| = \frac{n}{\sqrt{2n+1}} \left(1 - \frac{1}{2n+1}\right)^n \to \infty,$$

since

$$\left(1 - \frac{1}{2n+1}\right)^n = \left(1 + \frac{1}{2n}\right)^{-n} \to \frac{1}{\sqrt{e}}$$

and $n/\sqrt{2n+1} \to \infty$. Thus the convergence $f_n \to 0$ on $[0, 1]$ is not uniform.

4. In Example 9.4 the sequence (f_n) converges pointwise to f on $\mathbb{R}$, where f_n and f are defined in (9.5) and (9.6) respectively, but here also the convergence is not uniform, since

$$\sup_{x \in \mathbb{R}} |f_n(x) - f(x)| \geq |f_n(1/2n) - f(1/2n)|$$

$$= 1 - \sin \pi/4 \not\to 0.$$

5. The sequence

$$f_n(x) = \frac{\sin nx}{n}, \quad x \in \mathbb{R},$$

converges pointwise to 0 on $\mathbb{R}$. Here we see that

$$\sup_{x \in \mathbb{R}} |f_n(x) - f(x)| = \sup_{x \in \mathbb{R}} \left| \frac{\sin nx}{n} \right| \leq \frac{1}{n} \to 0,$$

so we conclude that $f_n \xrightarrow{u} 0$ on $\mathbb{R}$.

Example 9.6 Given
$$f_n(x) = nxe^{-nx},$$
it is clear that $f_n \to 0$ if and only if $x \geq 0$. To test this sequence for uniform convergence on $[a, \infty)$, where $a \geq 0$, we need to evaluate the limit of $\sup\{f_n(x) : x \geq 0\}$ as $n \to \infty$. Now $f'_n(x) = ne^{-nx}(1 - nx)$, hence $x = 1/n$ is the only critical point of f_n in $[0, \infty)$. We consider two cases:

(i) $a = 0$. Here the maximum value of f_n is attained at $1/n$ and equals $1/e$. Consequently,
$$\sup_{x \in [0,\infty)} f_n(x) = 1/e \not\to 0,$$
and the sequence (f_n) does not converge uniformly on $[0, \infty)$.

(ii) $a > 0$. In this case we can always find a positive integer N such that $a > 1/N$. For each $n \geq N$ the function f_n does not have any critical points in the interval (a, ∞). Since f_n is a decreasing function on this interval for all $n \geq N$, its maximum value is attained at $x = a$. Hence
$$\sup_{x \in [a,\infty)} f_n(x) = f_n(a) = nae^{-na} \to 0,$$
and we conclude that f_n converges uniformly to 0 on $[a, \infty)$ for any $a > 0$.

A Cauchy criterion is also available for testing uniform convergence.

Theorem 9.2 (Cauchy Criterion for Uniform Convergence) *The sequence of functions $f_n : D \to \mathbb{R}$ is uniformly convergent if, and only if, for every $\varepsilon > 0$, there is a positive integer N such that*
$$m, n \geq N \Rightarrow |f_n(x) - f_m(x)| < \varepsilon \quad \text{for all } x \in D, \tag{9.7}$$
or, equivalently,
$$m, n \geq N \Rightarrow \sup_{x \in D} |f_n(x) - f_m(x)| < \varepsilon.$$

Proof. Suppose $f_n \xrightarrow{u} f$ on D, and let $\varepsilon > 0$. Choose $N \in \mathbb{N}$ such that
$$n \geq N \Rightarrow |f_n(x) - f(x)| < \varepsilon/2, \ x \in D.$$
If $m, n \geq N$ then
$$|f_n(x) - f_m(x)| \leq |f_n(x) - f(x)| + |f(x) - f_n(x)| < \varepsilon, \ x \in D.$$

Now suppose (f_n) satisfies the criterion stated in the theorem; then, for every $x \in D$, the real sequence $(f_n(x))$ is clearly a Cauchy sequence, and hence convergent. Let $f(x)$ be its (pointwise) limit, that is,

$$f(x) = \lim_{n \to \infty} f_n(x), \quad x \in D.$$

Now hold $n \geq N$ fixed and let $m \to \infty$ in (9.7) to conclude, from the continuity of the absolute value, that

$$n \geq N \Rightarrow |f_n(x) - f(x)| \leq \varepsilon \text{ for all } x \in D,$$

which implies $f_n \overset{u}{\to} f$. □

EXERCISES 9.1

In Exercises 1 through 10 determine the pointwise limit of (f_n), then decide whether the convergence is uniform or not.

1. $f_n(x) = \dfrac{x}{n}$, $x \in \mathbb{R}$.

2. $f_n(x) = \dfrac{\sin(nx)}{nx}$, $x \in (0,1)$.

3. $f_n(x) = \dfrac{1}{nx+1}$, $x \in (0,1)$.

4. $f_n(x) = \dfrac{x}{nx+1}$, $x \in [0,1]$.

5. $f_n(x) = \dfrac{nx}{1+n^2 x^2}$, $x \in \mathbb{R}$.

6. $f_n(x) = \dfrac{nx^3}{1+nx}$, $x \in [0,1]$.

7. $f_n(x) = x^n(1-x)$, $x \in [0,1]$.

8. $f_n(x) = x^n(1-x^n)$, $x \in [0,1]$.

9. $f_n(x) = \begin{cases} nx, & x \in [0, 1/n] \\ 0, & x \in (1/n, 1]. \end{cases}$

10. $f_n(x) = \begin{cases} \sqrt{nx}, & x \in [0, 1/n] \\ 0, & x \in (1/n, 1]. \end{cases}$

11. If $f_n \xrightarrow{u} f$ on D and $f_n \xrightarrow{u} f$ on E, prove that $f_n \xrightarrow{u} f$ on $D \cup E$.

12. If $f_n \xrightarrow{u} f$ and $g_n \xrightarrow{u} g$ on D, prove that, for all $\alpha, \beta \in \mathbb{R}$, $\alpha f_n + \beta g_n \xrightarrow{u} \alpha f + \beta g$.

13. Suppose $f_n \xrightarrow{u} f$ and $g_n \xrightarrow{u} g$ on D. If each f_n and each g_n is bounded on D, prove that $f_n g_n \xrightarrow{u} fg$. Give an example where $f_n \xrightarrow{u} f$ and $g_n \xrightarrow{u} g$ on a set D and yet $(f_n g_n)$ does not converge uniformly.

14. Suppose (f_n) converges uniformly on D. If each f_n is bounded on D, prove that (f_n) is *uniformly bounded* on D, in the sense that there is a constant K such that $|f_n(x)| \leq K$ for all $x \in D$, $n \in \mathbb{N}$.

15. Let D be a compact set in $\mathbb{R}$. Suppose (f_n) is a sequence of continuous functions that *decreases pointwise* to f, in the sense that, for any $x \in D$, $f_{n+1}(x) \leq f_n(x)$ for all $n \in \mathbb{N}$ and $f_n(x) \to f(x)$. If f is continuous on D, prove that $f_n \xrightarrow{u} f$. Use the sequence $f_n(x) = x^n$ on $(0,1)$ to show that the compactness condition cannot be dropped. This is known as *Dini's Theorem*.

    Hint: Given $\varepsilon > 0$, show that, for each $c \in D$, there is a positive integer N_c and a $\delta_c > 0$ (which depend on c and ε) such that $f_n(x) - f(x) < \varepsilon$ for all $n \geq N_c$ and $x \in (c - \delta_c, c + \delta_c)$.

9.2 Properties of Uniform Convergence

Having demonstrated some of the shortcomings of pointwise convergence in the last section, we now examine the situation under uniform convergence.

Theorem 9.3
Let $x \in \hat{D}$ and suppose $f_n \xrightarrow{u} f$ on $D \setminus \{x\}$. If
$$\lim_{t \to x} f_n(t) = \ell_n$$
exists for each $n \in \mathbb{N}$, then
(i) (ℓ_n) is convergent,
(ii) $\lim_{t \to x} f(t)$ exists and coincides with $\lim_{n \to \infty} \ell_n$, i.e.,
$$\lim_{n \to \infty} \lim_{t \to x} f_n(t) = \lim_{t \to x} \lim_{n \to \infty} f_n(t).$$

Proof. Let $\varepsilon > 0$. Since (f_n) converges uniformly, it satisfies the Cauchy criterion, so there is an $N \in \mathbb{N}$ such that

$$m, n \geq N \Rightarrow |f_n(t) - f_m(t)| < \varepsilon \text{ for all } t \in D \setminus \{x\}. \tag{9.8}$$

Now let $t \to x$ in (9.8), and recall that the absolute value is a continuous function to conclude that

$$m, n \geq N \Rightarrow |\ell_n - \ell_m| \leq \varepsilon, \tag{9.9}$$

which implies that (ℓ_n) is a Cauchy sequence, and hence convergent. This proves (i).

To prove (ii), let $\ell = \lim_{n \to \infty} \ell_n$. Setting $n = N$ and letting $m \to \infty$ in (9.8) yields

$$|f_N(t) - f(t)| \leq \varepsilon \text{ for all } t \in D \setminus \{x\}, \tag{9.10}$$

and in (9.9), it gives

$$|\ell_N - \ell| \leq \varepsilon. \tag{9.11}$$

Since $f_N(t) \to \ell_N$ as $t \to x$, there is a $\delta > 0$ such that

$$t \in D \setminus \{x\}, |t - x| < \delta \Rightarrow |f_N(t) - \ell_N| < \varepsilon. \tag{9.12}$$

Using (9.10), (9.11), and (9.12), we obtain

$$|f(t) - \ell| \leq |f(t) - f_N(t)| + |f_N(t) - \ell_N| + |\ell_N - \ell| < 3\varepsilon$$

for all $t \in D \setminus \{x\}$ such that $|t - x| < \delta$, which means $\lim_{t \to x} f(t) = \ell$. □

Corollary 9.3 *Suppose $f_n \xrightarrow{u} f$ on D. If, for each $n \in \mathbb{N}$, f_n is continuous at $c \in D$, then so is f.*

Proof. We need only consider $c \in \hat{D} \cap D$. In view of Theorem 9.3, we have

$$\lim_{t \to c} f(t) = \lim_{n \to \infty} \lim_{t \to c} f_n(t)$$
$$= \lim_{n \to \infty} f_n(c)$$
$$= f(c),$$

which proves that f is continuous at c. □

Thus, if a sequence of continuous functions converges to a discontinuous limit function, then we can conclude, based on this corollary, that the convergence is not uniform. Example 9.1 demonstrates such a sequence. But Corollary 9.3 gives a necessary, not a sufficient, condition for uniform convergence; for it may happen that a sequence of continuous functions converges to a continuous limit function without the convergence being uniform. Example 9.3 provides an example of such a sequence.

Theorem 9.4
Suppose $f_n \in R(a,b)$ for each $n \in \mathbb{N}$. If $f_n \xrightarrow{u} f$ on $[a,b]$, then $f \in R(a,b)$ and
$$\int_a^b f(x)dx = \lim_{n \to \infty} \int_a^b f_n(x)dx. \tag{9.13}$$

Proof. Let $\varepsilon > 0$. To prove that $f \in \mathcal{R}(a,b)$, we need only exhibit a partition P such that
$$U(f,P) - L(f,P) < C\varepsilon,$$
where C is independent of ε.

Since f_n converges uniformly to f, we can find a positive integer N such that
$$n \geq N \Rightarrow f_n(x) - \varepsilon < f(x) < f_n(x) + \varepsilon \text{ for all } x \in [a,b].$$

Since $f_N \in \mathcal{R}(a,b)$, there is a partition $P = \{x_0, x_1, \cdots, x_n\}$ of $[a,b]$ such that
$$U(f_N, P) - L(f_N, P) < \varepsilon.$$
But since $f_N(x) - \varepsilon < f(x) < f_N(x) + \varepsilon$ for all $x \in D$, we have
$$L(f_N, P) - \varepsilon(b-a) \leq L(f,P),$$
$$U(f,P) \leq U(f_N, P) + \varepsilon(b-a).$$

Hence
$$U(f,P) - L(f,P) \leq U(f_N, P) - L(f_N, P) + 2\varepsilon(b-a)$$
$$< \varepsilon + 2\varepsilon(b-a)$$
$$= C\varepsilon,$$
where $C = 1 + 2(b-a)$, so $f \in \mathcal{R}(a,b)$.

Furthermore,

$$\left| \int_a^b f_n(x)dx - \int_a^b f(x)dx \right| \le \int_a^b |f_n(x) - f(x)|\,dx$$

$$\le \int_a^b \sup_{x \in [a,b]} |f_n(x) - f(x)|\,dx$$

$$\le (b-a) \sup_{x \in [a,b]} |f_n(x) - f(x)|.$$

The uniform convergence of (f_n) now ensures that

$$\sup_{x \in [a,b]} |f_n(x) - f(x)| \to 0 \quad \text{as } n \to \infty. \quad \square$$

Equation (9.13) can actually hold in the absence of uniform convergence, as may be seen by taking $f_n(x) = x^n$ (see Example 9.1), where

$$\int_0^1 f_n(x)dx = \int_0^1 x^n dx = \frac{1}{n+1} \to 0 = \int_0^1 f(x)dx,$$

so the conditions on f_n, as stated in Theorem 9.4, are not necessary for the validity of (9.13). In fact, these conditions will be significantly relaxed when we take up the Lebesgue integral in Chapter 11.

We may be tempted to expect an analogous result for differentiation, perhaps along the following lines: if $f_n \xrightarrow{u} f$ on $[a,b]$ and f_n is differentiable at $c \in [a,b]$, then so is f and $f'_n(c) \to f'(c)$. But this statement is false. Consider the following examples:

Example 9.7 The sequence

$$f_n(x) = \sqrt{x^2 + 1/n^2}, \quad x \in [-1, 1],$$

converges (pointwise) to the function $f(x) = \sqrt{x^2} = |x|$. The convergence is uniform on $[-1, 1]$, since

$$|f_n(x) - f(x)| = \sqrt{x^2 + 1/n^2} - \sqrt{x^2}$$

$$= \frac{1/n^2}{\sqrt{x^2 + 1/n^2} + \sqrt{x^2}}$$

$$\le \frac{1}{n}, \quad x \in [-1, 1].$$

Clearly each f_n is differentiable, with $f'_n(x) = x/\sqrt{x^2 + 1/n^2}$, but f is not differentiable at 0.

In Example 9.5 we saw that the sequence

$$f_n(x) = \frac{\sin nx}{n}, \quad x \in \mathbb{R},$$

converges uniformly to 0. However, the sequence $f'_n(x)$ does not converge at any x, except integral multiples of 2π. Even if the limit function f is differentiable and the sequence f'_n is convergent, we cannot be certain of the equality $f' = \lim_{n\to\infty} f'_n$, as the next example demonstrates.

Example 9.8 The sequence (f_n) defined by

$$f_n(x) = \frac{x^n}{n}, \quad x \in [0,1],$$

clearly converges uniformly to $f = 0$. However,

$$\lim_{n\to\infty} f'_n(1) = 1 \neq f'(1).$$

The following theorem gives sufficient conditions for the convergence $f'_n \to f'$.

Theorem 9.5
Suppose (f_n) is a differentiable sequence of functions on $[a,b]$ which converges at some point $x_0 \in [a,b]$. If the sequence (f'_n) is uniformly convergent on $[a,b]$, then (f_n) is also uniformly convergent on $[a,b]$ to a function f which is differentiable on $[a,b]$, and $f'_n \to f'$.

Proof. Let $\varepsilon > 0$. From the convergence of $(f_n(x_0))$ and the uniform convergence of (f'_n), we deduce that there is an $N \in \mathbb{N}$ such that

$$m, n \geq N \Rightarrow |f'_n(x) - f'_m(x)| < \varepsilon \text{ for all } x \in [a,b],$$

and

$$m, n \geq N \Rightarrow |f_n(x_0) - f_m(x_0)| < \varepsilon.$$

Given any two points $x, t \in [a,b]$, we know from the mean value theorem, applied to $f_n - f_m$, that we can find a point c between x and t such that

$$f_n(x) - f_m(x) - [f_n(t) - f_m(t)] = (x-t)[f'_n(c) - f'_m(c)].$$

If $m, n \geq N$ then

$$|f_n(x) - f_m(x) - [f_n(t) - f_m(t)]| < \varepsilon |x-t|, \tag{9.14}$$

hence
$$|f_n(x) - f_m(x)| \le |f_n(x) - f_m(x) - [f_n(x_0) - f_m(x_0)]|$$
$$+ |f_n(x_0) - f_m(x_0)|$$
$$< \varepsilon |x - x_0| + \varepsilon$$
$$\le \varepsilon(b - a + 1) = C\varepsilon, \ x \in [a, b].$$

This proves that (f_n) satisfies Cauchy's criterion for uniform convergence, and must therefore converge uniformly to some limit f.

For any fixed point $x \in [a, b]$, define
$$g_n(t) = \frac{f_n(t) - f_n(x)}{t - x}, \ t \in [a, b]\setminus\{x\},$$
$$g(t) = \frac{f(t) - f(x)}{t - x}, \ t \in [a, b]\setminus\{x\}.$$

Clearly $g_n \to g$ as $n \to \infty$, and we shall now show that this convergence is uniform. If $m, n \ge N$ then, using (9.14),
$$|g_n(t) - g_m(t)| < \varepsilon \ \text{for all} \ t \in [a, b]\setminus\{x\},$$
hence, by Cauchy's criterion, $g_n \overset{u}{\to} g$ on $[a, b]\setminus\{x\}$. Since
$$\lim_{t \to x} g_n(t) = f'_n(x) \ \text{for all} \ n \in \mathbb{N},$$
we can use Theorem 9.3 to conclude that
$$\lim_{t \to x} g(t) = \lim_{n \to \infty} f'_n(x).$$
But since $\lim_{t \to x} g(t) = f'(x)$, this proves that f is differentiable at x, and that
$$f'(x) = \lim_{n \to \infty} f'_n(x). \ \square$$

That the uniform convergence of (f'_n) is not necessary for the conclusions of this theorem may be seen by considering the sequence
$$f_n(x) = \frac{x^{n+1}}{n + 1}, \ x \in (0, 1),$$
which converges uniformly to 0, while $f'_n(x) = x^n$ converges also to 0, but not uniformly (see Example 9.1).

We end this section with a celebrated theorem of classical analysis due to Karl Weierstrass (1815-1897). According to Corollary 9.3, the uniform limit of a sequence of polynomials on a compact interval $[a, b]$ is necessarily continuous. The converse statement, that each continuous function on $[a, b]$ is such a limit, is known as the *Weierstrass approximation theorem*.

Theorem 9.6
If the function $f : [a, b] \to \mathbb{R}$ is continuous, then, for each $\varepsilon > 0$, there is a polynomial p such that $|f(x) - p(x)| < \varepsilon$ for all $x \in [a, b]$.

Proof. We first show that it is sufficient to prove the theorem for the special case $[a, b] = [0, 1]$. This follows from the observation that $x = (b - a)t + a$ is a continuous mapping of $[0, 1]$ onto $[a, b]$, hence the function
$$g(t) = f(x) = f((b - a)t + a)$$
is continuous on $[0, 1]$. If the theorem is valid on $[0, 1]$ and q is a polynomial which satisfies
$$|g(t) - q(t)| < \varepsilon \text{ for all } t \in [0, 1],$$
then we obtain
$$|f(x) - p(x)| < \varepsilon \text{ for all } x \in [a, b],$$
where p is the polynomial defined by
$$p(x) = q\left(\frac{x - a}{b - a}\right), \quad x \in [a, b].$$

Secondly, we may assume, without loss of generality, that $f(0) = f(1) = 0$. Indeed, if the theorem is proved in this case, then for the function
$$h(x) = f(x) - f(0) - [f(1) - f(0)]x, \quad x \in [0, 1],$$
which satisfies $h(0) = h(1) = 0$, there would exist a polynomial r such that
$$|h(x) - r(x)| < \varepsilon \text{ for all } x \in [0, 1].$$
Setting $p(x) = r(x) + f(0) + [f(1) - f(0)]x$, we now have
$$|f(x) - p(x)| < \varepsilon \text{ for all } x \in [0, 1].$$

Finally, we extend f as a continuous function to $\mathbb{R}$ by defining $f(x) = 0$ on $\mathbb{R} \setminus [0, 1]$.

For each $n \in \mathbb{N}$, let

$$\varphi_n(x) = \begin{cases} c_n(1-x^2)^n, & |x| \leq 1 \\ 0, & |x| > 1, \end{cases}$$

with c_n chosen so that $\int_{-1}^{1} \varphi_n(x)\,dx = 1$. That is,

$$c_n = \left[\int_{-1}^{1}(1-x^2)^n\,dx\right]^{-1}.$$

Since

$$\int_{-1}^{1}(1-x^2)^n\,dx = 2\int_{0}^{1}(1-x^2)^n\,dx$$
$$= 2\int_{0}^{1}(1-x)^n(1+x)^n\,dx$$
$$\geq 2\int_{0}^{1}(1-x)^n\,dx$$
$$= \frac{2}{n+1},$$

we arrive at the estimate

$$0 < c_n \leq \frac{n-1}{2}.$$

Define

$$p_n(x) = \int_{-1}^{1} f(x-t)\varphi_n(t)\,dt, \quad x \in \mathbb{R}.$$

Since f vanishes outside $[0,1]$, we have

$$p_n(x) = \int_{x-1}^{x} f(x-t)\varphi_n(t)\,dt.$$

The substitution $s = x - t$ yields

$$p_n(x) = \int_{0}^{1} f(s)\varphi_n(x-s)\,ds,$$

which is clearly a polynomial in x for all $x \in [0,1]$.

Now let ε be any positive number. Since f is continuous on the compact interval $[0,1]$, it is bounded and uniformly continuous there.

Therefore there is a number M such that $|f(x)| \le M$ for all $x \in [0,1]$ and a number $0 < \delta < 1$ such that

$$x, t \in [0,1], \ |x-t| < \delta \ \Rightarrow \ |f(x) - f(t)| < \varepsilon/2.$$

Thus, for all $x \in [0,1]$,

$$|f(x) - p_n(x)| = \left| \int_{-1}^{1} f(x)\varphi_n(t)dt - \int_{-1}^{1} f(x-t)\varphi_n(t)dt \right|$$

$$\le \int_{-1}^{1} |f(x) - f(x-t)| \varphi_n(t)dt$$

$$\le \left(\int_{-1}^{-\delta} + \int_{-\delta}^{\delta} + \int_{\delta}^{1} \right) |f(x) - f(x-t)| \varphi_n(t)dt$$

$$\le 2M \int_{-1}^{-\delta} \varphi_n(t)dt + \frac{\varepsilon}{2} \int_{-\delta}^{\delta} \varphi_n(t)dt + 2M \int_{\delta}^{1} \varphi_n(t)dt.$$

Using the upper bound on c_n, we have

$$\int_{\delta}^{1} \varphi_n(t)dt \le \frac{n+1}{2} \int_{\delta}^{1} (1-t^2)^n dt$$

$$\le \frac{n+1}{2} \int_{\delta}^{1} (1-t)^n dt$$

$$= \frac{1}{2}(1-\delta)^{n+1}.$$

By symmetry, we also have

$$\int_{-1}^{-\delta} \varphi_n(t)dt \le \frac{1}{2}(1-\delta)^{n+1}.$$

Since $(1-\delta)^{n+1} \to 0$ as $n \to \infty$, there is a positive integer N such that, for all $n \ge N$,

$$0 \le 2M \int_{-1}^{-\delta} \varphi_n(t)dt + 2M \int_{\delta}^{1} \varphi_n(t)dt < \frac{\varepsilon}{2}.$$

With $\int_{-\delta}^{\delta} \varphi_n(t)dt \le \int_{-1}^{1} \varphi_n(t)dt \le 1$, we finally conclude that

$$|f(x) - p_n(x)| < \varepsilon \text{ for all } t \in [0,1]. \ \Box$$

We have actually shown that, for any continuous function f on $[a, b]$, the sequence of polynomials p_n converges *uniformly* to f on $[a, b]$. By analogy with the density of $\mathbb{Q}$ in $\mathbb{R}$, in the sense that any real number is the limit of a sequence of rational numbers, this amounts to the statement that the polynomials on $[a, b]$ are *dense* in the set of continuous functions on $[a, b]$.

EXERCISES 9.2

1. Find the limit of the sequence $f_n(x) = x^n/(1 + x^n)$ on $[0, 2]$, and determine whether the convergence is uniform.

2. Define a sequence $f_n : [0, 1] \to \mathbb{R}$ such that each f_n is discontinuous at each point in $[0, 1]$, and yet (f_n) converges uniformly to a continuous function on $[0, 1]$.

3. Let φ be a continuous function on $[0, 1]$, and suppose $f_n(x) = \varphi(x)x^n$, $x \in [0, 1]$. Show that (f_n) converges uniformly if and only if $\varphi(1) = 0$. Deduce from this that $nx(1 - x)^n \xrightarrow{u} 0$ on $[0, 1]$.

4. Suppose $f_n \xrightarrow{u} f$ on D, where each f_n is continuous at $c \in D$. If (x_n) is a sequence of points in D which converges to c, prove that $f_n(x_n) \to f(c)$.

5. Let $f_n : [0, 1] \to \mathbb{R}$ be defined as follows:

$$f_n(x) = \begin{cases} n^2 x, & x \in [0, 1/n] \\ -n^2(x - 2/n), & x \in (1/n, 2/n] \\ 0, & x \in (2/n, 1]. \end{cases}$$

Prove that $\lim f_n = f = 0$. By comparing $\int_0^1 f$ to $\lim \int_0^1 f_n$, show that the convergence of f_n is not uniform.

6. Let $p > 0$ and suppose

$$f_n(x) = \frac{nx}{1 + n^2 x^p}, \quad x \in [0, 1].$$

For what values of p does (f_n) converge uniformly on $[0, 1]$ to a limit f? Does $\int_0^1 f_n \to \int_0^1 f$ if $p = 2$?

7. Evaluate the limit of the sequence
$$f_n(x) = \frac{x}{nx+1}$$
on $[0,1]$. Show that the convergence is not uniform, but that $\int_0^1 f_n \to \int_0^1 f$.

8. Theorem 9.4 is not, in general, valid for improper integrals. Let f_n and g_n be defined on $[1, \infty)$ as follows:
$$f_n(x) = \begin{cases} 1/x, & x \le n \\ 0, & x > n, \end{cases}$$
$$g_n(x) = \begin{cases} 1/n, & x \le n \\ 0, & x > n. \end{cases}$$

   (a) Show that $f_n \overset{u}{\to} f$ on $[1, \infty)$ and $\int_1^\infty f_n$ exists for all n, but $\int_1^\infty f$ does not exist.

   (b) Show that $g_n \overset{u}{\to} g$ on $[1, \infty)$, $\int_1^\infty g_n$ and $\int_1^\infty g$ exist, but $\int_1^\infty g_n \not\to \int_1^\infty g$.

9. The functions f_n on $[-1, 1]$ are defined by
$$f_n(x) = \frac{x}{1+n^2x^2}.$$
Show that (f_n) converges uniformly and that its limit f is differentiable, but that the equality $f'(x) = \lim f'_n(x)$ does not hold for all $x \in [-1, 1]$.

10. Let
$$f_n(x) = (1+x^n)^{1/n}, \quad x \in [0, 2].$$
Prove that f_n is differentiable, but that it converges uniformly to a limit which is not differentiable at $x = 1$.

11. Let
$$f_n(x) = \frac{n^2 x^3}{1+n^2 x^3}, \quad x \in [0, 1].$$
Show that (f_n) does not satisfy all the conditions of Theorem 9.5, but that the derivative of the limit function is the limit of (f'_n).

12. Suppose f_n is continuous on $[0,1]$ and $f_n \xrightarrow{u} f$ on $[0,1]$. Prove that

$$\lim_{n\to\infty} \int_0^{1-1/n} f_n(x)dx = \int_0^1 f(x)dx.$$

Hint: Use Exercise 9.1.14.

13. Let $f_n \to f$ on $[a,b]$. If f_n' is continuous on $[a,b]$ for each $n \in \mathbb{N}$ and $f_n' \xrightarrow{u} g$, prove that $\int_a^x g(t)dt = f(x) - f(a)$ for all $x \in [a,b]$ and that $g(x) = f'(x)$ for all $x \in [a,b]$.

14. Prove that

$$\int_a^\pi \lim_{n\to\infty} \frac{\sin nx}{nx} dx = 0 \text{ for all } a > 0.$$

Deduce the value of

$$\lim_{n\to\infty} \int_0^\pi \frac{\sin nx}{nx} dx.$$

15. If f is a continuous function on $[0,1]$ and $\int_0^1 x^n f(x)dx = 0$ for all $n \in \mathbb{N}_0$, use the Weierstrass approximation theorem to prove that $f = 0$ on $[0,1]$.

9.3 Series of Functions

Consider a sequence (f_n) of functions defined on a set $D \subseteq \mathbb{R}$. By analogy with numerical series, we define the n-th partial sum of the sequence as the function

$$S_n(x) = \sum_{k=1}^n f_k(x), \ x \in D.$$

The sequence (S_n) is called an *infinite series* (of functions) and is denoted by $\sum f_n$. The series is said to *converge* on D if the sequence (S_n) converges pointwise on D. In this case the limit of the sequence is called the *sum* of the series and is denoted by

$$\lim_{n\to\infty} S_n(x) = \sum_{k=1}^\infty f_k(x), \ x \in D.$$

A series which does not converge at a point is said to *diverge* at that point. The series $\sum f_k$ is said to *converge absolutely* on D if the series

$\sum |f_k|$ converges on D. If the sequence (S_n) converges uniformly on D, then $\sum f_k$ is also said to *converge uniformly* on D.

Applying Theorems 9.3, 9.4, and 9.5 to the sequence (S_n), and using the linearity of the limiting process, integration, and differentiation, we obtain the following parallel theorems:

Theorem 9.7
Let $x \in \hat{D}$ and suppose $\lim_{t \to x} f_n(t)$ exists for each $n \in \mathbb{N}$. If $\sum f_n$ converges uniformly on $D \setminus \{x\}$, then

$$\lim_{t \to x} \sum_{n=1}^{\infty} f_n(t) = \sum_{n=1}^{\infty} \lim_{t \to x} f_n(t).$$

Consequently, if each f_n is continuous at x, then so is the sum $\sum_{n=1}^{\infty} f_n$.

Theorem 9.8
Suppose $f_n \in R(a,b)$ for all $n \in \mathbb{N}$. If $\sum f_n$ converges uniformly on $[a,b]$, then $\sum_{n=1}^{\infty} f_n \in R(a,b)$ and

$$\int_a^b \sum_{n=1}^{\infty} f_n(x)\,dx = \sum_{n=1}^{\infty} \int_a^b f_n(x)\,dx.$$

Theorem 9.9
Let f_n be a differentiable function on $[a,b]$ for each $n \in \mathbb{N}$, and suppose the series $\sum f_n(x_0)$ converges at some point $x_0 \in [a,b]$. If the series $\sum f_n'$ is uniformly convergent on $[a,b]$, then $\sum f_n$ is also uniformly convergent on $[a,b]$, its sum $\sum_{n=1}^{\infty} f_n$ is differentiable on $[a,b]$, and

$$\left(\sum_{n=1}^{\infty} f_n \right)'(x) = \sum_{n=1}^{\infty} f_n'(x) \quad \text{for all } x \in [a,b].$$

Applied to series, the Cauchy criterion for uniform convergence (Theorem 9.2) takes the following form:

Theorem 9.10
The series $\sum f_n$ is uniformly convergent on D if, and only if, for every $\varepsilon > 0$, there is an $N \in \mathbb{N}$ such that

$$n > m \geq N \Rightarrow |S_n(x) - S_m(x)| = \left| \sum_{k=m+1}^{n} f_k(x) \right| < \varepsilon \text{ for all } x \in D.$$

The importance of this theorem derives from the fact that it gives a sufficient and necessary condition for the uniform convergence of series. It is obvious that the convergence of the series $\sum f_n$ implies the pointwise convergence $f_n \to 0$ on D. Setting $n = m + 1$ in Theorem 9.10, we see that $f_n \xrightarrow{u} 0$ on D is a necessary condition for the convergence of $\sum f_n$ to be uniform on D. This theorem can also be used to prove a more practical test for uniform convergence, known as the *Weierstrass M-test*.

Theorem 9.11 (Weierstrass M-Test)
Let (f_n) be a sequence of functions defined on D, and suppose that there is a sequence of non-negative numbers (M_n) such that

$$|f_n(x)| \leq M_n \quad for\ all\ x \in D,\ n \in \mathbb{N}.$$

If the series $\sum M_n$ converges, then both $\sum f_n$ and $\sum |f_n|$ converge uniformly on D.

Proof. Let $\varepsilon > 0$. Since $\sum M_n$ is convergent, there is an $N \in \mathbb{N}$ such that

$$n > m \geq N \Rightarrow \sum_{k=m+1}^{n} M_k < \varepsilon.$$

But

$$\left| \sum_{k=m+1}^{n} f_k(x) \right| \leq \sum_{k=m+1}^{n} |f_k(x)| \leq \sum_{k=m+1}^{n} M_n \quad \text{for all } x \in D.$$

Therefore

$$n > m \geq N \Rightarrow \left| \sum_{k=m+1}^{n} f_k(x) \right| \leq \sum_{k=m+1}^{n} |f_k(x)| < \varepsilon \quad \text{for all } x \in D,$$

and, by Cauchy's criterion, both the series $\sum f_n$ and $\sum |f_n|$ are uniformly convergent. $\square$

Example 9.9 Discuss the uniform convergence of the series $\sum f_n$, where

$(i)\ f_n(x) = \sin\left(\dfrac{x}{n^2}\right),\ (ii)\ f_n(x) = \dfrac{1}{n^2 x^2},\ x \neq 0,\ (iii)\ f_n(x) = \dfrac{\sin(3^n x)}{2^n}.$

Solution

(i) Suppose D is a subset of $\mathbb{R}$ bounded by K, i.e., $|x| \leq K$ for all $x \in D$. Then, using Example 7.11,

$$\left|\sin\left(\frac{x}{n^2}\right)\right| \leq \frac{|x|}{n^2} \leq \frac{K}{n^2} \quad \text{for all } x \in D.$$

Taking $M_n = K/n^2$ and noting that $\sum M_n$ is convergent, we conclude that $\sum f_n$ is uniformly (and absolutely) convergent on any bounded subset of $\mathbb{R}$.

Since, however, $f_n \to 0$ pointwise on $\mathbb{R}$ and

$$\sup_{x \in \mathbb{R}} |f_n(x)| \geq \left|\sin\left(\frac{n^2 \pi/2}{n^2}\right)\right| = 1 \not\to 0,$$

we see that (f_n) does not converge uniformly to 0 on $\mathbb{R}$. Consequently, the series $\sum f_n$ does not converge uniformly on $\mathbb{R}$.

(ii) The series $\sum 1/n^2 x^2$ clearly converges pointwise on the open set $\mathbb{R}\setminus\{0\}$. Now let $r > 0$. For all $x \in \mathbb{R}$ such that $|x| > r$, we have

$$|f_n(x)| \leq \frac{1}{n^2 r^2} \quad \text{for all } n \in \mathbb{N}.$$

Since $\sum 1/n^2 r^2$ is convergent, the series $\sum f_n$ converges uniformly, by the M-test, on the closed set $\mathbb{R}\setminus(-r,r) = (-\infty, r] \cup [r, \infty)$ for all $r > 0$. But, though $f_n(x) \to 0$ pointwise on $\mathbb{R}\setminus\{0\}$,

$$\sup_{x \neq 0} |f_n(x)| \geq |f_n(1/n)| = 1 \not\to 0.$$

Hence (f_n) does not converge uniformly to 0 on $\mathbb{R}\setminus\{0\}$, and $\sum f_n$ therefore fails to converge uniformly on $\mathbb{R}\setminus\{0\}$.

(iii) The series $\sum 2^{-n} \sin(3^n x)$ is easily seen to be uniformly convergent on $\mathbb{R}$ by the M-test, using $M_n = 2^{-n}$. Since f_n is continuous on $\mathbb{R}$ for all n, its sum

$$f(x) = \sum_{n=1}^{\infty} 2^{-n} \sin(3^n x)$$

is also a continuous function on $\mathbb{R}$, by Theorem 9.7. But, surprisingly, f is *nowhere differentiable* (see [KRA]). This is an example of a function, first pointed out by Weierstrass, which is continuous everywhere but nowhere differentiable.

Lemma 9.1 (Abel's Partial Summation Formula)

Let (a_n) and (b_n) be two real sequences. If $A_n = \sum_{k=1}^{n} a_k$, then

$$\sum_{k=1}^{n} a_k b_k = A_n b_{n+1} + \sum_{k=1}^{n} A_k (b_k - b_{k+1}). \tag{9.15}$$

Proof. Set $A_0 = 0$. Then

$$a_k = A_k - A_{k-1} \quad \text{for all } k = 1, \cdots, n.$$

Therefore

$$\begin{aligned}
\sum_{k=1}^{n} a_k b_k &= \sum_{k=1}^{n} (A_k - A_{k-1}) b_k \\
&= \sum_{k=1}^{n} A_k b_k - \sum_{k=1}^{n} A_{k-1} b_k \\
&= \sum_{k=1}^{n} A_k b_k - \sum_{k=0}^{n-1} A_k b_{k+1} \\
&= \sum_{k=1}^{n} A_k b_k - \sum_{k=1}^{n} A_k b_{k+1} + A_n b_{n+1} \quad \square
\end{aligned}$$

Theorem 9.12 (Dirichlet Test for Uniform Convergence)

Let (u_n) and (v_n) be two sequences of real valued functions on a set $D \subseteq \mathbb{R}$ which satisfy the following conditions:

(i) The sequence (U_n) of partial sums of (u_n) is uniformly bounded, in the sense that there is a constant K such that

$$|U_n(x)| = \left| \sum_{k=1}^{n} u_{k(x)} \right| \leq K \quad \text{for all } x \in D, n \in \mathbb{N}.$$

(ii) The sequence (v_n) is monotonically decreasing on D; that is, $v_{n+1}(x) \leq v_n(x)$ for all $n \in \mathbb{N}$ and $x \in D$.
(iii) $v_n \xrightarrow{u} 0$ on D.
Then the series $\sum u_n v_n$ is uniformly convergent on D.

Proof. Let $S_n = \sum_{k=1}^{n} u_k v_k$. From Lemma 9.1, we have

$$S_n = U_n v_{n+1} + \sum_{k=1}^{n} U_k (v_k - v_{k+1}).$$

If $n > m$ and $x \in D$, noting that $v_k(x) - v_{k+1}(x) \geq 0$, we obtain

$$|S_n(x) - S_m(x)| \leq K |v_{n+1}(x)| + K |v_{m+1}(x)|$$

$$+ K \sum_{k=m+1}^{n} [v_k(x) - v_{k+1}(x)]$$

$$= K |v_{n+1}(x)| + K |v_{m+1}(x)| + K [v_{m+1}(x) - v_{n+1}(x)].$$

Properties (ii) and (iii) imply $v_k(x) \geq 0$ for all k, hence

$$|S_n(x) - S_m(x)| \leq 2K v_{m+1}(x).$$

Since $v_n \to 0$ uniformly on D, we see that (S_n) satisfies the Cauchy criterion for uniform convergence. □

Example 9.10 Discuss the convergence properties of the two series

$$(i) \ \sum \frac{\sin nx}{n^2}, \quad (ii) \ \sum \frac{\sin nx}{n}.$$

Solution

(i) Since $|\sin nx| \leq 1$ for all $x \in \mathbb{R}$ and $\sum \frac{1}{n^2}$ is convergent, the series $\sum \frac{\sin nx}{n^2}$ is uniformly convergent on $\mathbb{R}$ by the M-test.

(ii) If $\left|\frac{\sin nx}{n}\right| \leq M_n$ for all x in some interval I, then $M_n \geq \frac{c}{n}$, where $c = \sup_{x \in I} |\sin nx| > 0$. Since $\sum \frac{1}{n}$ is divergent, we see that the M-test is not applicable on any interval in $\mathbb{R}$. We shall use the Dirichlet test to prove uniform convergence on any interval $[a, b] \subseteq (2m\pi, (2m+1)\pi)$, $m \in \mathbb{Z}$. The proof relies on the equality

$$\sum_{k=1}^{n} \sin kx = \frac{\cos \frac{1}{2}x - \cos(n + \frac{1}{2})x}{2 \sin \frac{1}{2}x}, \quad x \neq 2m\pi, m \in \mathbb{Z}, \quad (9.16)$$

which can be proved by induction on n. Setting $u_k(x) = \sin kx$ and $U_n(x) = \sum_{k=1}^{n} u_k(x)$, we see that

$$|U_n(x)| \leq \frac{|\cos \frac{1}{2}x| + |\cos(n + \frac{1}{2})x|}{2|\sin \frac{1}{2}x|}$$

$$\leq \frac{1}{|\sin \frac{1}{2}x|}$$

$$\leq \max\left\{\frac{1}{|\sin \frac{1}{2}a|}, \frac{1}{|\sin \frac{1}{2}b|}\right\} \quad \text{for all } n \in \mathbb{N}, \ x \in [a, b].$$

With $v_n(x) = 1/n$ on $[a, b]$, we obtain a decreasing sequence which converges (uniformly) to 0. By Dirichlet's test the series $\sum \frac{\sin nx}{n}$ converges uniformly on $[a, b]$.

Using Fourier expansions (see [CAR], for example), it can be shown that $\sum \frac{\sin nx}{n}$ converges pointwise on $\mathbb{R}$ to a function which is periodic in 2π, and which is defined on $[-\pi, \pi]$ by

$$S(x) = \begin{cases} -\frac{\pi + x}{2}, & x \in [-\pi, 0) \\ 0, & x = 0 \\ \frac{\pi - x}{2}, & x \in (0, \pi]. \end{cases}$$

Note that $x = 2m\pi$, $m \in \mathbb{Z}$, are points of (jump) discontinuity for the function S. Hence the series $\sum \frac{\sin nx}{n}$ of continuous terms cannot converge uniformly in any interval containing such points.

Theorem 9.13 (Abel's Test for Uniform Convergence)
Suppose the sequences (u_n) and (v_n) of real valued functions defined on $D \subseteq \mathbb{R}$ satisfy the following conditions:
(i) $\sum u_n$ converges uniformly on D.
(ii) The sequence (v_n) is uniformly bounded and monotonically decreasing on D.
Then the series $\sum u_n v_n$ is uniformly convergent on D.

Proof. Suppose $|v_n(x)| \leq M$ for all $n \in \mathbb{N}$, $x \in D$. If $n > m$ then, applying the equality (9.15) to the sequences $\tilde{u}_k = u_{k+m}$ and $\tilde{v}_k = v_{k+m}$, we obtain

$$\sum_{k=1}^{n-m} \tilde{u}_k \tilde{v}_k = \tilde{U}_{n-m} \tilde{v}_{n-m+1} + \sum_{k=1}^{n-m} \tilde{U}_k (\tilde{v}_k - \tilde{v}_{k+1}).$$

$$\Rightarrow \sum_{k=m+1}^{n} u_k v_k = v_{n+1} \sum_{k=m+1}^{n} u_k + \sum_{k=m+1}^{n} (v_k - v_{k+1}) \sum_{j=m+1}^{k+m} u_j.$$

Now let ε be an arbitrary positive number. Relying on condition (i), we can choose N such that

$$n > m \geq N \Rightarrow \left| \sum_{j=m+1}^{n} u_j(x) \right| < \varepsilon \text{ for all } x \in D.$$

Therefore, if $n > m \geq N$, then for all $x \in D$ we have

$$\left| \sum_{k=m+1}^{n} u_k(x) v_k(x) \right| < \varepsilon |v_{n+1}(x)| + \varepsilon \sum_{k=m+1}^{n} [v_k(x) - v_{k+1}(x)]$$

$$= \varepsilon |v_{n+1}(x)| + \varepsilon v_{m+1}(x) - \varepsilon v_{n+1}(x)$$

$$\leq 3M\varepsilon.$$

By the Cauchy criterion, the series $\sum u_n v_n$ converges uniformly on D. □

Example 9.11 The series $\sum a_n x^n$ is uniformly convergent on $[0,1]$ if $\sum a_n$ is convergent. To see that, set $u_n(x) = a_n$ and $v_n(x) = x^n$ for all $n \in \mathbb{N}$, $x \in [0,1]$. With this choice of u_n and v_n the conditions of Abel's test are clearly satisfied, hence its conclusion applies. If $\sum a_n$ is absolutely convergent, the same conclusion follows from the M-test.

EXERCISES 9.3

1. Determine where the series $\sum f_n$ converges pointwise and where it converges uniformly, if $f_n(x)$ is defined as

    (a) $\dfrac{1}{x^2 + n^2}$

    (b) $\dfrac{1}{x^n + 1}$, $x \neq -1$

    (c) $\dfrac{x^n}{x^n + 1}$, $x \neq -1$

    (d) $\dfrac{(-1)^n}{n + |x|}$

    (e) $\dfrac{x}{n(1 + nx^2)}$.

2. If $\sum a_n$ is absolutely convergent, prove that $\sum a_n \cos nx$ and $\sum a_n \sin nx$ are uniformly convergent on $\mathbb{R}$.

3. Let (a_n) be a decreasing sequence of non-negative numbers. If $\sum a_n \sin nx$ is uniformly convergent on $\mathbb{R}$, prove that $na_n \to 0$.

4. Determine the domain of (i) absolute convergence and (ii) uniform convergence for the series $\sum \dfrac{1}{1 + n^2 x}$.

5. Prove that the series $\sum \dfrac{x^p}{1 + n^2 x^2}$, where $p > 0$, is uniformly convergent on $[0,1]$ if and only if $p > 1$. Hint: For $p \leq 1$ consider the sum from n to $2n$ evaluated at $x = 1/n$.

6. Prove that $\sum (-1)^n (x^2 + n^2)/n^3$ converges uniformly on any bounded interval, but that it does not converge absolutely at any $x \in \mathbb{R}$.

7. Use induction to prove the identity 9.16.

8. Discuss the convergence of $\sum \dfrac{\cos nx}{n^p}$ and $\sum \dfrac{\sin nx}{n^p}$, where $p > 0$.
   Hint: Use Dirichlet's test.

9. Let $p > 0$. Prove that $\sum \dfrac{x^n \sin nx}{n^p}$ is uniformly convergent on $[0,1]$. Hint: Use Abel's test and Exercise 9.3.8.

10. Prove that $\sum \dfrac{x^n(1-x)}{\log(n+1)}$ is uniformly convergent on $[0,1]$. If
    $$M_n = \sup_{x \in [0,1]} \dfrac{x^n(1-x)}{\log(n+1)},$$
    show that $\sum M_n$ is divergent.
    This is an example of a uniformly convergent series to which the M-test does not apply.

11. Prove that the series $\sum 1/n^x$ is uniformly convergent on $[a, \infty)$ for all $a > 1$. If $S(x) = \sum_{n=1}^{\infty} 1/n^x$, prove that
    $$S'(x) = -\sum_{n=1}^{\infty} \dfrac{\log n}{n^x} \quad \text{for all } x > 1.$$
    Use the definition $n^x = e^{x \log n}$.

12. By dividing the interval $[0, \infty)$ into the subintervals $[(n-1)\pi, n\pi]$, prove the existence of the improper integral $\int_0^{\infty} \dfrac{\sin x}{x}\, dx$. Does the integral $\int_0^{\infty} \left| \dfrac{\sin x}{x} \right| dx$ converge?

13. Prove that the series $\sum \dfrac{x^2}{n^2 + x^2}$ converges uniformly on $[-1, 1]$ to a differentiable function S. Determine $\lim_{x \to 0} \dfrac{S(x)}{x^2}$ and $S''(0)$.

9.4 Power Series

A series $\sum f_n$ is called a *power series* if the function f_n has the form
$$f_n(x) = a_n(x-c)^n, \ n \in \mathbb{N}_0,$$
where c and a_n are real numbers and $f_0(x) = a_0$. Since the translation $t = x - c$ reduces the series $\sum a_n(x-c)^n$ to $\sum a_n t^n$, there is no loss of generality, and more convenience, in taking $c = 0$. When a_n is not defined at $n = 0$, as in $\sum x^n/n$, it is tacitly assumed that the summation is over $\mathbb{N}$ rather than $\mathbb{N}_0$.

The power series $\sum a_n x^n$ obviously converges at $x = 0$ to a_0 regardless of the values of $a_1, a_2, a_3, \cdots$. For other values of x we should expect convergence to depend on these values.

Example 9.12 Consider the series

$$(i) \ \sum n! x^n, \quad (ii) \ \sum x^n, \quad (iii) \ \sum \frac{x^n}{n!}.$$

Applying the ratio test (Corollary 4.9.2) leads to the following results:

$(i) \ \dfrac{(n+1)! \, |x^{n+1}|}{n! \, |x^n|} = (n+1)\,|x| \to \infty \ \text{ for all } x \neq 0.$

$(ii) \ \dfrac{|x^{n+1}|}{|x^n|} = |x| < 1 \ \text{ for all } x \in (-1, 1).$

$(iii) \ \dfrac{|x^{n+1}| \, n!}{(n+1)! \, |x^n|} = \dfrac{|x|}{n+1} \to 0 \ \text{ for all } x \in \mathbb{R}.$

Thus the first series converges only at $x = 0$. The second series converges on $(-1, 1)$ and diverges for $|x| > 1$. Since the n-th term does not converge to 0 at $x = \pm 1$, the series diverges at these points also. The third series converges on $\mathbb{R}$. Note how the domain of convergence of $\sum a_n x^n$ grows larger as the sequence (a_n) decreases more rapidly. This is to be expected in light of the fact that convergence ultimately depends on how rapidly the product $a_n x^n$ tends to 0.

For any power series $\sum a_n x^n$, we define the extended number
$$\rho = \limsup |a_n|^{1/n}. \tag{9.17}$$
Clearly $0 \leq \rho \leq \infty$. Now we set
$$R = \frac{1}{\rho}, \ 0 < \rho < \infty, \tag{9.18}$$
and define $R = \infty$ if $\rho = 0$ and $R = 0$ when $\rho = \infty$.

Theorem 9.14 (Cauchy-Hadamard)
The series $\sum a_n x^n$ is absolutely convergent if $|x| < R$ and divergent if $|x| > R$.

Proof. Applying the root test (Theorem 4.8), we obtain
$$\limsup |a_n x^n|^{1/n} = |x| \limsup |a_n|^{1/n} = \frac{|x|}{R}.$$
Hence the series converges absolutely when $\frac{|x|}{R} < 1$ and diverges when $\frac{|x|}{R} > 1$. □

When $0 < R < \infty$, Theorem 9.14 asserts that the series $\sum a_n x^n$ converges on the open interval $(-R, R)$ and diverges on $\mathbb{R}\setminus[-R, R] = (-\infty, -R) \cup (R, \infty)$. For that reason R is called the *radius of convergence* of the series $\sum a_n x^n$. Note that the theorem says nothing about what happens when $x = \pm R$. In other words, the interval of convergence for the series is either $(-R, R)$, $[-R, R)$, $(-R, R]$, or $[-R, R]$.

According to Corollary 4.9.2, if
$$\lim_{n \to \infty} \left| \frac{a_{n+1}}{a_n} \right|$$
exists, then it equals $\limsup |a_n|^{1/n}$, in which case
$$R = \lim_{n \to \infty} \left| \frac{a_n}{a_{n+1}} \right|.$$
In many examples, this limit is easier to evaluate than the limit in (9.17).

Example 9.13 Test the following series for convergence:

(i) $\sum \frac{x^n}{n}$ (ii) $\sum \frac{x^n}{n^2}$ (iii) $\sum \left(\sin \frac{1}{n} \right) x^n$

(iv) $\sum a_n x^n$, $a_n = \begin{cases} 2^{-n} & n = 2k \\ 3^{-n} & n = 2k+1. \end{cases}$

Solution
(i) With $a_n = 1/n$, we have
$$R = \lim \left| \frac{a_n}{a_{n+1}} \right| = \lim \frac{n+1}{n} = 1.$$

At $x = 1$, we obtain the harmonic series $\sum 1/n$, which we know is divergent. $x = -1$ yields the alternating series $\sum (-1)^n/n$, which is convergent. Thus the series converges on $[-1, 1)$ and diverges on $\mathbb{R}\setminus[-1, 1)$.
(ii) In this case $a_n = 1/n^2$ and
$$R = \lim \frac{(n+1)^2}{n^2} = 1.$$
The series converges at $x = \pm 1$, hence it converges on the closed interval $[-1, 1]$ and diverges outside it.
(iii)
$$R = \lim \frac{\sin(1/n)}{\sin(1/(n+1))} = 1.$$
When $x = -1$, the alternating series $\sum (-1)^n \sin(1/n)$ results. Since the sequence $\sin(1/n)$ decreases monotonically to 0, the series converges at this point. Using the inequality $\sin x \geq 2x/\pi$ for all $x \in [0, \pi/2]$, which was proved in example 7.11, we obtain
$$\sin \frac{1}{n} \geq \frac{2}{n\pi} \quad \text{for all } n \in \mathbb{N}.$$
Since the series $\sum 2/n\pi$ diverges, so does $\sum \sin(1/n)$ by the comparison test. The interval of convergence is therefore $[-1, 1)$.
(iv) Since
$$\lim_{k \to \infty} a_{2k}^{1/2k} = \frac{1}{2}, \quad \lim_{k \to \infty} a_{2k+1}^{1/(2k+1)} = \frac{1}{3},$$
we have $\limsup a_n^{1/n} = 1/2$, hence $R = 2$. At $x = \pm 2$, the n-th term $a_n x^n$ does not converge to 0, so this series converges on $(-2, 2)$ only.

Uniform convergence for power series is characterized by the following simple result.

Theorem 9.15
Let R be the radius of convergence of the power series $\sum a_n x^n$. If $0 < r < R$, then the series converges uniformly on $[-r, r]$.

Proof. For any $x \in [-r, r]$,
$$|a_n x^n| \leq |a_n| r^n, \quad n \in \mathbb{N}_0.$$
Since $r \in (-R, R)$, the (numerical) series $\sum |a_n| r^n$ is convergent. By the M-test, with $M_n = |a_n| r^n$, the series $\sum a_n x^n$ is uniformly convergent on $[-r, r]$. □

Remarks 9.2

1. If R is the radius of convergence of the power series $\sum a_n x^n$, then the series is uniformly convergent on any compact subset of $(-R, R)$. Indeed, if $D \subseteq (-R, R)$ is compact, then $\sup D$ and $\inf D$ both exist and belong to D, hence $D \subseteq [-r, r]$, where

$$0 \le r = \max\{\sup D, -\inf D\} < R.$$

2. In general, $\sum a_n x^n$ may not converge uniformly on $(-R, R)$. Consider $\sum x^n$, whose radius of convergence is 1. Since

$$\sup_{x \in (-1,1)} |x^n| = 1 \not\to 0,$$

the n-th term does not converge uniformly to 0 on $(-1, 1)$, and hence the series cannot converge uniformly on $(-1, 1)$.

3. Since each function $x \mapsto a_n x^n$ is continuous, the sum

$$S(x) = \sum_{n=0}^{\infty} a_n x^n$$

is continuous on $[-r, r]$ for all $r \in (0, R)$ (by Theorem 9.7). Consequently S is continuous at every $x \in (-R, R)$. In view of Remark 9.2.2 we cannot assume continuity at $\pm R$, even if the series converges there. A related result in this connection, known as *Abel's continuity theorem*, states that if $\sum a_n$ is convergent, then

$$\lim_{x \to 1^-} \sum_{n=0}^{\infty} a_n x^n = \sum_{n=0}^{\infty} a_n.$$

This follows from the observation that, by Abel's test (Theorem 9.13), the series $\sum a_n x^n$ is uniformly convergent on $[0, 1]$ (see Example 9.11). Thus the sum S is continuous (from the left) at $x = 1$, and we arrive at the desired result.

Lemma 9.2 *If $b_n \to b > 0$, then, for any sequence (a_n),*

$$\limsup a_n b_n = b \limsup a_n. \qquad (9.19)$$

Proof. We may assume, without loss of generality, that $b_n > 0$ for all $n \in \mathbb{N}$. Choose a subsequence of (a_n), call it (a_{n_k}), such that $a_{n_k} \to \limsup a_n$. Since $b_{n_k} \to b$, we have

$$\lim_{k \to \infty} a_{n_k} b_{n_k} = b \limsup a_n.$$

But, by Theorem 3.16, $\limsup a_n b_n$ is the largest limit attained by a subsequence of $(a_n b_n)$, hence

$$\limsup a_n b_n \geq b \limsup a_n.$$

Replacing b_n in this inequality by $1/b_n$, and a_n by $a_n b_n$, we obtain

$$\limsup a_n = \limsup \frac{1}{b_n}(a_n b_n) \geq \frac{1}{b} \limsup a_n b_n,$$

which proves (9.19). □

Theorem 9.16
Let R be the radius of convergence of the series $\sum a_n x^n$. Then the sum $S(x) = \sum_{n=0}^{\infty} a_n x^n$ is differentiable on $(-R, R)$, and its derivative is given by the series

$$S'(x) = \sum_{n=1}^{\infty} n a_n x^{n-1}, \quad (9.20)$$

whose radius of convergence is also R.

Proof. Since $\lim n^{1/n} = 1$, we can use Lemma 9.2 to obtain

$$\limsup |n a_n|^{1/n} = \limsup |a_n|^{1/n} = 1/R.$$

By Theorem 9.15 both series $\sum a_n x^n$ and $\sum n a_n x^{n-1}$ are uniformly convergent on $[-r, r]$ for all $0 < r < R$. Since any $x \in (-R, R)$ lies in $[-r, r]$ for some $r \in (0, R)$, Equation (9.20) follows from Theorem 9.9. □

By applying this theorem a number of times, we obtain

Corollary 9.16.1 *If R is the radius of convergence of the series $\sum a_n x^n$, then its sum $S(x) = \sum_{n=0}^{\infty} a_n x^n$ is differentiable any number of times on $(-R, R)$, with its m-th derivative given by*

$$S^{(m)}(x) = \sum_{n=m}^{\infty} n(n-1)\cdots(n-m+1) a_n x^{n-m} \quad \text{for all } m \in \mathbb{N}. \quad (9.21)$$

In particular

$$S^{(m)}(0) = m! a_m. \quad (9.22)$$

Suppose a function f is represented by a power series

$$f(x) = \sum_{n=0}^{\infty} a_n x^n$$

on the interval $(-R, R)$. It follows from (9.22) that the coefficients a_n in this representation are uniquely defined by

$$a_n = f^{(n)}(0)/n!.$$

In other words, a function that is expressible on an open interval by a power series can be so represented in only one way. Equivalently, we can say

Corollary 9.16.2 *If the series $\sum a_n x^n$ and $\sum b_n x^n$ have the same sum on an interval $(-\varepsilon, \varepsilon)$, where $\varepsilon > 0$, then $a_n = b_n$ for all $n \in \mathbb{N}_0$.*

According to Corollary 9.16.1, if we have a power series $\sum a_n(x - c)^n$ about a point c, the function which represents its sum, $S(x) = \sum_{n=0}^{\infty} a_n(x - c)^n$, is differentiable any number of times on the interval $(c - R, c + R)$. This raises the question: When can a function be represented as a power series on an open interval (a, b)? Clearly the function has to be differentiable any number of times, or *infinitely differentiable*, on (a, b). But that is not enough. We have already seen in Exercise 7.4.6 that the function

$$f(x) = \begin{cases} e^{-1/x^2}, & x \neq 0 \\ 0, & x = 0 \end{cases}$$

is infinitely differentiable in a neighborhood of $x = 0$, with $f^{(k)}(0) = 0$ for all $k \in \mathbb{N}$. If $f(x)$ were representable by a power series

$$\sum_{n=0}^{\infty} a_n x^n$$

in a neighborhood of 0, then

$$a_k = \frac{f^{(k)}(0)}{k!} = 0 \text{ for all } k \in \mathbb{N},$$

which would imply that $f(x) = 0$ on that neighborhood, and this is clearly false. For a complete answer to the question raised here, we have to go back to Taylor's theorem in Chapter 7.

According to Theorem 7.17, if f is infinitely differentiable on an open interval I and $c \in I$, then we can write

$$f(x) = p_n(x) + R_n(x),$$

where p_n is the polynomial

$$p_n(x) = f(c) + f'(c)(x - c) + \cdots + \frac{f^{(n)}(c)}{n!}(x - c)^n$$

and $R_n(x)$, the remainder term, can be expressed in a variety of ways. Two forms of R_n were given in Section 7.4: The Lagrange form

$$R_n(x) = \frac{f^{(n+1)}(d)}{(n+1)!}(x-c)^{n+1},$$

where d lies between x and c, and the Cauchy form

$$R_n(x) = \frac{f^{(n+1)}(c)}{n!}(x-u)(x-c)^{n+1},$$

where u is a point in I which lies between x and c. A third form, given in Exercise 8.4.10, has the integral representation

$$R_n(x) = \frac{1}{n!}\int_c^x (x-t)^n f^{(n+1)}(t)\,dt.$$

If $\lim_{n\to\infty} R_n(x) = 0$ for all $x \in I$, then $p_n(x) \to f(x)$ on I, so that

$$f(x) = \sum_{n=0}^{\infty} \frac{f^{(n)}(c)}{n!}(x-c)^n, \quad x \in I. \tag{9.23}$$

The series on the right-hand side of equation (9.23) is called *Taylor's series* for f about c, or the *Taylor series expansion* of f about c. A function is called *analytic* on an open interval I if it can be expressed by its Taylor series at every point of I, that is, if it is infinitely differentiable on I and its remainder term $R_n(x)$ tends to 0 on I as $n \to \infty$. Here are some examples:

1. Polynomials

If f is a polynomial of degree n on $\mathbb{R}$, then $R_m(x) = 0$ for all $m > n$ and all $x \in \mathbb{R}$. Hence its Taylor expansion about $c = 0$ (for example) is

$$f(x) = p_n(x) = a_0 + a_1 x + \cdots + a_n x^n, \quad x \in \mathbb{R}.$$

2. The exponential function

We have already used the term *exponential function* to denote the function $x \mapsto \exp x = e^x$ on $\mathbb{R}$, which was defined in Section 2.4 and some of whose algebraic and analytical properties were derived in the following chapters. We need only assume that e^x is an increasing function whose derivative is e^x. This implies that its derivative to any order is also e^x. By Taylor's theorem, e^x is represented about $c = 0$ by

$$e^x = \sum_{k=0}^{n} \frac{x^k}{k!} x^k + R_n(x),$$

and
$$R_n(x) = \frac{e^d}{(n+1)!}x^{n+1},$$
where d lies between 0 and x. Since e^x is increasing, we have
$$|R_n(x)| \leq \frac{e^{|x|}}{(n+1)!}|x|^{n+1} \to 0, \text{ for all } x \in \mathbb{R},$$
as $n \to \infty$. This follows from the fact that the convergence of $\sum x^n/n!$ on $\mathbb{R}$ (Example 9.12) implies $x^n/n! \to 0$. Thus the exponential function is analytic on $\mathbb{R}$ and has the following power series representation about $c = 0$:
$$e^x = \sum_{n=0}^{\infty} \frac{x^n}{n!}, \quad x \in \mathbb{R}, \tag{9.24}$$

3. The sine and cosine functions

Using the known properties of $\sin : \mathbb{R} \to [-1, 1]$ and $\cos : \mathbb{R} \to [-1, 1]$, namely that
$$\sin' x = \cos x, \quad \cos' x = -\sin x,$$
we see that both of these functions are infinitely differentiable and satisfy $|\sin x| \leq 1$ and $|\cos x| \leq 1$. This immediately implies that $\lim_{n \to \infty} R_n(x) = 0$, and leads to the following Taylor series representation about $c = 0$ of these functions:
$$\sin x = \sum_{n=0}^{\infty} \frac{(-1)^n}{(2n+1)!} x^{2n+1}, \tag{9.25}$$
$$\cos x = \sum_{n=0}^{\infty} \frac{(-1)^n}{(2n)!} x^{2n}, \quad x \in \mathbb{R}. \tag{9.26}$$

4. The binomial function $x \mapsto \dfrac{1}{1-x}$

This function is also infinitely differentiable on $(-1, 1)$, and we saw in Example 7.20 that
$$\frac{1}{1-x} = 1 + x + x^2 + \cdots + x^n + R_n(x), \quad x \in (-1, 1),$$
where the remainder term in Cauchy's form is given by
$$R_n(x) = \frac{(n+1)x^{n+1}(1-\theta)^n}{(1-\theta x)^{n+2}}, \quad 0 < \theta < 1,$$

which tends to 0 as $n \to \infty$ for all $x \in (-1, 1)$, as was shown in Example 7.20. Hence
$$\frac{1}{1-x} = \sum_{n=0}^{\infty} x^n$$
is the Taylor expansion of this analytic function on $(-1, 1)$.

Example 9.14
$$\pi = 4 \sum_{n=0}^{\infty} \frac{(-1)^n}{2n+1}.$$

Proof. Using the Taylor series for the binomial function, we have, for all $x \in (-1, 1)$,
$$\frac{1}{1-x} = \sum_{n=0}^{\infty} x^n,$$
$$\Rightarrow \frac{1}{1+x} = \frac{1}{1-(-x)} = \sum_{n=0}^{\infty} (-1)^n x^n,$$
$$\Rightarrow \frac{1}{1+x^2} = \sum_{n=0}^{\infty} (-1)^n x^{2n}.$$

Integrating both sides of the last equation, and using Theorem 9.8, we obtain
$$\arctan u = \sum_{n=0}^{\infty} \frac{(-1)^n u^{2n+1}}{2n+1}, \quad u \in (0, 1).$$

Since the alternating series $\sum (-1)^n/(2n+1)$ converges, we may apply Abel's continuity theorem to obtain
$$\lim_{u \to 1^-} \arctan u = \sum_{n=0}^{\infty} \frac{(-1)^n}{2n+1},$$
noting that the left-hand side of this equation is $\pi/4$.

The power series expressions for the exponential and trigonometric functions, given by (9.24), (9.25), and (9.26), were obtained by employing some old definitions of these functions which have certain drawbacks. The definition of the exponential function, as given in Section 2.4 for example, is somewhat awkward to use in actual calculations, while the sine and cosine functions were defined in geometrical, rather than analytical, terms. Now we can turn things around and *redefine* these functions in terms of their power series representations. This has the advantage of placing the definitions on a firm analytical basis, which is supported by a well developed theory, that of power series. We then

Sequences and Series of Functions

have to show that all the known properties of these so-called *transcendental functions* can be derived solely from the new definitions.

Definition 9.3 The *exponential function* $\exp : \mathbb{R} \to \mathbb{R}$ is defined by

$$\exp x = \sum_{n=0}^{\infty} \frac{x^n}{n!}.$$

Since the radius of convergence of the series is infinite (Example 9.12), this function is infinitely differentiable on $\mathbb{R}$, and

$$\exp' x = \sum_{n=1}^{\infty} \frac{n x^{n-1}}{n!} = \sum_{n=0}^{\infty} \frac{x^n}{n!} = \exp x.$$

This is the fundamental property which characterizes the exponential function.

To prove the relation

$$\exp(x+t) = \exp x \cdot \exp t, \tag{9.27}$$

we define $f(x) = \exp x \exp(c - x)$ on $\mathbb{R}$ with c a fixed real number. Differentiating this function, we get

$$f'(x) = \exp x \cdot \exp(c-x) - \exp x \cdot \exp(c-x) = 0 \quad \text{for all } x \in \mathbb{R}.$$

By Theorem 7.8, f is a constant on $\mathbb{R}$, which can be evaluated by setting $x = 0$ and noting that

$$\exp 0 = 1.$$

Thus

$$f(x) = f(0) = \exp 0 \cdot \exp c = \exp c, \quad x \in \mathbb{R}.$$

Setting $t = c - x$ now yields (9.27). Setting $t = -x$ in (9.27), we get

$$\exp x \cdot \exp(-x) = \exp 0 = 1, \tag{9.28}$$

which shows that $\exp x \neq 0$ and that $\exp(-x) = 1/\exp x$ for any $x \in \mathbb{R}$. Since the exponential function is continuous, it must have the same sign on $\mathbb{R}$ (see Corollary 6.6.3), and since $\exp(0) = 1$, that sign is positive, i.e.,

$$\exp x > 0 \quad \text{for all } x \in \mathbb{R}.$$

Thus the exponential function, being its own derivative, is strictly increasing. Furthermore, since $\exp x > x$ for all $x > 0$, we have

$$\lim_{x \to \infty} \exp x = \infty, \tag{9.29}$$

and hence, in view of (9.28),
$$\lim_{x \to -\infty} \exp x = 0. \tag{9.30}$$

The exponential function being continuous, its range is necessarily an interval, and we conclude, based on (9.29) and (9.30), that it maps $\mathbb{R}$ onto $(0, \infty)$. By defining $\exp 1 = e$, we see that the constant e is given by
$$e = 1 + 1 + \frac{1}{2!} + \frac{1}{3!} + \frac{1}{4!} \cdots,$$
which can be shown to agree with its previous definition as $\lim(1 + 1/n)^n$ (see Exercise 9.4.17). Now we define
$$e^x = \exp x,$$
and this allows us to use the more common notation e^x to denote the exponential function. In Exercises 9.4.18 and 9.4.19 the reader is asked to show that this definition is consistent with the definition of the exponential function in Section 2.4.

Thus we have shown that $\exp : \mathbb{R} \to (0, \infty)$ is a differentiable bijection. It therefore has a differentiable inverse function, which is the *logarithmic function* $\log : (0, \infty) \to \mathbb{R}$, or, more specifically, *the logarithmic function to the base e*, since
$$y = \log x \Leftrightarrow e^y = x.$$

From the inverse function theorem (Theorem 7.11), we have
$$\log' x = \frac{1}{\exp(\log x)} = \frac{1}{x}, \quad x \in (0, \infty).$$

Since $\log 1 = \exp^{-1} 1 = 0$, we see that
$$\log x = \int_1^x \frac{1}{t} dt. \tag{9.31}$$

This is the definition of the logarithmic function given in many calculus books.

Since $x = \exp(\log x)$, the identity (9.27) implies
$$x^n = \exp(n \log x), \quad x > 0, \ n \in \mathbb{N}.$$

Similarly, by taking roots, we obtain
$$x^{1/m} = \exp\left(\frac{1}{m}\log x\right), \quad x > 0, \ m \in \mathbb{N}.$$
Consequently, using (9.28),
$$x^r = \exp(r \log x), \quad x > 0, \ r \in \mathbb{Q}. \tag{9.32}$$
Based on (9.32), we now define
$$x^\alpha = \exp(\alpha \log x), \quad x > 0, \ \alpha \in \mathbb{R}. \tag{9.33}$$
Differentiating (9.33), and using the chain rule, we arrive at the formula
$$\begin{aligned}(x^\alpha)' &= \exp(\alpha \log x) \cdot \frac{\alpha}{x} \\ &= \alpha[\exp(\alpha \log x) \cdot \exp(-\log x)] \\ &= \alpha[\exp((\alpha - 1)\log x] \\ &= \alpha x^{\alpha-1},\end{aligned}$$
which agrees with the differentiation rules for powers of x. Other properties of exponents can be similarly derived.

Definition 9.4 The sine and cosine functions are defined on $\mathbb{R}$ by the power series
$$\sin x = \sum_{n=0}^{\infty} \frac{(-1)^n}{(2n+1)!} x^{2n+1}, \tag{9.34}$$
$$\cos x = \sum_{n=0}^{\infty} \frac{(-1)^n}{(2n)!} x^{2n}. \tag{9.35}$$

Using the ratio test, it is straightforward to show that the series in (9.34) and (9.35) converge on $\mathbb{R}$, hence the sine and cosine functions are infinitely differentiable on $\mathbb{R}$. Their first derivatives are given, according to Theorem 9.16, by
$$\sin' x = \cos x, \quad \cos' x = -\sin x.$$
Using Definition 9.4, we immediately see that
$$\sin(-x) = -\sin x, \quad \cos(-x) = \cos x, \quad x \in \mathbb{R},$$
$$\sin 0 = 0, \ \cos 0 = 1, \ \lim_{x \to 0} \frac{\sin x}{x} = 1.$$

The other familiar properties of these functions, such as

$$\sin^2 x + \cos^2 x = 1,$$
$$\sin(x+y) = \sin x \cos y + \cos x \sin y,$$
$$\cos(x+y) = \cos x \cos y - \sin x \sin y,$$

can also be obtained in ways not too different from those used in the case of the exponential function.

EXERCISES 9.4

1. Determine the radius of convergence of the series $\sum a_n x^n$, where a_n is given by

   (a) $\dfrac{1}{n^n}$,

   (b) $\dfrac{n^n}{n!}$,

   (c) $\dfrac{(n!)^2}{(2n)!}$,

   (d) $n^{-\sqrt{n}}$,

   (e) 1 if $n = k^2$ and 0 if $n \neq k^2$, $k \in \mathbb{N}$.

2. Given that $0 < b \leq |a_n| < c$, find the radius of convergence of $\sum a_n x^n$.

3. Give an example of a power series which converges on $(-1, 1]$ and diverges at -1.

4. If the function $f(x) = \sum_{n=0}^{\infty} a_n x^n$ is even on $(-R, R)$, show that $a_n = 0$ for all odd n, and if f is odd then $a_n = 0$ for all even n.

5. For what values of c is each of the following series uniformly convergent on $[-c, c]$?

   (a) $\sum \dfrac{x^n}{n^{1/n}}$,

   (b) $\sum n x^n$,

   (c) $\sum \dfrac{2^n}{\sqrt{n!}} x^n$,

(d) $\sum \dfrac{(-1)^n x^n}{5^n(n+1)}$,

(e) $\sum \dfrac{x^2}{\sqrt{n}(1+nx^2)}$,

(f) $\sum x^n(1-x)$.

6. If $0 < R < \infty$ is the radius of convergence of $\sum a_n x^n$ and k is a fixed positive integer, find the radius of convergence of

   (a) $\sum a_n^k x^n$,
   (b) $\sum a_n x^{kn}$,
   (c) $\sum a_n x^{n^k}$.

7. Suppose $\sum a_n x^n$ and $\sum b_n x^n$ have radii of convergence R_1 and R_2, respectively.

   (a) If $R_1 \neq R_2$, prove that $\sum(a_n + b_n)x^n$ has radius of convergence $\min\{R_1, R_2\}$. What can happen if $R_1 = R_2$?
   (b) Show that the radius of convergence of $\sum a_n b_n x^n$ is not less than $R_1 R_2$.

8. (a) If $\sum a_n$ converges, prove that $\sum \dfrac{a_n}{n+1}$ converges and that
   $$\int_0^1 \sum_{n=0}^{\infty} a_n x^n \, dx = \sum_{n=0}^{\infty} \dfrac{a_n}{n+1}.$$

   (b) If $\sum \dfrac{a_n}{n+1}$ converges, show that the above equation still holds, even though the integral may be improper.

9. Prove that the function $f(x) = (1+x)^\alpha$ is analytic on $(-1, 1)$ for all $\alpha \in \mathbb{R}$, and that
   $$(1+x)^\alpha = \sum_{n=0}^{\infty} \binom{\alpha}{n} x^n,$$
   where $\binom{\alpha}{0} = 1$, $\binom{\alpha}{n} = \dfrac{\alpha(\alpha-1)(\alpha-2)\cdots(\alpha-n+1)}{n!}$, $n \in \mathbb{N}$.
   Use this to prove that x^α is analytic on $(0, \infty)$ by writing $x^\alpha = c^\alpha[1 + (x/c - 1)]^\alpha$.

10. Use the Taylor expansion of $\dfrac{1}{1-x}$ to prove that

$$\log(1-x) = -\sum_{n=1}^{\infty} \frac{x^n}{n}, \quad x \in (-1,1).$$

Deduce that

$$\log 2 = \sum_{n=1}^{\infty} \frac{1}{n 2^n}.$$

11. Use the Taylor expansion of $\dfrac{1}{1-x}$ about $x = 0$ to prove that

$$\frac{1}{(1-x)^2} = \sum_{n=1}^{\infty} n x^{n-1}, \quad |x| < 1.$$

12. Find the Taylor expansions of $\dfrac{1}{2-x}$ about $x = 0$ and $x = 1$. What is the radius of convergence in each case?

13. Determine the Taylor expansion of the function $\dfrac{1}{x(2-x)}$ about $x = 1$ and its radius of convergence.

14. Find the power series representing the so-called *error function*

$$f(x) = \int_0^x e^{-t^2}\, dt, \quad x \in \mathbb{R}.$$

Evaluate $\lim_{x \to \infty} f(x)$.

15. Let R be the radius of convergence of $\sum a_n x^n$.

   (a) If $f(x) = \sum_{n=0}^{\infty} a_n x^n = 0$ for all $x \in (-R, R)$, prove that $a_n = 0$ for all $n \in \mathbb{N}_0$.

   (b) If there is a sequence of distinct points (c_k) in $(-R, R)$ such that $c_k \to c \in (-R, R)$, and $f(c_k) = 0$ for all $k \in \mathbb{N}$, prove that $a_n = 0$ for all $n \in \mathbb{N}_0$.

16. Suppose $f : \mathbb{R} \to \mathbb{R}$ satisfies $f(x+y) = f(x)f(y)$ for all $x, y \in \mathbb{R}$. If f is continuous, prove that $f(x) = e^{cx}$ for some constant c.

17. Prove that

$$e = \lim_{n \to \infty} \left(1 + \frac{1}{n}\right)^n = \sum_{n=0}^{\infty} \frac{1}{n!}.$$

Hint: Recall from Example 3.12 that the sequence $x_n = (1+n^{-1})^n$ is monotonically increasing and bounded above by $\sum 1/n!$.

18. Prove that, for any rational number r, $e^r = \exp r$.

19. Using the definition of a^x for any real number x (see Section 2.4), prove that $e^x = \exp x$.

10

Lebesgue Measure

In Chapter 11 we shall present a theory of integration more general than the Riemann theory of Chapter 8, called *Lebesgue integration*, which is due to the French mathematician *Henri Lebesgue* (1875-1941). From among the standard approaches to this theory, we shall follow the path based on first extending the notion of length of a real interval to define the so-called *Lebesgue measure* of certain subsets of $\mathbb{R}$, which forms the subject of this chapter, then using this to define the *Lebesgue integral* in Chapter 11. This approach has the advantage of giving prominence to *measure* as a fundamental mathematical concept in its own right with wide applications in probability and stochastic theory, besides providing the cornerstone for building a more complete integration theory, in a sense which will be clarified in the next chapter.

So far we have followed the usual convention of denoting sets, such as intervals, by capital letters, like A and B, and their elements by small letters, such as a and b. In this chapter we shall deal extensively with sets composed of subsets of $\mathbb{R}$, and these will usually be denoted by calligraphic letters, such as $\mathcal{A}$ and $\mathcal{B}$. If $E \subseteq \mathbb{R}$, we have already used the symbol $\mathcal{P}(E)$ to denote the set of subsets of E, or the *power set* of E (see Exercise 1.1.7). Thus

$$\mathcal{P}(\varnothing) = \{\varnothing\},$$
$$\mathcal{P}(\{1\}) = \{\varnothing, \{1\}\},$$
$$\mathcal{P}(\{1,2\}) = \{\varnothing, \{1\}, \{2\}, \{1,2\}\}.$$

We shall also have occasion to deal with the extended real numbers $\bar{\mathbb{R}} = \mathbb{R} \cup \{-\infty, \infty\}$, where the operations of addition and multiplication are extended according to the rules

$$a \pm \infty = \pm\infty \quad \text{for all } a \in \mathbb{R},$$
$$a \cdot \infty = \infty \quad \text{for all } a \in (0, \infty),$$
$$a \cdot \infty = -\infty \quad \text{for all } a \in (-\infty, 0),$$
$$\infty + \infty = \infty,$$
$$\infty \cdot (\pm\infty) = \pm\infty.$$

The form $\infty - \infty$ is indeterminate, but we shall set $0 \cdot \infty = 0$.

As in all significant concepts in mathematics, the notion of *measure* is based on capturing the essence of a simple intuitive idea and extending it by a mathematical procedure to a more general setting. The intuitive idea in our case is the *length* of a real interval I, denoted by $l(I)$, which is the (non-negative) difference between its end-points. In order to extend this idea to *any* subset of $\mathbb{R}$, we should like to define a non-negative function, or *measure*, m on $\mathcal{P}(\mathbb{R})$ which satisfies the following conditions:

1. The function m is defined on $\mathcal{P}(\mathbb{R})$.

2. For any interval I in $\mathbb{R}$, $m(I) = l(I)$.

3. If E and F are disjoint subsets of $\mathcal{P}(\mathbb{R})$, then $m(E \cup F) = m(E) + m(F)$. More generally, if (E_n) is a sequence of pairwise disjoint elements of $\mathcal{P}(\mathbb{R})$, in the sense that $E_i \cap E_j = \emptyset$ for all $i \neq j$, then we would like to have

$$m\left(\bigcup_{i=1}^{\infty} E_i\right) = \sum_{i=1}^{\infty} m(E_i).$$

This property is known as *countable additivity*.

4. m is invariant under translation, in the sense that

$$m(E + x) = m(E) \quad \text{for all } x \in \mathbb{R}, E \in \mathcal{P}(\mathbb{R}),$$

where $E + x = \{y + x : y \in E\}$.

But, as it turns out, there is no function which satisfies all four conditions (see [HAL], for example), so we have to drop some of them. Since the last three conditions are basic for any definition of measure, as they capture the essence of the notion of length, we shall drop the first condition, and thereby accept the existence of subsets of $\mathbb{R}$ on which m is not defined, that is, which are not measurable. But if we cannot define m on all of $\mathcal{P}(\mathbb{R})$, we should define it on as large a subset of $\mathcal{P}(\mathbb{R})$ as possible while preserving properties 2,3 and 4. In order to avoid cumbersome additional conditions for applying these properties, such a subset, or *class,* should naturally be closed under the formation of unions and complementation. We shall therefore start by considering such classes of $\mathcal{P}(\mathbb{R})$.

10.1 Classes of Subsets of $\mathbb{R}$

In all that follows, Ω will denote a non-empty subset of $\mathbb{R}$.

Definition 10.1 A non-empty class $\mathcal{R}$ of subsets of Ω is called a *Boolean ring*, or simply a *ring*, on Ω if

$$A, B \in \mathcal{R} \Rightarrow A \cup B, A \backslash B \in \mathcal{R}.$$

A ring, in other words, is characterized by its closure under the operations of forming finite unions and differences.

Example 10.1 Let Ω be a non-empty subset of $\mathbb{R}$, and suppose that $A, B \subseteq \Omega$. Each of the following classes is a ring on Ω:

(i) $\mathcal{P}(\Omega)$, the power set of Ω.

(ii) $\{\varnothing, A\}$.

(iii) $\{\varnothing, A, A^c, \Omega\}$.

(iv) $\{\varnothing, A, B, A \cap B, A \cup B, A \backslash B, B \backslash A, A \Delta B\}$. See Exercise 1.1.5 for the definition of $A \Delta B$.

(v) The class of all finite subsets of Ω.

(vi) The class of all finite subsets of Ω and their complements in Ω.

(vii) The class of all countable subsets of Ω.

(viii) The class of all countable subsets of Ω and their complements in Ω.

Definition 10.2 A class $\mathcal{D} \subseteq \mathcal{P}(\mathbb{R})$ is said to be *pairwise disjoint*, or simply *disjoint*, if

$$A, B \in \mathcal{D}, A \neq B \Rightarrow A \cap B = \varnothing.$$

The union $\cup \{A : A \in \mathcal{D}\}$ is then described as a *disjoint union*.

The symbol $\mathcal{I}$ will be used to denote the class of semi-open intervals $[a, b)$ where $a, b \in \mathbb{R}$ and $a < b$. If $a \geq b$ then $[a, b)$ should be read as the empty set $\varnothing$. The class of all finite, disjoint unions of members of of $\mathcal{I}$ will be denoted by $\mathcal{E}$.

Lemma 10.1 For all $I_1, I_2 \in \mathcal{I}$, we have

$$(i)\ I_1 \cap I_2 \in \mathcal{I}, \quad (ii)\ I_1 \backslash I_2 \in \mathcal{E}.$$

Proof. Let $I_k = [a_k, b_k)$, $k = 1, 2$. Using the notation $\alpha \wedge \beta = \min\{\alpha, \beta\}$, $\alpha \vee \beta = \max\{\alpha, \beta\}$ for any $\alpha, \beta \in \mathbb{R}$, we have

$$I_1 \cap I_2 = [a_1 \vee a_2, b_1 \wedge b_2) \in \mathcal{I},$$
$$I_1 \backslash I_2 = [a_1, b_1 \wedge a_2) \cup [a_1 \vee b_2, b_1) \in \mathcal{E}. \quad \square$$

We are now in a position to present a more substantial example of a ring.

Theorem 10.1
$\mathcal{E}$ *is a ring on* $\mathbb{R}$.

Proof. $\mathcal{E}$ is by definition closed under disjoint finite unions. We begin by verifying its closure under finite intersections and differences. Let

$$E = \bigcup_{i=1}^{n} I_i, \quad F = \bigcup_{j=1}^{m} J_j,$$

where $I_i, J_j \in \mathcal{I}$, and each union is disjoint. Then

$$E \cap F = \bigcup_{i=1}^{n} \bigcup_{j=1}^{m} I_i \cap J_j.$$

Since each $I_i \cap J_j \in \mathcal{I}$, by Lemma 10.1, and the class $\{I_i \cap J_j : 1 \leq i \leq n, 1 \leq j \leq m\}$ is disjoint, we see that $E \cap F \in \mathcal{E}$.

Furthermore,

$$E \backslash F = \bigcap_{j=1}^{m} \bigcup_{i=1}^{n} I_i \backslash J_j.$$

Here we have $I_i \backslash J_j \in \mathcal{E}$, again by Lemma 10.1, and the class $\{I_i \backslash J_j \in \mathcal{E} : 1 \leq i \leq n\}$ is disjoint for each j. Hence $\bigcup_{i=1}^{n} I_i \backslash J_j \in \mathcal{E}$ for all $1 \leq j \leq m$. Since $\mathcal{E}$ has been shown to be closed under the formation of finite intersections, we conclude that $E \backslash F \in \mathcal{E}$.

To prove that $\mathcal{E}$ is a ring, it remains to show that $E \cup F \in \mathcal{E}$. But since $E \cup F = (E \backslash F) \cup F$ is a disjoint union of elements of $\mathcal{E}$, $E \cup F \in \mathcal{E}$. $\quad \square$

Corollary 10.1 $\mathcal{E}$ *is the class of all finite unions of elements of* $\mathcal{I}$.

Remarks 10.1

1. Since a ring is non-empty and closed under differences, it necessarily includes the empty set $\varnothing$.

2. A ring is closed under the binary operations $\cap$ and $\triangle$. This is seen from the equalities

$$E \cap F = E \backslash (E \backslash F),$$
$$E \triangle F = (E \backslash F) \cup (F \backslash E).$$

The converse is also true. Suppose $\mathcal{R}$ is a non-empty class of subsets of Ω which is closed under $\cap$ and Δ. If $E, F \in \mathcal{R}$, then

$$E \cup F = (E \Delta F) \Delta (F \cap E) \in \mathcal{R},$$
$$E \backslash F = E \cap (E \Delta F) \in \mathcal{R},$$

which means $\mathcal{R}$ is a ring.

Definition 10.3 A class $\mathcal{A}$ of subsets of Ω is called an *algebra* on Ω if it is a ring and $\Omega \in \mathcal{A}$.

It follows immediately that a non-empty class $\mathcal{A}$ is an algebra if and only if

$$E, F \in \mathcal{A} \Rightarrow E \cup F \in \mathcal{A},$$
$$E \in \mathcal{A} \Rightarrow E^c = \Omega \backslash E \in \mathcal{A}.$$

We leave the details of the proof as an exercise.

Remarks 10.2

1. Amongst the rings described in Example 10.1, (i), (iii), (vi) and (viii) are algebras; (ii) is an algebra if and only if $A = \Omega$; (v) is an algebra if and only if Ω is finite; (vii) is an algebra if and only if Ω is countable.

2. Since $\mathbb{R}$ cannot be expressed as a finite union of bounded intervals, $\mathcal{E}$ is not an algebra on $\mathbb{R}$.

Definition 10.4

(i) A non-empty class $\mathcal{R}$ of subsets of Ω is called a σ-*ring* on Ω if it is closed under the formation of differences and countable unions, i.e.,

$$E, F \in \mathcal{R} \Rightarrow E \backslash F \in \mathcal{R},$$
$$E_i \in \mathcal{R} \Rightarrow \bigcup_{i=1}^{\infty} E_i \in \mathcal{R}.$$

(ii) A σ-ring on Ω that contains Ω is called a σ-*algebra* on Ω.

Equivalently, a σ-algebra is a non-empty class which is closed under complementation in Ω and the formation of countable unions.

Remarks 10.3

1. Every σ-ring (σ-algebra) is a ring (algebra). Every finite ring (algebra) is a σ-ring (σ-algebra).

2. In Example 10.1, the rings (i), (iii), and (viii) are σ-algebras, while (ii), (iv) and (vii) are σ-rings that may not be σ-algebras.

3. $\mathcal{E}$ is not a σ-ring.

4. A σ-ring is closed under the formation of countable intersections. This follows from the observation that, if $E_i \in \mathcal{R}$ for all $i \in \mathbb{N}$, where $\mathcal{R}$ is a σ-ring, then $E = \bigcup_{i=1}^{\infty} E_i \in \mathcal{R}$, and therefore

$$\bigcap_{i=1}^{\infty} E_i = E \setminus \bigcup_{i=1}^{\infty} (E \setminus E_i) \in \mathcal{R}.$$

The next theorem is basic for the construction of σ-algebras.

Theorem 10.2
Let $\mathcal{D}$ be any class of subsets of Ω. Then there is a smallest σ-algebra $\mathcal{A}(\mathcal{D})$ on Ω that contains $\mathcal{D}$.

Proof. Let $\mathcal{S}$ be the family of all σ-algebras containing $\mathcal{D}$. $\mathcal{S}$ is not empty as $\mathcal{P}(\Omega) \in \mathcal{S}$, so we can define $\mathcal{A}(\mathcal{D})$ to be the (non-empty) intersection of all the σ-algebras in $\mathcal{S}$. Since each σ-algebra in $\mathcal{S}$ contains Ω and is closed under the formation of countable unions and complementation, the same is true of $\mathcal{A}(\mathcal{D})$. Thus $\mathcal{A}(\mathcal{D})$ is a σ-algebra on Ω and is clearly the smallest such σ-algebra, in the sense that any σ-algebra on Ω that contains $\mathcal{D}$ also contains $\mathcal{A}(\mathcal{D})$. □

$\mathcal{A}(\mathcal{D})$ is called the σ-algebra *generated* by $\mathcal{D}$. Note, in this context, that the term *class* and *family* are used as synonyms of *set*. A simple modification of the proof yields a smallest ring, a smallest algebra, or a smallest σ-ring containing $\mathcal{D}$, or *generated* by $\mathcal{D}$.

Example 10.2
(i) Let $E \subseteq \Omega$ and $\mathcal{D} = \{E\}$. Then $\mathcal{A}(\mathcal{D}) = \{E, E^c, \Omega, \varnothing\}$.
(ii) Let $E, F \subseteq \Omega$ and $\mathcal{D} = \{E, F\}$. Then the σ-ring generated by $\mathcal{D}$ is $\{E, F, E \cup F, E \cap F, E \setminus F, F \setminus E, E \Delta F, \varnothing\}$.
(iii) Let Ω be a non-countable subset of $\mathbb{R}$, and suppose $\mathcal{D}$ is the class of all *singleton sets*. A singleton set is a set which contains exactly one element. Clearly $\mathcal{A}(\mathcal{D})$ contains all countable sets in Ω and their complements. Since the class of all such sets is a σ-algebra, it coincides with $\mathcal{A}(\mathcal{D})$.

While the ring generated by $\mathcal{I}$ is $\mathcal{E}$, the σ-algebra generated by $\mathcal{I}$ is important enough to warrant a formal definition, as it plays a central role in this study.

Definition 10.5 The σ-algebra generated by $\mathcal{I}$ is called the *Borel σ-algebra* on $\mathbb{R}$. It is denoted by $\mathcal{B}$ and its elements are called *Borel sets*.

The next theorem shows that $\mathcal{B}$ can be generated by intervals other than the semi-open type $[a, b)$.

Theorem 10.3
Let
$$\mathcal{I}_0 = \{(a, b) : a, b \in \mathbb{R}\},$$
$$\mathcal{I}_1 = \{[a, b] : a, b \in \mathbb{R}\},$$
$$\mathcal{I}_2 = \{(a, \infty) : a \in \mathbb{R}\},$$
$$\mathcal{I}_3 = \{[a, \infty) : a \in \mathbb{R}\}.$$

Then $\mathcal{B} = \mathcal{A}(\mathcal{I}_0) = \mathcal{A}(\mathcal{I}_1) = \mathcal{A}(\mathcal{I}_2) = \mathcal{A}(\mathcal{I}_3)$.

Proof. It suffices to prove that
$$\mathcal{B} \supseteq \mathcal{A}(\mathcal{I}_0) \supseteq \mathcal{A}(\mathcal{I}_1) \supseteq \mathcal{A}(\mathcal{I}_2) \supseteq \mathcal{A}(\mathcal{I}_3) \supseteq \mathcal{B}.$$

For the first inclusion, note that
$$(a, b) = \bigcup_{n=1}^{\infty} [a + 1/n, b),$$
which proves that $\mathcal{B} \supseteq \mathcal{I}_0$, and hence $\mathcal{B} \supseteq \mathcal{A}(\mathcal{I}_0)$. The second inclusion follows from the observation that
$$[a, b] = \bigcap_{n=1}^{\infty} (a - 1/n, b + 1/n)$$
$$\Rightarrow \mathcal{A}(\mathcal{I}_0) \supseteq \mathcal{I}_1$$
$$\Rightarrow \mathcal{A}(\mathcal{I}_0) \supseteq \mathcal{A}(\mathcal{I}_1).$$

Similarly, the equalities
$$(a, \infty) = \bigcup_{n=1}^{\infty} [a + 1/n, n]$$
$$[a, \infty) = \bigcap_{n=1}^{\infty} (a - 1/n, \infty)$$
$$[a, b) = [a, \infty) \setminus [b, \infty)$$

lead to the remaining inclusions. □

Corollary 10.3 $\mathcal{B}$ *is generated by the class of open subsets of* $\mathbb{R}$, *and also by the class of closed subsets of* $\mathbb{R}$.

Proof

(i) Let $\mathcal{G}$ denote the class of open sets in $\mathbb{R}$. Since $\mathcal{G} \supseteq \mathcal{I}_0$, it follows that $\mathcal{A}(\mathcal{G}) \supseteq \mathcal{A}(\mathcal{I}_0) = \mathcal{B}$. On the other hand, every open subset of $\mathbb{R}$ is a countable union of open intervals (Theorem 3.18), hence $\mathcal{G} \subseteq \mathcal{B}$, and therefore $\mathcal{A}(\mathcal{G}) \subseteq \mathcal{B}$.

(ii) Since a σ-algebra is closed under complementation and countable intersections, $\mathcal{B}$ is also generated by the class of all closed subsets of $\mathbb{R}$. □

Corollary 10.3 indicates that the σ-algebra of Borel sets is large enough to include all the "interesting" subsets of $\mathbb{R}$ from our point of view, such as the open sets, closed sets, compact sets, intervals, and countable sets. In fact, this corollary allows us to extend the notion of Borel σ-algebras to abstract spaces: If X is any set and τ is a topology on X, we define the Borel σ-algebra of (X, τ) to be $\mathcal{A}(\tau)$.

We conclude this section with a useful result which we shall have occasion to fall back on before the end of this chapter: If $f : \Omega \to \Gamma$ is any function and $\mathcal{D}$ is a class of subsets of Γ, we define

$$f^{-1}(\mathcal{D}) = \{f^{-1}(D) : D \in \mathcal{D}\},$$

where, as usual,

$$f^{-1}(D) = \{x \in \Omega : f(x) \in D\}.$$

Theorem 10.4

Let $f : \Omega \to \Gamma$ and suppose $\mathcal{D}$ is a class of subsets of Γ. Then

$$\mathcal{A}(f^{-1}(\mathcal{D})) = f^{-1}(\mathcal{A}(\mathcal{D})).$$

Proof. As

$$f^{-1}(\cup_{i=1}^{\infty} E_i) = \bigcup_{i=1}^{\infty} f^{-1}(E_i), \quad f^{-1}(E^c) = [f^{-1}(E)]^c,$$

by (1.21) and (1.23), $f^{-1}(\mathcal{A}(\mathcal{D}))$ is easily seen to be a σ-algebra; and since $f^{-1}(\mathcal{A}(\mathcal{D})) \supseteq f^{-1}(\mathcal{D})$, we conclude that

$$f^{-1}(\mathcal{A}(\mathcal{D})) \supseteq \mathcal{A}(f^{-1}(\mathcal{D})).$$

To prove the inclusion in the other direction, let

$$\mathcal{H} = \{E \in \mathcal{A}(\mathcal{D}) : f^{-1}(E) \in \mathcal{A}(f^{-1}(\mathcal{D}))\}.$$

Clearly, $\mathcal{H} \supseteq \mathcal{D}$. Furthermore, $\mathcal{H}$ is a σ-algebra; for suppose $E \in \mathcal{H}$, then $f^{-1}(E) \in \mathcal{A}(f^{-1}(\mathcal{D}))$ and therefore

$$f^{-1}(E^c) = [f^{-1}(E)]^c \in \mathcal{A}(f^{-1}(\mathcal{D})).$$

Hence $E^c \in \mathcal{H}$. Also, if $E_i \in \mathcal{H}$ for all $i \in \mathbb{N}$, then

$$f^{-1}\left(\bigcup_{i=1}^{\infty} E_i\right) = \bigcup_{i=1}^{\infty} f^{-1}(E_i) \in \mathcal{A}(f^{-1}(\mathcal{D})),$$

which implies $\bigcup_{i=1}^{\infty} E_i \in \mathcal{H}$. Being a σ-algebra, $\mathcal{H}$ necessarily includes $\mathcal{A}(\mathcal{D})$. Therefore, if $F \in f^{-1}(\mathcal{A}(\mathcal{D}))$ then $F \in f^{-1}(\mathcal{H}) = \mathcal{A}(f^{-1}(\mathcal{D}))$. Hence

$$f^{-1}(\mathcal{A}(\mathcal{D})) \subseteq \mathcal{A}(f^{-1}(\mathcal{D})). \quad \square$$

EXERCISES 10.1

1. Prove that a non-empty class of subsets is an algebra if, and only if, it is closed under complementation and the formation of finite unions.

2. (a) When is Example 10.1(iv) an algebra?
   (b) When is Example 10.1(v) a σ-ring?
   (c) When is Example 10.1(vii) a σ-algebra?

3. Let $E, F \subseteq \Omega$. Determine the algebra generated by $\{E, F\}$.

4. Let $\Omega = [0,1] \cap \mathbb{Q}$, and suppose $\mathcal{D}$ consists of all subsets of Ω of the form $[a, b) \cap \mathbb{Q}$. Show that $\mathcal{A}(\mathcal{D}) = \mathcal{P}(\Omega)$.

5. Suppose $\mathcal{D}$ is a class of subsets of $\mathbb{R}$. Let $\mathcal{R}(\mathcal{D})$ be the ring generated by $\mathcal{D}$. Show that each of the elements of $\mathcal{R}(\mathcal{D})$ can be covered by a finite union of elements of $\mathcal{D}$. State a parallel result for the σ-ring generated by $\mathcal{D}$.

6. Let $\mathcal{R}$ be a Boolean ring of subsets of Ω. If the binary operations $+$ and $\cdot$ are defined on $\mathcal{R}$ by $A + B = A \Delta B$ and $A \cdot B = A \cap B$, show that $(\mathcal{R}, +, \cdot)$ is an algebraic ring.

7. Make the necessary modifications to Theorem 10.2 to verify the existence of a smallest ring, algebra, and σ-ring which contains $\mathcal{D}$.

8. If $\mathcal{S}$ is any σ-ring on Ω, show that $\mathcal{A}(\mathcal{S}) = \mathcal{S} \cup \{E^c : E \in \mathcal{S}\}$.

9. Give an example of a class of subsets of $\mathbb{R}$ that is closed under Δ and the formation of countable intersections, but is not a σ-ring.

10. Show that $\mathbb{N}, \mathbb{Z}, \mathbb{Q}, \mathbb{Q}^c$ are all Borel sets.

11. If $\Gamma \subseteq \Omega$ and $\mathcal{D} \subseteq \mathcal{P}(\Omega)$, we write $\Gamma \cap \mathcal{D}$ to mean $\{\Gamma \cap D : D \in \mathcal{D}\}$. Show that $\mathcal{A}(\Gamma \cap \mathcal{D}) = \Gamma \cap \mathcal{A}(\mathcal{D})$. Hint: Consider the function $i : \Gamma \to \Omega$, $i(x) = x$.

10.2 Lebesgue Outer Measure

In order to extend the concept of "length" from intervals in $\mathcal{I}$ to all of $\mathcal{P}(\mathbb{R})$, we shall define the *Lebesgue outer measure* of any subset of $\mathbb{R}$. This is achieved by first making the extension to $\mathcal{E}$. The remainder of this section is then devoted to examining the basic properties of outer measure.

Definition 10.6 For any real, bounded interval I with end-points a and b, where $a \leq b$, we define the *length* of I as

$$l(I) = b - a.$$

If I is unbounded, we define $l(I) = \infty$.

Lemma 10.2 *The function $l : \mathcal{I} \to [0, \infty)$ satisfies the following properties:*
(i) $l(\varnothing) = 0$.
(ii) *Monotonicity, i.e.,*

$$I, J \in \mathcal{I}, I \subseteq J \Rightarrow l(I) \leq l(J).$$

(iii) *Finite additivity, in the sense that, if $\{I_1, I_2, \cdots, I_n\}$ is a finite set of disjoint intervals in $\mathcal{I}$ whose union lies in $\mathcal{I}$, then*

$$l\left(\bigcup_{i=1}^n I_i\right) = \sum_{i=1}^n l(I_i).$$

Proof. Properties (i) and (ii) easily follow from Definition 10.6, so we shall only prove finite additivity. Let $I_i = [a_i, b_i)$ and $\bigcup_{i=1}^n I_i = [a, b)$.

By rearranging the intervals, if necessary, we can assume that $b_1 < b_2 < \cdots < b_n$. Since the set $\{I_1, I_2, \cdots I_n\}$ is disjoint, we have

$$a_1 < b_1 \leq a_2 < b_2 \leq \cdots \leq a_n < b_n.$$

Now the equality $\bigcup_{i=1}^{n} I_i = [a, b)$ implies

$$a_1 = a,$$
$$b_i = a_{i+1}, \ 1 \leq i \leq n-1,$$
$$b_n = b.$$

Hence

$$\sum_{i=1}^{n} l(I_i) = (b_1 - a_1) + (b_2 - a_2) + \cdots + (b_n - a_n)$$
$$= b_n - a_1$$
$$= b - a$$
$$= l\left(\bigcup_{i=1}^{n} I_i\right). \quad \square$$

Definition 10.7 If $E \in \mathcal{E}$ is represented by a finite union $\bigcup_{i=1}^{n} I_i$ of disjoint intervals $I_i \in \mathcal{I}$, the *length* of E is defined as

$$l(E) = \sum_{i=1}^{n} l(I_i).$$

For this definition to make sense, we have to show that $l(E)$ does not depend on how E is represented as a union of disjoint intervals in $\mathcal{I}$. To that end, let $E = \bigcup_{j=1}^{m} J_j$ be any other representation, where $J_1, \cdots, J_m$ are disjoint intervals in $\mathcal{I}$, and we have to show that

$$\sum_{j=1}^{m} l(J_j) = \sum_{i=1}^{n} l(I_i).$$

Observe that, for all $i \in \{1, \cdots, n\}$,

$$I_i = I_i \cap E = \bigcup_{j=1}^{m} (I_i \cap J_j).$$

Since $\{I_i \cap J_j : 1 \leq j \leq m\}$ is a disjoint set of intervals, we can use Lemma 10.2 to write

$$\sum_{i=1}^{n} l(I_i) = \sum_{i=1}^{n} \sum_{j=1}^{m} l(I_i \cap J_j).$$

Similarly, by writing $J_j = \bigcup_{i=1}^{n} J_j \cap I_i$, we obtain

$$\sum_{j=1}^{m} l(J_j) = \sum_{j=1}^{m} \sum_{i=1}^{n} l(J_j \cap I_i) = \sum_{i=1}^{n} l(I_i). \quad \square$$

It is not difficult to verify that the function $l : \mathcal{E} \to [0, \infty)$ satisfies the three properties of $l|_{\mathcal{I}}$ which were derived in Lemma 10.2. l enjoys an additional property, known as *finite subadditivity*, expressed by the following lemma.

Lemma 10.3 *If $E_i \in \mathcal{E}$ for all $1 \leq i \leq n$, then*

$$l\left(\bigcup_{i=1}^{n} E_i\right) \leq \sum_{i=1}^{n} l(E_i).$$

Proof. In order to express the $\bigcup_{i=1}^{n} E_i$ as a disjoint union, we define

$$F_1 = E_1,$$
$$F_i = E_i \setminus \bigcup_{j=1}^{i-1} E_j, \ i = 2, \cdots, n.$$

Clearly, the sets $F_i \in \mathcal{E}$ are pairwise disjoint and $\bigcup_{i=1}^{n} F_i = \bigcup_{i=1}^{n} E_i$. By the finite additive property of l on $\mathcal{E}$, we can write

$$l\left(\bigcup_{i=1}^{n} E_i\right) = l\left(\bigcup_{i=1}^{n} F_i\right) = \sum_{i=1}^{n} l(F_i).$$

Since $F_i \subseteq E_i$ for all $1 \leq i \leq n$, the monotonicity property of l implies

$$l\left(\bigcup_{i=1}^{n} E_i\right) = \sum_{i=1}^{n} l(F_i) \leq \sum_{i=1}^{n} l(E_i). \quad \square$$

Let us now expand our underlying set from $\mathcal{E}$ to $\mathcal{P}(\mathbb{R})$. In Section 6.5 we defined an open cover of a set $E \subseteq \mathbb{R}$ as a collection of open sets $\{G_\lambda : \lambda \in \Lambda\}$ whose union contains E. Similarly, if $\{I_i : i \in \mathbb{N}\}$ is a (countable) collection of intervals in $\mathcal{I}$ such that $E \subseteq \bigcup_{i=1}^{\infty} I_i$, then the collection $\{I_i\}$ is referred to as a countable cover of E by intervals in $\mathcal{I}$, or a countable $\mathcal{I}$-*cover* of E. Since $\mathbb{R} = \bigcup_{n=1}^{\infty} [-n, n)$, the collection $\{[-n, n) : n \in \mathbb{N}\}$ is an $\mathcal{I}$-cover of any $E \subseteq \mathbb{R}$.

The following definition is a first attempt at defining a measure on the subsets of $\mathbb{R}$.

Definition 10.8 The *Lebesgue outer measure*, or simply the *outer measure*, of any set $E \subseteq \mathbb{R}$ is defined as

$$m^*(E) = \inf \left\{ \sum_{i=1}^{\infty} l(I_i) : I_i \in \mathcal{I}, E \subseteq \bigcup_{i=1}^{\infty} I_i \right\}. \quad (10.1)$$

In other words, the outer measure of E is $\inf \sum l(I_i)$ over all countable $\mathcal{I}$-covers $\{I_i : i \in \mathbb{N}\}$ of E.

The next theorem indicates that m^* is an extension of l from $\mathcal{I}$ to $\mathcal{P}(\mathbb{R})$.

Theorem 10.5
For all $I \in \mathcal{I}$, $m^*(I) = l(I)$.

Proof. Let $I = [a, b)$. Since $\{I, \varnothing, \varnothing, \cdots\}$ is a countable $\mathcal{I}$-cover of the interval I, $m^*(I) \leq l(I)$, and it remains to prove that $l(I) \leq m^*(I)$. Let $\varepsilon > 0$. Using the defining property of the infimum, there is a cover $\{I_i \in \mathcal{I} : i \in \mathbb{N}\}$ of I such that

$$m^*(I) + \varepsilon > \sum_{i=1}^{\infty} l(I_i). \quad (10.2)$$

Suppose $I_i = [a_i, b_i)$ and let $J_i = (a_i - \varepsilon/2^i, b_i)$. The collection $\{J_i : i \in \mathbb{N}\}$ constitutes an open cover of the compact interval $[a, b - \varepsilon]$. Therefore, by the Heine-Borel theorem, there is a positive integer n such that

$$[a, b - \varepsilon] \subseteq \bigcup_{i=1}^{n} J_i$$

$$\Rightarrow [a, b - \varepsilon) \subseteq \bigcup_{i=1}^{n} (a_i - \varepsilon/2^i, b_i).$$

By the subadditive property of l (Lemma 10.3),

$$b - \varepsilon - a \leq \sum_{i=1}^{n} (b_i - a_i + \varepsilon/2^i)$$

$$\leq \sum_{i=1}^{\infty} (b_i - a_i + \varepsilon/2^i)$$

$$\leq \sum_{i=1}^{\infty} l(I_i) + \varepsilon,$$

which implies

$$l(I) = b - a \leq \sum_{i=1}^{\infty} l(I_i) + 2\varepsilon.$$

Using (10.2) we obtain

$$l(I) < m^*(I) + 3\varepsilon,$$

and since $\varepsilon > 0$ is arbitrary, we conclude that $l(I) \leq m^*(I)$. □

The next theorem presents the basic properties of outer measure.

Theorem 10.6
The function $m^* : \mathcal{P}(\mathbb{R}) \to [0, \infty]$ satisfies the following properties:
(i) $m^*(\varnothing) = 0$.
(ii) Monotonicity: $E \subseteq F \Rightarrow m^*(E) \subseteq m^*(F)$.
(iii) Countable subadditivity: $m^* \left(\bigcup_{i=1}^\infty E_i \right) \leq \sum_{i=1}^\infty m^*(E_i)$ for any collection $\{E_i : n \in \mathbb{N}\}$ of subsets of $\mathbb{R}$.

Proof

(i) follows directly from the definition of m^*, so we shall only prove (ii) and (iii).

(ii) Suppose $\{I_i : I_i \in \mathcal{I}, i \in \mathbb{N}\}$ covers F, then it also covers E, and we therefore have

$$m^*(E) \leq \sum_{i=1}^\infty l(I_i).$$

This means that $m^*(E)$ is a lower bound of the set of positive numbers

$$\left\{ \sum_{i=1}^\infty l(I_i) : I_i \in \mathcal{I},\ F \subseteq \bigcup_{i=1}^\infty I_i \right\}. \tag{10.3}$$

Since $m^*(F)$ is the infimum of the set (10.3), we have $m^*(E) \leq m^*(F)$.

(iii) Let $\varepsilon > 0$. For every $i \in \mathbb{N}$ there is an $\mathcal{I}$-cover $\{I_{ij} : j \in \mathbb{N}\}$ of E_i such that

$$m^*(E_i) + \frac{\varepsilon}{2^i} \geq \sum_{j=1}^\infty l(I_{ij}).$$

Since the collection $\{I_{ij} : i, j \in \mathbb{N}\}$ is countable and covers $\bigcup_{i=1}^\infty E_i$,

$$m^* \left(\bigcup_{i=1}^\infty E_i \right) \leq \sum_{i=1}^\infty \sum_{j=1}^\infty l(I_{ij}) \leq \sum_{i=1}^\infty m^*(E_i) + \varepsilon,$$

and since this true for all $\varepsilon > 0$,

$$m^* \left(\bigcup_{i=1}^\infty E_i \right) \leq \sum_{i=1}^\infty m^*(E_i). \quad \square$$

Example 10.3 Since $[0, n] \subseteq [0, \infty)$ for all $n \in \mathbb{N}$, the monotonicity of m^* implies

$$m^*([0, \infty)) \geq m^*([0, n])$$
$$\Rightarrow m^*([0, \infty)) \geq l([0, n]) = n \quad \text{for all } n \in \mathbb{N}$$
$$\Rightarrow m^*([0, \infty)) = \infty.$$

Corollary 10.6.1 *If E is a countable subset of $\mathbb{R}$, then $m^*(E) = 0$.*

Proof. Assume first that $E = \{x\}$, where x is a real number. Then $E \subseteq [x, x + 1/n)$ for all $n \in \mathbb{N}$, and therefore

$$m^*(E) \leq l([x, x + 1/n)) = 1/n \quad \text{for all } n \in \mathbb{N},$$

which implies $m^*(E) = 0$. If $E = \{x_i : i \in \mathbb{N}\}$, where $x_i \in \mathbb{R}$, then we can write $E = \bigcup_{i=1}^{\infty} \{x_i\}$ and use the countable subadditive property of m^* to conclude that

$$m^*(E) \leq \sum_{i=1}^{\infty} m^*(\{x_i\}) = 0. \quad \square$$

Corollary 10.6.2 *The outer measure of any real interval is its length.*

Proof. By Theorem 10.5, this statement is true when the interval is in $\mathcal{I}$, and we have to check its validity for other types of intervals. We shall only prove the theorem for compact intervals $I = [a, b]$ and leave the other cases as an exercise. Since $[a, b) \subseteq I = [a, b) \cup \{b\}$, we have from monotonicity and countable subadditivity

$$m^*([a, b)) \leq m^*(I) \leq m^*([a, b)) + m^*(\{b\}).$$

Using Theorem 10.5 and Corollary 10.6.1, we obtain

$$b - a \leq m^*(I) \leq b - a,$$

which means $m^*(I) = l(I)$. $\square$

Corollary 10.6.3 *The set $[0, 1]$ is not countable.*

The next theorem addresses the effects of translation, reflection, and contraction of a set on its outer measure. As would be expected, m^* is invariant under translation and reflection, and preserves contraction. For the sake of convenience, we shall use the notation $cE = \{cx : x \in E\}$, where E is a subset of $\mathbb{R}$ and c is a real number.

Theorem 10.7
For any $E \subseteq \mathbb{R}$ and all $c \in \mathbb{R}$,
(i) $m^*(E+c) = m^*(E)$,
(ii) $m^*(cE) = |c|\, m^*(E)$.

Proof
(i) Suppose $\{I_i = [a_i, b_i] : i \in \mathbb{N}\}$ is an $\mathcal{I}$-cover of E. If $J_i = [a_i+c, b_i+c]$, then clearly $\{J_i : i \in \mathbb{N}\}$ is an $\mathcal{I}$-cover of $E+c$, and we therefore have

$$m^*(E+c) \leq \sum_{i=1}^{\infty} l(J_i) = \sum_{i=1}^{\infty} l(I_i),$$

which implies $m^*(E+c) \leq m^*(E)$. Using $E+c$ instead of E, and $-c$ instead of c, we obtain

$$m^*(E) = m^*(E+c-c) \leq m^*(E+c).$$

(ii) If $c = 0$ then $cE = \{0\}$ and we obtain the desired result by using Corollary 10.6.1. If $c > 0$, let $\{I_i = [a_i, b_i]\}$ be an $\mathcal{I}$-cover of E, in which case the intervals $\{J_i = [ca_i, cb_i]\}$ cover cE. Consequently

$$m^*(cE) \leq \sum_{i=1}^{\infty} l(J_i) = c \sum_{i=1}^{\infty} l(I_i),$$

and we obtain the inequality $m^*(cE) \leq cm^*(E)$. If E is replaced by cE and c by $1/c$ in this inequality, then $m^*(E) \leq c^{-1}m^*(cE)$, hence $m^*(cE) = cm^*(E)$.

To complete the proof, we have to show that $m^*(-E) = m^*(E)$, that is, the outer measure of a set is not affected by reflection. Once again we assume that the intervals $\{I_i = [a_i, b_i]\}$ cover E, so that $\{-I_i = (-b_i, -a_i]\}$ cover $-E$, i.e.,

$$-E \subseteq \bigcup_{i=1}^{\infty} (-b_i, -a_i]$$

$$\Rightarrow m^*(-E) \leq \sum_{i=1}^{\infty} m^*(-b_i, -a_i] \leq \sum_{i=1}^{\infty} l(I_i).$$

This means that $m^*(-E)$ is a lower bound of the sum $\sum_{i=1}^{\infty} l(I_i)$, hence $m^*(-E) \leq m^*(E)$. By interchanging E and $-E$ we obtain this inequality in the reverse direction, which proves $m^*(-E) = m^*(E)$. □

The reader may already have noticed the higher dose of set theory used in the proofs involving outer measure, with the supremum property of $\mathbb{R}$ playing a central role. This interplay between "covering intervals" and "epsilons" will feature prominently in much of our work for the remainder of this chapter, thereby giving it a distinctive flavour.

EXERCISES 10.2

1. Verify that the properties of $l|_{\mathcal{I}}$, as spelled out in Lemma 10.2, are satisfied by $l|_{\mathcal{E}}$.

2. Complete the proof of Corollary 10.6.2 by treating the cases where the interval is $(a, b]$, (a, b), $[a, \infty)$, (a, ∞), $(-\infty, b]$, $(-\infty, b)$, or $(-\infty, \infty)$.

3. If $\{I_1, I_2, \cdots, I_n\}$ is a finite set of intervals in $\mathcal{I}$ which covers $\mathbb{Q} \cap [0, 1]$, prove that $\sum_{i=1}^{n} l(I_i) \geq 1$.

4. Given $m^*(E) = 0$, prove that $m^*(E \cup F) = m^*(F)$ for all $F \subseteq \mathbb{R}$.

5. Given any $E \subseteq \mathbb{R}$, prove that, for all $\varepsilon > 0$, there is an open set G which contains E and satisfies $m^*(G) \leq m^*(E) + \varepsilon$.

6. In Definition 10.8, show that taking the intervals I_i in $\mathcal{I}_0$ or $\mathcal{I}_1$, instead of $\mathcal{I}$, does not change the definition of m^*.

10.3 Lebesgue Measure

The results we obtained in Section 10.2 would seem to imply that outer measure constitutes a reasonable extension of the length function l to $\mathcal{P}(\mathbb{R})$. But we were not able to prove that m^* is *countably additive*, in the sense that

$$m^*\left(\bigcup_{i=1}^{\infty} E_i\right) = \sum_{i=1}^{\infty} m^*(E_i)$$

for any disjoint countable collection $\{E_i : i \in \mathbb{N}\}$ of subsets of $\mathbb{R}$. We were only able to prove that m^* is *countably subadditive*, i.e.,

$$m^*\left(\bigcup_{i=1}^{\infty} E_i\right) \leq \sum_{i=1}^{\infty} m^*(E_i).$$

As pointed out at the beginning of this chapter, countable additivity is indispensable for any definition of measure, and we shall have to reduce the domain of definition of m^* in order to achieve it. We start with a general definition of measure, which does not necessarily extend l. Suppose Ω is any non-empty subset of $\mathbb{R}$.

Definition 10.9 Let $\mathcal{A}$ be a σ-algebra on Ω. A function $\mu : \mathcal{A} \to [0, \infty]$ is called a *measure* on $\mathcal{A}$ if

(i) $\mu(\varnothing) = 0$.
(ii) μ is *countably additive*, i.e., for any countable class $\{E_i : i \in \mathbb{N}\}$ of pairwise disjoint sets in $\mathcal{A}$,

$$\mu\left(\bigcup_{i=1}^{\infty} E_i\right) = \sum_{i=1}^{\infty} \mu(E_i).$$

The triplet $(\Omega, \mathcal{A}, \mu)$ is then called a *measure space*. When $\mathcal{A}$ is understood, we often refer to μ simply as a *measure on* Ω.

Examples 10.4

(i) For every $E \subseteq \mathbb{R}$, we define $\mu(E)$ to be the number of elements in E if E is finite, and to be ∞ otherwise. It is straightforward to verify that μ is a measure on $\mathcal{P}(\mathbb{R})$, known as the *counting measure*.

(ii) Let $c \in \mathbb{R}$ and define μ_c on $\mathcal{P}(\mathbb{R})$ by

$$\mu_c(E) = \begin{cases} 1 & \text{if } c \in E \\ 0 & \text{otherwise.} \end{cases}$$

This is the so-called *Dirac measure* at c.

(iii) Let (c_i) be a sequence in $\mathbb{R}$ and suppose that (p_i) is a corresponding sequence of non-negative numbers. Define

$$\mu(E) = \sum_{\{i : c_i \in E\}} p_i.$$

Then μ is a measure on $\mathcal{P}(\mathbb{R})$ which, in fact, is related to the Dirac measure by the equality $\mu = \sum_{i=1}^{\infty} p_i \mu_{c_i}$.

(iv) Let μ be defined on $\mathcal{P}(\mathbb{R})$ by

$$\mu(E) = \begin{cases} 0 & \text{if } E \text{ is finite} \\ \infty & \text{otherwise.} \end{cases}$$

Clearly μ is finitely additive. But it is not countably additive, for $\mathbb{N} = \bigcup_{i=1}^{\infty} \{i\}$, and yet

$$\mu(\mathbb{N}) = \infty \neq \sum_{i=1}^{\infty} \mu(\{i\}) = 0.$$

Thus μ is not a measure.

The next theorem indicates the properties of a general measure which derive from Definition 10.9 above. It is significant to note that these include monotonicity and countable subadditivity, but not invariance under translation, reflection, or preservation of contraction which, according to Theorem 10.7, apply to outer measure. This should come as no surprise in as much as m^*, being an extension of l, inherits the homogeneous and isotropic properties of "length".

Theorem 10.8
Let μ be a measure on a σ-algebra $\mathcal{A}$. Then
(i) μ is monotonic, i.e.,

$$E, F \in \mathcal{A}, \ E \subseteq F \Rightarrow \mu(E) \leq \mu(F).$$

(ii) μ is subtractive, in the sense that

$$E, F \in \mathcal{A}, \ E \subseteq F, \ \mu(E) < \infty \Rightarrow \mu(F \backslash E) = \mu(F) - \mu(E).$$

(iii) μ is countably subadditive, so that if $E_i \in A$ for all $i \in \mathbb{N}$, then

$$\mu\left(\bigcup_{i=1}^{\infty} E_i\right) \leq \sum_{i=1}^{\infty} \mu(E_i).$$

Proof. Since $F = E \cup (F \backslash E)$ represents F as a disjoint union of elements of $\mathcal{A}$, we have

$$\mu(F) = \mu(E) + \mu(F \backslash E).$$

Now (i) follows by noting that $\mu(F \backslash E) \geq 0$. Furthermore, if $\mu(E) < \infty$, we obtain (ii).

To prove (iii), we shall first express $\bigcup_{i=1}^{\infty} E_i$ as a disjoint union. Define

$$F_1 = E_1, \quad F_i = E_i \backslash \bigcup_{j=1}^{i-1} E_j, \ i \geq 2.$$

Clearly, the collection $\{F_i : i \in \mathbb{N}\}$ is pairwise disjoint and $\bigcup_{i=1}^{\infty} F_i = \bigcup_{i=1}^{\infty} E_i$. Consequently

$$\mu\left(\bigcup_{i=1}^{\infty} E_i\right) = \sum_{i=1}^{\infty} \mu(F_i),$$

and since $F_i \subseteq E_i$, we have $\mu(F_i) \leq \mu(E_i)$ by monotonicity, which proves (iii). □

Another property of measure which follows from Definition 10.9 is continuity in a sense which will be defined below.

Definition 10.10 Let (E_i) be a sequence of subsets of $\Omega \subseteq \mathbb{R}$. We define
$$\limsup E_i = \bigcap_{i=1}^{\infty} \bigcup_{j=i}^{\infty} E_j, \quad \liminf E_i = \bigcup_{i=1}^{\infty} \bigcap_{j=i}^{\infty} E_j.$$
Denote $\limsup E_i$ by E^*, and $\liminf E_i$ by E_*. The sequence (E_i) is said to be *convergent* with limit E if $E^* = E_* = E$.

Example 10.5
(i) If (E_i) is *increasing*, in the sense that $E_{i+1} \supseteq E_i$ for all $i \in \mathbb{N}$, then (E_i) converges to $\bigcup_{i=1}^{\infty} E_i$. This follows from the observation that $\bigcup_{j=i}^{\infty} E_j = \bigcup_{j=1}^{\infty} E_j$ and $\bigcap_{j=i}^{\infty} E_j = E_i$ for all $i \in \mathbb{N}$.
(ii) If (E_i) is *decreasing*, in the sense that $E_{i+1} \subseteq E_i$ for all $i \in \mathbb{N}$, then $\bigcup_{j=i}^{\infty} E_j = E_i$ and $\bigcap_{j=i}^{\infty} E_j = \bigcap_{j=1}^{\infty} E_j$ for all $i \in \mathbb{N}$. Therefore (E_i) converges to $\bigcap_{i=1}^{\infty} E_i$.

Theorem 10.9
Let $\mathcal{A}$ be a σ-algebra on Ω, μ a measure on $\mathcal{A}$, and (E_i) a sequence in $\mathcal{A}$.
(i) If (E_i) is increasing, then $\mu(\lim E_i) = \lim \mu(E_i)$, and μ is said to be continuous from below on $\mathcal{A}$.
(ii) If (E_i) is decreasing and there is an N such that $\mu(E_N) < \infty$, then $\mu(\lim E_i) = \lim \mu(E_i)$, and μ is said to be continuous from above on $\mathcal{A}$.

Proof
(i) Let $E = \lim E_i$. From Example 10.5 we know that $E = \bigcup_{i=1}^{\infty} E_i$. As we did in the proof of Theorem 10.8, we first express E as a disjoint union by defining
$$F_1 = E_1, \quad F_i = E_i \setminus \bigcup_{j=1}^{i-1} E_j, \quad i \geq 2.$$
$F_i \in \mathcal{A}$ for all $i \in \mathbb{N}$ and the unions
$$E_i = \bigcup_{j=1}^{i} F_j, \quad E = \bigcup_{j=1}^{\infty} F_j$$
are disjoint. By countable additivity,
$$\mu(E) = \sum_{i=1}^{\infty} \mu(F_i)$$
$$= \lim_{n \to \infty} \sum_{i=1}^{n} \mu(F_i) = \lim_{n \to \infty} \mu(E_n).$$

(ii) In this case, set $E = \lim E_i = \bigcap_{i=1}^{\infty} E_i$. Since (E_i) is decreasing, $G_i = E_N \setminus E_i$ is an increasing sequence in $\mathcal{A}$ with limit $E_N \setminus E$. From part (i) we therefore have

$$\mu(E_N \setminus E) = \lim \mu(E_N \setminus E_i).$$

The monotonicity of μ implies that $\mu(E) < \infty$ and that $\mu(E_i) < \infty$ for all $i \geq N$. Now the subtractive property gives

$$\mu(E_N) - \mu(E) = \lim[\mu(E_N) - \mu(E_i)] = \mu(E_N) - \lim \mu(E_i)$$

and the desired result. □

Remark 10.4 The requirement that $\mu(E_N) < \infty$ for some integer N cannot be dropped. This can be seen by taking μ to be the counting measure on $\mathcal{P}(\mathbb{R})$, and $E_i = \{j \in \mathbb{N} : j > i\}$. Clearly (E_i) is decreasing, with limit $\varnothing$ (by the theorem of Archimedes). Thus $\mu(\lim E_i) = \mu(\varnothing) = 0$, while $\mu(E_i) = \infty$ for all i, so that $\lim \mu(E_i) = \infty$.

Definition 10.9 of the measure μ and the subsequent variety of examples given in Example 10.4 indicate the scope and richness of this concept. Our interest here, however, is in a measure which extends the function $l : \mathcal{I} \to [0, \infty]$ to a suitably large σ-algebra. We shall achieve this by restricting m^* to a sub-σ-algebra of $\mathcal{P}(\mathbb{R})$ on which m^* is countably additive, and thereby ensuring that it satisfies the conditions of a measure according to Definition 10.9.

Definition 10.11 A set $E \subseteq \mathbb{R}$ is said to be *Lebesgue measurable* if

$$m^*(A) = m^*(A \cap E) + m^*(A \cap E^c) \text{ for all } A \subseteq \mathbb{R}. \qquad (10.4)$$

The Lebesgue measurable sets in $\mathcal{P}(\mathbb{R})$ will be denoted by $\mathcal{M}$.

Remarks 10.5

1. We shall often refer to the elements of $\mathcal{M}$ simply as *measurable sets* with the understanding that they are measurable in the sense of Lebesgue, that is, they satisfy equation (10.4), also known as the *Carathéodory condition*.

2. Clearly equation (10.4) is satisfied by $E = \varnothing$ and $E = \mathbb{R}$, and therefore both of these sets are measurable.

3. By the subadditivity of m^* we always have

$$m^*(A) \leq m^*(A \cap E) + m^*(A \cap E^c),$$

so a set E belongs to $\mathcal{M}$ if and only if

$$m^*(E) \geq m^*(A \cap E) + m^*(A \cap E^c) \quad \text{for all } A \subseteq \mathbb{R}. \tag{10.5}$$

4. Suppose $m^*(E) = 0$ and let $A \subseteq \mathbb{R}$. By the monotonicity of m^*, $m^*(A \cap E) = 0$ and

$$m^*(A) \geq m^*(A \cap E^c) = m^*(A \cap E) + m^*(A \cap E^c),$$

so we conclude that $E \in \mathcal{M}$. In other words, a set whose outer measure is 0 is measurable.

Lemma 10.4 $\mathcal{M}$ *is an algebra on* $\mathbb{R}$.

Proof. Since interchanging E and E^c does not alter the right-hand side of equation (10.4), it follows that

$$E \in \mathcal{M} \Leftrightarrow E^c \in \mathcal{M}.$$

Now suppose $E, F \in \mathcal{M}$. We shall prove that $E \cup F \in \mathcal{M}$. Let A be any subset of $\mathbb{R}$. Since $E \in \mathcal{M}$, equation (10.4) remains valid with A replaced by $A \cap (E \cup F)$. Thus

$$m^*(A \cap (E \cup F)) = m^*(A \cap (E \cup F) \cap E) + m^*(A \cap (E \cup F) \cap E^c)$$
$$= m^*(A \cap E) + m^*(A \cap F \cap E^c).$$

Noting that

$$m^*(A \cap (E \cup F)^c) = m^*(A \cap E^c \cap F^c),$$

and adding these two equations, we obtain

$$m^*(A \cap (E \cup F)) + m^*(A \cap (E \cup F)^c) =$$
$$m^*(A \cap E) + m^*(A \cap E^c \cap F) + m^*(A \cap E^c \cap F^c).$$

But since F is measurable, we have

$$m^*(A \cap E^c) = m^*(A \cap E^c \cap F) + m^*(A \cap E^c \cap F^c).$$

Hence

$$m^*(A \cap (E \cup F)) + m^*(A \cap (E \cup F)^c) = m^*(A \cap E) + m^*(A \cap E^c)$$
$$= m^*(A),$$

since E is measurable. This implies $E \cup F$ is measurable. Thus $\mathcal{M}$ is closed under complementation and the formation of unions, and is therefore an algebra on $\mathbb{R}$. □

Theorem 10.10
$\mathcal{M}$ is a σ-algebra on $\mathbb{R}$.

Proof. In view of Lemma 10.4, we need only prove that $\mathcal{M}$ is closed under the formation of countable unions. Suppose $E_i \in \mathcal{M}$ for every $i \in \mathbb{N}$. We may assume that the sets $E_1, E_2, E_3, \cdots$ are pairwise disjoint, otherwise we apply the method used in the proof of Lemma 10.3 to define a disjoint class in $\mathcal{M}$ having the same union. Let $A \subseteq \mathbb{R}$. Using induction, we shall prove that

$$m^*(A \cap \bigcup_{i=1}^n E_i) = \sum_{i=1}^n m^*(A \cap E_i). \tag{10.6}$$

Equation (10.6) clearly holds for $n = 1$. Assume it is true for $n = k$. Since $E_{k+1} \in \mathcal{M}$,

$$m^*\left(A \cap \bigcup_{i=1}^{k+1} E_i\right) = m^*\left(A \cap \bigcup_{i=1}^{k+1} E_i \cap E_{k+1}\right)$$
$$+ m^*\left(A \cap \bigcup_{i=1}^{k+1} E_i \cap E_{k+1}^c\right)$$
$$= m^*(A \cap E_{k+1}) + m^*\left(A \cap \bigcup_{i=1}^{k} E_i\right)$$
$$= \sum_{i=1}^{k+1} m^*(A \cap E_i),$$

using the inductive hypothesis. This means (10.6) is true for $n = k+1$, and hence for all $n \in \mathbb{N}$.

Define

$$F_n = \bigcup_{i=1}^n E_i, \quad F = \bigcup_{i=1}^\infty E_i.$$

Since $F_n \in \mathcal{M}$,

$$m^*(A) = m^*(A \cap F_n) + m^*(A \cap F_n^c)$$
$$= \sum_{i=1}^n m^*(A \cap E_i) + m^*(A \cap F_n^c)$$
$$\geq \sum_{i=1}^n m^*(A \cap E_i) + m^*(A \cap F^c)$$

for all $n \in \mathbb{N}$. In the limit as $n \to \infty$, we obtain

$$m^*(A) \geq \sum_{i=1}^{\infty} m^*(A \cap E_i) + m^*(A \cap F^c)$$
$$\geq m^*(A \cap F) + m^*(A \cap F^c), \tag{10.7}$$

where we use the subadditive property of m^* in the last inequality. From (10.7) we conclude that F is measurable. $\square$

Corollary 10.10.1 *If $\{E_i : i \in \mathbb{N}\}$ is a disjoint family of measurable sets, then*

$$m^*\left(A \cap \bigcup_{i=1}^{\infty} E_i\right) = \sum_{i=1}^{\infty} m^*(A \cap E_i) \ \text{for all } A \subseteq \mathbb{R}. \tag{10.8}$$

Proof. For all n,

$$m^*(A \cap \bigcup_{i=1}^{\infty} E_i) \geq m^*(A \cap \bigcup_{i=1}^{n} E_i)$$
$$= \sum_{i=1}^{n} m^*(A \cap E_i)$$

by (10.6). Therefore

$$m^*(A \cap \bigcup_{i=1}^{\infty} E_i) \geq \sum_{i=1}^{\infty} m^*(A \cap E_i).$$

But $m^*(A \cap \bigcup_{i=1}^{\infty} E_i) \leq \sum_{i=1}^{\infty} m^*(A \cap E_i)$ by countable subadditivity, hence the equality (10.8). $\square$

Corollary 10.10.2 $m^*|_{\mathcal{M}}$, *the restriction of outer measure to the class $\mathcal{M}$, is a measure on $\mathcal{M}$.*

Proof. To prove this result, it suffices to verify that m^* is countably additive on $\mathcal{M}$. Let $\{E_i : i \in \mathbb{N}\}$ be a disjoint collection of members of $\mathcal{M}$. Setting $A = \mathbb{R}$ in (10.8) yields

$$m^*\left(\bigcup_{i=1}^{\infty} E_i\right) = \sum_{i=1}^{\infty} m^*(E_i). \ \square$$

The measure $m^*|_{\mathcal{M}}$ is known as the *Lebesgue measure*, and is denoted by m. Thus $m : \mathcal{M} \to [0, \infty]$ is the natural extension of l from $\mathcal{I}$ to the σ-algebra $\mathcal{M}$ which, in accordance with Definition 10.9, yields the measure space $(\mathbb{R}, \mathcal{M}, m)$. We shall now see that restricting the outer measure function m^* from $\mathcal{P}(\mathbb{R})$ to $\mathcal{M}$ does not really restrict our ability to handle practically all sets of interest in $\mathbb{R}$, for $\mathcal{M}$ contains all Borel sets.

Theorem 10.11
Every Borel set is measurable, that is, $\mathcal{B} \subseteq \mathcal{M}$.

Proof. We have already proved that $\mathcal{B} = \mathcal{A}(\mathcal{I}_3)$, $\mathcal{I}_3$ being the class of intervals of the form $[a, \infty)$. Since $\mathcal{M}$ is a σ-algebra, all we need to show is that $[a, \infty) \in \mathcal{M}$ for every $a \in \mathbb{R}$. Let $A \subseteq \mathbb{R}$, and set

$$A_1 = A \cap [a, \infty), \quad A_2 = A \cap (-\infty, a).$$

Let ε be any positive number. By Definition 10.8, there is an $\mathcal{I}$-cover $\{I_i : i \in \mathbb{N}\}$ of A such that

$$m^*(A) + \varepsilon \geq \sum_{i=1}^{\infty} l(I_i). \tag{10.9}$$

Setting $J_i = I_i \cap [a, \infty)$ and $J_i' = I_i \cap (-\infty, a)$, it is clear that J_i and J_i' both lie in $\mathcal{I}$, and that

$$A_1 \subseteq \bigcup_{i=1}^{\infty} J_i, \quad A_2 \subseteq \bigcup_{i=1}^{\infty} J_i'.$$

Therefore

$$m^*(A_1) \leq \sum_{i=1}^{\infty} m^*(J_i) = \sum_{i=1}^{\infty} l(J_i), \tag{10.10}$$

$$m^*(A_2) \leq \sum_{i=1}^{\infty} m^*(J_i') = \sum_{i=1}^{\infty} l(J_i'). \tag{10.11}$$

Since $I_i = J_i \cup J_i'$ and $J_i \cap J_i' = \varnothing$,

$$l(I_i) = l(J_i) + l(J_i'). \tag{10.12}$$

Now inequalities (10.9), (10.10), (10.11) and equation (10.12) imply

$$m^*(A_1) + m^*(A_2) \leq \sum_{i=1}^{\infty} l(I_i) \leq m^*(A) + \varepsilon.$$

Since $\varepsilon > 0$ is arbitrary,

$$m^*(A) \geq m^*(A_1) + m^*(A_2)$$
$$= m^*(A \cap [a, \infty)) + m^*(A \cap [a, \infty)^c),$$

hence $[a, \infty) \in \mathcal{M}$. $\square$

374   Elements of Real Analysis

Thus we have the following inclusion relation between three important σ-algebras on $\mathbb{R}$:
$$\mathcal{B} \subseteq \mathcal{M} \subseteq \mathcal{P}(\mathbb{R}).$$
It can be shown that both inclusions are proper. In other words there are non-measurable subsets of $\mathbb{R}$, and there are measurable sets which are not Borel sets. But we shall not give examples of such sets, as their definitions are rather involved (see [ROY] or [HAL], for example), an indication that the subsets of $\mathbb{R}$ which we are likely to deal with all lie in $\mathcal{B}$.

Remarks 10.6

1. The fact that $\mathcal{P}(\mathbb{R}) \setminus \mathcal{M} \neq \varnothing$ means that m^* is not additive on $\mathcal{P}(\mathbb{R})$, since m^* cannot be additive if (10.4) is not satisfied.

2. The construction of a non-measurable subset of $\mathbb{R}$, cited in [ROY], shows that there is no translation invariant measure on $\mathcal{P}(\mathbb{R})$ which extends ℓ.

3. Since $\mathcal{B}$ is a σ-algebra, the restriction of m to $\mathcal{B}$ is also a measure, known as the *Borel measure*.

In view of Remark 10.6.3, and the fact that $\mathcal{B}$ includes all the interesting subsets of $\mathbb{R}$, it is natural to ask why we bother with $\mathcal{M}$. The reason is that m enjoys a property that $m|_{\mathcal{B}}$ does not share, the so-called *completeness* property, which states: If $F \in \mathcal{M}$ and $m(F) = 0$, then every $E \subseteq F$ is measurable (and, of course, $m(E) = 0$). To see this, note that, since m^* is monotonic,
$$0 \leq m^*(E) \leq m^*(F) = m(F) = 0.$$
Thus $m^*(E) = 0$ and, by Remark 10.5.4, E is measurable. A set whose measure is 0 is called a *null set*. To see why Borel measure is not complete, refer to Exercise 10.4.9.

We saw in Corollary 10.6.1 that every countable set has outer measure 0 and is therefore measurable. The next example exhibits a non-countable set whose Lebesgue measure is 0.

Example 10.6 Recalling the definition of the Cantor set $F = \bigcap_{n=1}^{\infty} F_n$ in Example 3.17, we found that F is uncountable. As F is closed, it is a Borel set and hence measurable. Since F_i is obtained by dropping the middle third of each interval in F_{i-1}, we have

$$m(F_n) = \frac{2}{3}m(F_{n-1}) = \left(\frac{2}{3}\right)^n m(F_0) = \left(\frac{2}{3}\right)^n.$$

(F_n) is a decreasing sequence with limit F, so we can use continuity from above to conclude that

$$m(F) = \lim m(F_n) = \lim \left(\frac{2}{3}\right)^n = 0.$$

Theorem 10.12
If $E \in \mathcal{M}$ and $a \in \mathbb{R}$, then $E + a \in \mathcal{M}$, $aE \in \mathcal{M}$, and

$$m(E + a) = m(E), \quad m(aE) = |a| \, m(E).$$

Proof. In view of Theorem 10.7, we need only prove that $E + a$ and aE are measurable. For any $A \subseteq \mathbb{R}$, it is a simple matter to show that

$$A \cap (E + a) = [(A - a) \cap E] + a,$$
$$A \cap (E + a)^c = A \cap (E^c + a) = [(A - a) \cap E^c] + a.$$

Since $E \in \mathcal{M}$ and outer measure is preserved under translation,

$$m^*(A \cap (E + a)) + m^*(A \cap (E + a)^c) = m^*((A - a) \cap E)$$
$$+ m^*((A - a) \cap E^c)$$
$$= m^*(A - a)$$
$$= m^*(A),$$

which implies $E + a \in \mathcal{M}$.

A similar argument leads to the conclusion that $cE \in \mathcal{M}$. $\square$

The next theorem shows that the extension of the function $l : \mathcal{I} \to [0, \infty]$ to a measure on $\mathcal{B}$ is unique.

Theorem 10.13
If μ is any measure on $\mathcal{B}$ such that $\mu(I) = l(I)$ for all $I \in \mathcal{I}$, then $\mu = m$ on $\mathcal{B}$.

Proof. Let E be any Borel set and suppose $\varepsilon > 0$. By Definition 10.8, there is a collection $\{I_i \in \mathcal{I} : i \in \mathbb{N}\}$ such that $E \subseteq \bigcup_{i=1}^\infty I_i$ and

$$m(E) + \varepsilon \geq \sum_{i=1}^\infty l(I_i) = \sum_{i=1}^\infty \mu(I_i).$$

Since μ is monotonic and countably subadditive,

$$\sum_{i=1}^{\infty} \mu(I_i) \geq \mu(E),$$

hence, ε being arbitrary,

$$m(E) \geq \mu(E). \tag{10.13}$$

Suppose now that $m(E) < \infty$. Given $\varepsilon > 0$, choose $\{I_i : i \in \mathbb{N}\}$ as above and assume, without loss of generality, that the union $A = \bigcup_{i=1}^{\infty} I_i$ is disjoint. Since m and μ are both countably additive, $m(A) = \mu(A)$; and since $E \subseteq A$,

$$m(E) \leq m(A) = \mu(A) = \mu(E) + \mu(A \backslash E).$$

But $\mu(A \backslash E) \leq m(A \backslash E)$ by (10.13), and $m(A) \leq m(E) + \varepsilon$ by assumption, hence

$$\begin{aligned} m(E) &\leq \mu(E) + m(A \backslash E) \\ &= \mu(E) + m(A) - m(E) \\ &\leq \mu(E) + \varepsilon, \end{aligned}$$

which implies $m(E) \leq \mu(E)$. In view of (10.13), we conclude that $\mu(E) = m(E)$.

We have therefore proved that $\mu(E) = m(E)$ for every $E \in \mathcal{B}$ with $m(E) < \infty$. For a general E in $\mathcal{B}$, we can write

$$E = \bigcup_{n \in \mathbb{Z}} E \cap [n, n+1) = \bigcup_{n \in \mathbb{Z}} E_n,$$

where $E_n = E \cap [n, n+1)$. Since, for each n, $m(E_n) < \infty$, we have $\mu(E_n) = m(E_n)$. The sets E_n being pairwise disjoint, the countable additivity of μ and m implies

$$m(E) = \sum_{n \in \mathbb{Z}} m(E_n) = \sum_{n \in \mathbb{Z}} \mu(E_n) = \mu(E). \quad \square$$

Thus we have shown that m, the Lebesgue measure on $\mathbb{R}$, extends the notion of length from intervals to more general subsets of $\mathbb{R}$. However, to compute the measure of a set using Definition 10.8 is no simple matter, so the following approximation theorems (especially the second) is useful in that it gives us a manipulation grip on m.

Lebesgue Measure

Theorem 10.14 (First Approximation Theorem)
For any $E \subseteq \mathbb{R}$, the following statements are equivalent:
(i) $E \in \mathcal{M}$.
(ii) For every $\varepsilon > 0$ there is an open set $G \supseteq E$ such that
$$m^*(G \backslash E) < \varepsilon.$$
(iii) For every $\varepsilon > 0$ there is a closed set $F \subseteq E$ such that
$$m^*(E \backslash F) < \varepsilon.$$

Proof
(i)$\Rightarrow$(ii): Suppose $m(E) < \infty$, and let $\{I_i : i \in \mathbb{N}\}$ be an $\mathcal{I}$-cover of E which satisfies
$$m(E) + \frac{\varepsilon}{2} > \sum_{i=1}^{\infty} \ell(I_i).$$
Assuming $I_i = [a_i, b_i)$, we can choose $G = \bigcup_{i=1}^{\infty} J_i$ with $J_i = (a_i - \varepsilon/2^{i+1}, b_i)$. G is clearly open and contains E. Furthermore,
$$m(G) \leq \sum_{i=1}^{\infty} m(J_i) = \sum_{i=1}^{\infty} \ell(I_i) + \frac{\varepsilon}{2} < m(E) + \varepsilon.$$
Since both G and E are measurable,
$$m^*(G \backslash E) = m(G \backslash E) = m(G) - m(E) < \varepsilon.$$

If $m(E) = \infty$ then we set $E_n = E \cap (-n, n)$. Since $m(E_n) \leq 2n < \infty$ for all $n \in \mathbb{N}$, there is an open set G_n such that $G_n \supseteq E_n$ for each n and
$$m^*(G_n \backslash E_n) < \frac{\varepsilon}{2^n}.$$
Clearly, the set $G = \bigcup_{n=1}^{\infty} G_n$ is open, contains E, and satisfies
$$G \backslash E = \bigcup_{n=1}^{\infty} (G_n \backslash E) \subseteq \bigcup_{n=1}^{\infty} (G_n \backslash E_n).$$
Hence
$$m^*(G \backslash E) \leq \sum_{n=1}^{\infty} m^*(G_n \backslash E_n) < \varepsilon.$$
(ii)$\Rightarrow$(i): For every $n \in \mathbb{N}$, let $\varepsilon = 1/n$ and choose the open set $G_n \supseteq E$ so that $m^*(G_n \backslash E) < 1/n$. The intersection $G = \bigcap_{n=1}^{\infty} G_n$ is clearly measurable, contains E, and satisfies
$$G \backslash E = \bigcap_{n=1}^{\infty} (G_n \backslash E).$$

Consequently, $m^*(G\backslash E) < 1/n$ for every $n \in \mathbb{N}$, which implies $m^*(G\backslash E) = 0$. $G\backslash E$ is therefore measurable, hence so is $G\backslash(G\backslash E) = E$.

(i)$\Rightarrow$(iii): Since E is measurable, so is E^c. Given $\varepsilon > 0$, from the equivalence of (i) and (ii), we can find an open set $G \supseteq E^c$ such that $m^*(G\backslash E^c) < \varepsilon$. Setting $F = G^c$, we see that F is closed, contained in E, and satisfies

$$E\backslash F = E \cap G = G\backslash E^c$$
$$\Rightarrow m^*(E\backslash F) = m^*(G\backslash E^c) < \varepsilon.$$

(iii)$\Rightarrow$(i): Since $F^c\backslash E^c = E\backslash F$, it is clear that that the set E^c satisfies statement (ii) with $G = F^c$. E^c is therefore measurable, so E is also measurable. $\square$

Theorem 10.15 (Second Approximation Theorem)
Let $E \subseteq \mathbb{R}$. If $m^*(E) < \infty$ then $E \in \mathcal{M}$ if and only if, for every $\varepsilon > 0$, there is a set $A \in \mathcal{E}$ such that $m^*(E \Delta A) < \varepsilon$.

Proof. Suppose $E \in \mathcal{M}$ and $\varepsilon > 0$. Then, by Theorem 10.14, there is an open set $G \supseteq E$ such that

$$m(G\backslash E) < \frac{\varepsilon}{2}. \tag{10.14}$$

Since G is open, it can be represented by a union of disjoint open intervals $\{I_i : i \in \mathbb{N}\}$. Thus $G = \bigcup_{i=1}^{\infty} I_i$, and, since $m(E) < \infty$, we can use countable additivity and (10.14) to write

$$\sum_{i=1}^{\infty} m(I_i) = m(G) < \infty.$$

Therefore there is a positive integer n such that

$$m(G) - m\left(\bigcup_{i=1}^{n} I_i\right) < \frac{\varepsilon}{2}. \tag{10.15}$$

Setting $I_i = (a_i, b_i)$, $A_0 = \bigcup_{i=1}^{n} I_i$, and $A = \bigcup_{i=1}^{n} [a_i, b_i)$, we note that $A \in \mathcal{E}$ and differs from A_0 by a finite number of points. Hence $m(A) = m(A_0) < m(G) < \infty$. Now the inclusion relation $E\backslash A \subseteq G\backslash A_0$ and (10.15) imply

$$m(E\backslash A) \leq m(G\backslash A_0) = m(G) - m(A_0) < \frac{\varepsilon}{2},$$

while the inclusion $A_0\backslash E \subseteq G\backslash E$ and (10.14) imply

$$m(A\backslash E) = m(A_0\backslash E) \leq m(G\backslash E) < \frac{\varepsilon}{2}.$$

Thus $m(E\Delta A) < \varepsilon$.

Conversely, given $m^*(E) < \infty$, let $\varepsilon > 0$ and choose $A = \bigcup_{i=1}^n [a_i, b_i) \in \mathcal{E}$ so that $m^*(E\Delta A) < \varepsilon$. To prove that E is measurable, it is sufficient, by Theorem 10.14, to prove the existence of an open set $G \supseteq E$ such that $m^*(G\backslash E) < 3\varepsilon$. Let $A_1 = \bigcup_{i=1}^n (a_i - \varepsilon/2^i, b_i)$. Then A_1 is an open set which satisfies

$$m^*(E\Delta A_1) \leq m^*(E\backslash A_1) + m^*(A_1\backslash E)$$
$$< m^*(E\backslash A) + m^*(A\backslash E) + \varepsilon < 2\varepsilon. \tag{10.16}$$

Furthermore, from the definition of $m^*(E\backslash A_1)$, there is a cover $\{I_i \in \mathcal{I} : i \in \mathbb{N}\}$ of $E\backslash A_1$ such that

$$m^*(E\backslash A_1) + \frac{\varepsilon}{2} > \sum_{i=1}^{\infty} l(I_i).$$

Following the procedure in the proof of Theorem 10.14, we can construct an open set A_2 that contains $E\backslash A_1$ and satisfies

$$m^*(A_2) \leq \sum_{i=1}^{\infty} l(I_i) + \frac{\varepsilon}{2} < m^*(E\backslash A_1) + \varepsilon. \tag{10.17}$$

By defining $G = A_1 \cup A_2$, we see that $G \supseteq E$ and $G\backslash E \subseteq A_2 \cup (A_1\backslash E)$. Hence, using (10.16) and (10.17),

$$m^*(G\backslash E) \leq m^*(A_2) + m^*(A_1\backslash E)$$
$$< m^*(E\backslash A_1) + m^*(A_1\backslash E) + \varepsilon$$
$$= m^*(E\Delta A_1) + \varepsilon = 3\varepsilon. \quad \square$$

EXERCISES 10.3

1. Let $(x_i : i \in \mathbb{N})$ be a sequence of positive numbers. If $\mu(E)$ is defined for each $E \subseteq \mathbb{R}$ by $\mu(E) = \sum_{x_i \in E} x_i$, prove that μ is a measure on $\mathcal{P}(\mathbb{R})$.

2. For any $E, F \in \mathcal{M}$, prove that $m(E) + m(F) = m(E \cup F) + m(E \cap F)$.

3. Given $E \in \mathcal{M}$, prove that $m^*(E) + m^*(F) = m^*(E \cup F) + m^*(E \cap F)$ for all $F \subseteq \mathbb{R}$. Use this to prove the equality in Exercise 10.3.2.

4. Let E be a subset of $[0,1]$ such that 5 does not appear in the decimal expansion of any $x \in E$. Prove that $E \in \mathcal{M}$ and that $m(E) = 0$.

5. If (E_n) is a sequence in $\mathcal{M}$, prove that
$$m(\liminf E_n) \leq \liminf m(E_n).$$
Assuming that $m(\bigcup_{i=N}^{\infty} E_i) < \infty$ for some N, prove that
$$m(\limsup E_n) \geq \limsup m(E_n).$$

6. Let $\mathcal{G}$ be the collection of open sets in $\mathbb{R}$ and $\mathcal{F}$ the collection of closed sets in $\mathbb{R}$. $\mathcal{G}_\delta$ will denote the class of all intersections of all sequences in $\mathcal{G}$, that is, $G \in \mathcal{G}_\delta$ if and only if there is a sequence of open sets $(G_i : i \in \mathbb{N})$ such that $G = \bigcap_{i=1}^{\infty} G_i$. Similarly $\mathcal{F}_\sigma$ is made up of the unions of sequences in $\mathcal{F}$, so that $F \in \mathcal{F}_\sigma$ if and only if there is a sequence $(F_i : i \in \mathbb{N})$ in $\mathcal{F}$ such that $F = \bigcup_{i=1}^{\infty} F_i$.

   (a) Prove that all sets in $\mathcal{G}_\delta$ and $\mathcal{F}_\sigma$ are Borel sets.
   (b) Give an example of a set in $\mathcal{G}_\delta$, and one of a set in $\mathcal{F}_\sigma$, such that neither set belongs to $\mathcal{G} \cup \mathcal{F}$.
   (c) Prove that, for every $E \subseteq \mathbb{R}$, there is a set $G \in \mathcal{G}_\delta$ such that $E \subseteq G$ and $m^*(E) = m(G)$.
   (d) If $E \in \mathcal{M}$, prove that there is a $G \in \mathcal{G}_\delta$ and an $F \in \mathcal{F}_\sigma$ such that $F \subseteq E \subseteq G$ and $m(G \backslash F) = 0$.

7. If G is an open, non-empty subset of $\mathbb{R}$, prove that $m(G) > 0$.

10.4 Measurable Functions

We now turn our attention from sets to functions. In Section 6.5 we saw that a function $f : \Omega \to \mathbb{R}$ is continuous if, and only if, the inverse image under f of every open subset of $\mathbb{R}$ is open in Ω. For a function defined from one topological space to another, this idea whereby f^{-1} preserves the topologies of the underlying spaces serves as the appropriate definition of continuity. We define measurability for a function along the same lines, with the σ-algebras $\mathcal{B}$ and $\mathcal{M}$ replacing the topologies. Throughout this section, Ω is assumed to be a (Lebesgue) measurable set.

Lebesgue Measure

Definition 10.12 A function $f : \Omega \to \mathbb{R}$ is said to be *Lebesgue measurable*, or simply *measurable*, if the set $f^{-1}(B)$ is measurable for every Borel set B, that is, if

$$B \in \mathcal{B} \Rightarrow f^{-1}(B) \in \mathcal{M}. \tag{10.18}$$

In order to include extended real functions in this definition, we say that $f : \Omega \to \bar{\mathbb{R}}$ is measurable if, in addition to (10.18) being satisfied, the sets $f^{-1}(\{-\infty\})$ and $f^{-1}(\{\infty\})$ are also measurable.

Definition 10.13 A function $f : \Omega \to \bar{\mathbb{R}}$ is *Borel measurable* if $f^{-1}(B) \in \mathcal{B}$ for every $B \in \mathcal{B}$, and $f^{-1}(\{-\infty\}), f^{-1}(\{\infty\}) \in \mathcal{B}$.

Note that, in this definition, $\Omega = f^{-1}(\bar{\mathbb{R}})$ is necessarily a Borel set.

Lemma 10.5 *Suppose $\mathcal{B}$ is generated by a collection $\mathcal{D}$ of subsets of $\mathbb{R}$. A function $f : \Omega \to \bar{\mathbb{R}}$ is measurable if, and only if,*
(i) $f^{-1}(D) \in \mathcal{M}$ for every $D \in \mathcal{D}$,
(ii) $f^{-1}(\{-\infty\}), f^{-1}(\{\infty\}) \in \mathcal{M}$.

Proof. Suppose (i) and (ii) are satisfied. By Theorem 10.4,

$$\mathcal{A}(f^{-1}(\mathcal{D})) = f^{-1}(\mathcal{A}(\mathcal{D})) = f^{-1}(\mathcal{B}).$$

From (i) we have $f^{-1}(\mathcal{D}) \subseteq \mathcal{M}$, hence $f^{-1}(\mathcal{B}) \subseteq \mathcal{M}$, which means f is measurable. The converse is obvious. $\square$

Theorem 10.16
For a function $f : \Omega \to \bar{\mathbb{R}}$, the following statements are equivalent:
(i) f is measurable.
(ii) For all $\alpha \in \mathbb{R}$, $\{x \in \Omega : f(x) > \alpha\}$ is measurable.
(iii) For all $\alpha \in \mathbb{R}$, $\{x \in \Omega : f(x) \geq \alpha\}$ is measurable.
(iv) For all $\alpha \in \mathbb{R}$, $\{x \in \Omega : f(x) < \alpha\}$ is measurable.
(v) For all $\alpha \in \mathbb{R}$, $\{x \in \Omega : f(x) \leq \alpha\}$ is measurable.

Proof
(i)$\Rightarrow$(ii): We have

$$\{x \in \Omega : f(x) > \alpha\} = f^{-1}((\alpha, \infty)) \cup f^{-1}(\{\infty\}).$$

Since f is measurable and (α, ∞) is a Borel set, $f^{-1}(\{\infty\})$ and $f^{-1}((\alpha, \infty))$ are both measurable, hence so is $\{x \in \Omega : f(x) > \alpha\}$.
(ii)$\Rightarrow$(i): By Theorem 10.3, the intervals $\{(\alpha, \infty) : \alpha \in \mathbb{R}\}$ generate $\mathcal{B}$. Furthermore, the sets

$$f^{-1}(\{\infty\}) = \bigcap_{n=1}^{\infty} \{x \in \Omega : f(x) > n\},$$

$$f^{-1}(\{-\infty\}) = \bigcap_{n=1}^{\infty} \{x \in \Omega : f(x) \leq -n\} = \bigcap_{n=1}^{\infty} \{x \in \Omega : f(x) > -n\}^c$$

are both measurable. Since

$$f^{-1}((\alpha, \infty)) = \{x \in \Omega : f(x) > \alpha\} \setminus \{x \in \Omega : f(x) = \infty\}$$

is also measurable, we conclude that f is measurable by Lemma 10.5.

The equivalence of (i) and (iii) is established by choosing $\mathcal{D}$ to be composed of all sets of the form $[\alpha, \infty)$, and using Lemma 10.5. To prove the equivalence of (ii) and (v) and that of (iii) and (iv), we resort to the fact that a σ-algebra is closed under complementation. □

Remarks 10.7

1. Suppose $f : \Omega \to \bar{\mathbb{R}}$ is a measurable function. If Ω_0 is a measurable subset of Ω, then $f|_{\Omega_0}$ is also measurable. Conversely, if $\Omega_1, \Omega_2 \in \mathcal{M}$ and $f|_{\Omega_1}, f|_{\Omega_2}$ are measurable, then so is $f : \Omega_1 \cup \Omega_2 \to \bar{\mathbb{R}}$.

2. Since every singleton $\{\alpha\}$ is a Borel set, the set $\{x \in \Omega : f(x) = \alpha\}$ is measurable for all $\alpha \in \mathbb{R}$ whenever f is measurable. The converse, however, is not true (see Exercise 10.4.2).

3. If $\Omega \in \mathcal{B}$, Lemma 10.4 and Theorem 10.16 are valid for Borel measurability as well. All that is needed is to replace $\mathcal{M}$ by $\mathcal{B}$, as the proof depends only on the properties shared by all σ-algebras.

4. Recalling the definition of neighborhoods of ∞ and $-\infty$, a function

$$f : \Omega \to \bar{\mathbb{R}}$$

is continuous at $x \in \Omega$ if, for every neighborhood V of $f(x)$, there is a neighborhood U of x such that $f(U) \subseteq V$. Thus a straightforward modification of Theorem 6.17, whereby $\mathbb{R}$ is replaced by $\bar{\mathbb{R}}$, leads to the conclusion that f is continuous if and only if, whenever G is open in $\mathbb{R}$ or is a neighborhood of ∞ or $-\infty$, $f^{-1}(G)$ is open in Ω. Suppose $f : \Omega \to \bar{\mathbb{R}}$ is continuous. Then

$$f^{-1}(\{\infty\}) = \bigcap_{n=1}^{\infty} f^{-1}((n, \infty]) \in \mathcal{M},$$

$$f^{-1}(\{-\infty\}) = \bigcap_{n=1}^{\infty} f^{-1}([-\infty, -n)) \in \mathcal{M},$$

and for any open set $G \subseteq \mathbb{R}$, $f^{-1}(G)$ is open and therefore measurable. Since the open sets in $\mathbb{R}$ generate $\mathcal{B}$, the function f is measurable. It is precisely to achieve this result, namely that continuous functions be measurable, that Definition 10.12 was designed. The "neater" requirement that the inverse image of a measurable set be measurable would not have achieved that.

5. Suppose $f : \Omega \to \mathbb{R}$ is Lebesgue measurable and $\varphi : \Gamma \to \bar{\mathbb{R}}$, where $\Gamma \supseteq f(\Omega)$, is Borel measurable. Then, with $E \in \mathcal{B}$, $E = \{\infty\}$, or $E = \{-\infty\}$, we have

$$(\varphi \circ f)^{-1}(E) = f^{-1}(\varphi^{-1}(E)) \in \mathcal{M}$$

since $\varphi^{-1}(E) \in \mathcal{B}$. Consequently, the composite function $\varphi \circ f$ is measurable. In particular, if φ is continuous, $\Gamma \in \mathcal{B}$, and f is measurable, then $\varphi \circ f$ is measurable. In general, we cannot relax the requirement that φ be Borel measurable to Lebesgue measurable and ensure the measurability of $\varphi \circ f$ (see Exercise 10.4.8). Of course, if $\Omega \in \mathcal{B}$ and φ and f are both Borel measurable, then so is $\varphi \circ f$.

In the next theorem we verify the closure of the set of measurable functions under the usual algebraic operations. We have at hand a variety of criteria for measurability, and the choices we make below are aimed as much at illustrating this variety as they are dictated by convenience. The reader is encouraged to try alternative proofs.

Theorem 10.17
If the functions $f, g : \Omega \to \bar{\mathbb{R}}$ are measurable, then each of the following functions is also measurable:
(i) $f + g : \Omega_0 \to \bar{\mathbb{R}}$, where

$$\Omega_0 = \{x \in \Omega : (f(x), g(x)) \neq (\infty, -\infty), (-\infty, \infty)\}.$$

(ii) $cf : \Omega \to \bar{\mathbb{R}}$, where $c \in \mathbb{R}$.
(iii) $|f| : \Omega \to [0, \infty]$.
(iv) $f^n : \Omega \to \bar{\mathbb{R}}$, where $n \in \mathbb{N}$.
(v) $1/f : \Omega \to \bar{\mathbb{R}}$, where $(1/f)(x) = \infty$ if $f(x) = 0$.
(vi) $fg : \Omega \to \bar{\mathbb{R}}$.

Proof
(i) Note that Ω is replaced here by Ω_0 in order to avoid the indeterminate forms $\infty - \infty$ and $-\infty + \infty$. Since the sets $f^{-1}(\{\infty\})$, $f^{-1}(\{-\infty\})$,

$g^{-1}(\{\infty\})$, and $g^{-1}(\{-\infty\})$ are all measurable, Ω_0 is itself measurable. Let $\alpha \in \mathbb{R}$ and consider the set

$$A = \{x \in \Omega_0 : f(x) + g(x) < \alpha\}$$
$$= \{x \in \Omega_0 : f(x) < \alpha - g(x)\}.$$

Since $\mathbb{Q}$ is dense in $\mathbb{R}$, for every $x \in \Omega$ there is a $q \in \mathbb{Q}$ such that $f(x) < q < \alpha - g(x)$. Consequently

$$A = \bigcup_{q \in \mathbb{Q}} \{x \in \Omega_0 : f(x) < q\} \cap \{x \in \Omega_0 : g(x) < \alpha - q\}.$$

But for each $q \in \mathbb{Q}$, the set $\{x \in \Omega_0 : f(x) < q\} \cap \{x \in \Omega_0 : g(x) < \alpha - q\}$ is measurable, hence $A \in \mathcal{M}$.

(ii) If $c = 0$ then the function $cf = 0$ on Ω is measurable because it is continuous. Suppose, therefore, that $c \neq 0$. In this case, for all $\alpha \in \mathbb{R}$,

$$\{x \in \Omega : cf(x) > \alpha\} = \{x \in \Omega : f(x) > \alpha/c\} \quad \text{if } c > 0,$$
$$\{x \in \Omega : cf(x) > \alpha\} = \{x \in \Omega : f(x) < \alpha/c\} \quad \text{if } c < 0.$$

Since both of these sets are measurable, cf is measurable for all c.

(iii) For all $\alpha \geq 0$, the set

$$\{x \in \Omega : |f(x)| > \alpha\} = \{x \in \Omega : f(x) > \alpha\} \cup \{x \in \Omega : f(x) < -\alpha\}$$

is clearly measurable. If $\alpha < 0$, then $\{x \in \Omega : |f(x)| > \alpha\} = \Omega \in \mathcal{M}$.

(iv) Let $\Omega_1 = \{x \in \Omega : |f(x)| < \infty\}$ and $g = f|_{\Omega_1}$. Then $\Omega_1 \in \mathcal{M}$ and g is measurable. Since the function $\varphi(x) = x^n$ is continuous on $\mathbb{R}$, it follows from Remark 10.7.5 above that $g^n = \varphi \circ g$ is measurable. Also $f^n(x) = \infty$ if and only if $|f(x)| = \infty$, so $(f^n)^{-1}(\{\infty\}) \in \mathcal{M}$. On the other hand, $(f^n)^{-1}(\{-\infty\})$ is either empty or $f^{-1}(\{-\infty\})$, and is therefore measurable. Finally, noting that every Borel set B lies in $\mathbb{R}$, we have

$$(f^n)^{-1}(B) = (g^n)^{-1}(B) \in \mathcal{M},$$

so we conclude that f^n is measurable.

(v) Let $\Omega_2 = \{x \in \Omega : f(x) = 0\}$, $\Omega_3 = \{x \in \Omega : f(x) = -\infty\}$, and $\Omega_4 = \Omega \setminus (\Omega_2 \cup \Omega_3)$. We shall prove that the functions

$$g = \frac{1}{f}\bigg|_{\Omega_2}, \quad h = \frac{1}{f}\bigg|_{\Omega_3}, \quad k = \frac{1}{f}\bigg|_{\Omega_4}$$

are all measurable. By definition of $1/f$, we have $\{x \in \Omega_2 : g(x) < \alpha\} = \varnothing$ for all $\alpha \in \mathbb{R}$, so this set is measurable. Furthermore,

$$\{x \in \Omega_3 : h(x) < \alpha\} = \begin{cases} \Omega_3, & \alpha > 0 \\ \varnothing, & \alpha \leq 0, \end{cases}$$

and in either case this set is also measurable. Finally we can write $k = \varphi \circ f$, where φ is the continuous function defined on $\mathbb{R}\setminus\{0\}$ by $\varphi(x) = 1/x$. Hence k is measurable.

(vi) It is a simple matter to ascertain that the product fg restricted to any of the sets $\{x : f(x) = \infty\}$, $\{x : f(x) = -\infty\}$, $\{x : g(x) = \infty\}$, $\{x : g(x) = -\infty\}$ is measurable. Therefore we may assume that f and g are real valued on all of Ω. Since

$$fg = \frac{1}{2}\left[(f+g)^2 - f^2 - g^2\right],$$

we conclude from (i), (ii), and (iv) that fg is measurable. $\square$

Using the function $f - g$ in Theorem 10.17, we have

Corollary 10.17.1 *If $f, g : \Omega \to \bar{\mathbb{R}}$ are measurable functions, then the sets $\{x : f(x) < g(x)\}$, $\{x : f(x) \leq g(x)\}$, and $\{x : f(x) = g(x)\}$ are all measurable.*

For any two functions $f : D_f \to \bar{\mathbb{R}}$ and $g : D_g \to \bar{\mathbb{R}}$, we have already defined the functions $f \vee g$ and $f \wedge g$ on $D_f \cap D_g$ in Exercise 6.2.3 by

$$(f \vee g)(x) = f(x) \vee g(x) = \max\{f(x), g(x)\},$$
$$(f \wedge g)(x) = f(x) \wedge g(x) = \min\{f(x), g(x)\}.$$

We now define the *positive part* and the *negative part* of f, respectively, by

$$f^+ = f \vee 0 : D_f \to [0, \infty],$$
$$f^- = -(f \wedge 0) : D_f \to [0, \infty].$$

Noting that

$$f \vee g = \frac{1}{2}(f + g + |f - g|),$$
$$f \wedge g = \frac{1}{2}(f + g - |f - g|),$$

and using Theorem 10.17, we can state

Corollary 10.17.2 *If the functions $f, g : \Omega \to \bar{\mathbb{R}}$ are measurable, then $f \vee g$, $f \wedge g$, f^+, and f^- are all measurable.*

An important feature of measurability is its closure under limiting operations, whether we are dealing with a sequence of sets, as we have already seen, or a sequence of functions, as we shall now show. Let

$$f_n : D \to \bar{\mathbb{R}}, \ n \in \mathbb{N},$$

be a sequence of extended real functions. For every $x \in D$, we define

$$\sup f_n : x \mapsto \sup\{f_n(x) : n \in \mathbb{N}\},$$
$$\inf f_n : x \mapsto \inf\{f_n(x) : n \in \mathbb{N}\},$$
$$\limsup f_n : x \mapsto \limsup f_n(x),$$
$$\liminf f_n : x \mapsto \liminf f_n(x).$$

These functions are therefore well defined from D to $\bar{\mathbb{R}}$. As we already know from Theorem 3.16, the pointwise limit of f_n exists at every point $x \in D$ where $\limsup f_n(x) = \liminf f_n(x)$. The subset of D where this equality holds is the *domain of existence* of $\lim f_n$.

Theorem 10.18
For any sequence $f_n : \Omega \to \bar{\mathbb{R}}$ of measurable functions, the functions $\sup f_n$, $\inf f_n$, $\limsup f_n$, $\liminf f_n$ are all measurable on Ω, and $\lim f_n$ is measurable on its domain of existence.

Proof. Since f_n is measurable for every $n \in \mathbb{N}$, the sets

$$\{x : \sup f_n(x) > \alpha\} = \bigcup_{n=1}^{\infty} \{x : f_n(x) > \alpha\},$$
$$\{x : \inf f_n(x) < \alpha\} = \bigcup_{n=1}^{\infty} \{x : f_n(x) < \alpha\}$$

are both measurable, which means that $\sup f_n$ and $\inf f_n$ are measurable. Noting that

$$\limsup f_n = \inf_{n \in \mathbb{N}} g_n, \quad g_n = \sup_{k \geq n} f_k,$$
$$\liminf f_n = \sup_{n \in \mathbb{N}} h_n, \quad h_n = \inf_{k \geq n} f_k,$$

we conclude that the functions $g_n, h_n, \limsup f_n$, and $\liminf f_n$ are all measurable. The measurability of $\limsup f_n$ and $\liminf f_n$ implies, by Corollary 10.17.1, that

$$\Omega_0 = \{x \in \Omega : \limsup f_n(x) = \liminf f_n(x)\},$$

which is the domain of existence of $\lim f_n$, is measurable. Therefore $\lim f_n = \limsup f_n|_{\Omega_0}$ is measurable. ⌑

Definition 10.14
(i) The functions $f, g : \Omega \to \bar{\mathbb{R}}$ are said to be equal *almost everywhere* (a.e.) if they are equal except on a set of measure 0, that is, if
$$m(\{x \in \Omega : f(x) \neq g(x)\}) = 0,$$
and we express this symbolically by writing
$$f = g \quad (\text{a.e.}).$$
Similarly $f \geq g$ (a.e.) means $m\{x \in \Omega : f(x) < g(x)\} = 0$, and $f > g$ (a.e.), $f \leq g$ (a.e.), $f < g$ (a.e.) are defined in the same way.
(ii) The sequence of functions $f_n : \Omega \to \bar{\mathbb{R}}$ are said to converge to f *almost everywhere* if
$$m(\{x \in \Omega : \lim f_n(x) \neq f(x)\}) = 0,$$
in which case we write
$$\lim f_n = f \quad (a.e.) \quad \text{or} \quad f_n \to f \quad (\text{a.e.}).$$

Theorem 10.19
If the functions $f, g : \Omega \to \bar{\mathbb{R}}$ are equal almost everywhere and the function f is measurable, then g is also measurable.

Proof. Let
$$E = \{x \in \Omega : f(x) \neq g(x)\}.$$
Then, for any $\alpha \in \mathbb{R}$,
$$\{x : g(x) > \alpha\} = (\{x : f(x) > \alpha\} \cap E^c) \cup (\{x : g(x) > \alpha\} \cap E). \quad (10.19)$$
Since $m(E) = 0$, both E and E^c are measurable, hence, f being measurable,
$$\{x : f(x) > \alpha\} \cap E^c \in \mathcal{M}.$$
Since $m(E) = 0$ and $\{x : g(x) > \alpha\} \cap E \subseteq E$, the completeness of m implies that $\{x : g(x) > \alpha\} \cap E$ is measurable. By equation (10.19), the set $\{x : g(x) > \alpha\}$, and therefore g, is measurable. □

Since Borel measure is not complete (Exercise 10.4.9), this theorem is not valid if we use Borel measurability, and that is the main advantage of Lebesgue measurability.

Definition 10.15 For any set $E \subseteq \mathbb{R}$, the function χ_E defined on $\mathbb{R}$ by

$$\chi_E(x) = \begin{cases} 1, & x \in E \\ 0, & x \in E^c, \end{cases}$$

is called the *characteristic function* of E.

Clearly, χ_E is measurable if, and only if, E is measurable.

Definition 10.16 If $\{E_1, E_2, \cdots, E_n\}$ is a finite collection of disjoint, measurable sets, then any linear combination of the corresponding characteristic functions $\chi_{E_1}, \chi_{E_2}, \cdots, \chi_{E_n}$,

$$\varphi = \sum_{i=1}^{n} c_i \chi_{E_i}, \quad c_i \in \mathbb{R}, \tag{10.20}$$

is called a *simple function*.

Thus a simple function is a measurable function whose range is a finite set of real numbers. Clearly, the representation (10.20) of φ is not unique. It becomes unique if $c_i \neq c_j$ whenever $i \neq j$, in which case $E_i = \varphi^{-1}(\{c_i\})$, $D_\varphi = \bigcup_{i=1}^{n} E_i$, and $\varphi(D_\varphi) = \{c_1, c_2, \cdots, c_n\}$. Equation (10.20) is then said to be a *canonical representation* of φ. Algebraic combinations of simple functions, such as sums and products, are also simple functions; for if

$$\varphi = \sum_{i=1}^{n} c_i \chi_{E_i}, \quad \psi = \sum_{i=1}^{m} d_i \chi_{F_i},$$

then it is straightforward to verify that

$$k\varphi = \sum_{i=1}^{n} k c_i \chi_{E_i},$$

$$\varphi + \psi = \sum_{i=1}^{n} \sum_{j=1}^{m} (c_i + d_j) \chi_{E_i \cap F_j},$$

$$\varphi\psi = \sum_{i=1}^{n} \sum_{j=1}^{m} (c_i d_j) \chi_{E_i \cap F_j},$$

$$\varphi \vee \psi = \sum_{i=1}^{n} \sum_{j=1}^{m} (c_i \vee d_j) \chi_{E_i \cap F_j},$$

$$\varphi \wedge \psi = \sum_{i=1}^{n} \sum_{j=1}^{m} (c_i \wedge d_j) \chi_{E_i \cap F_j}.$$

We end this section with an important density theorem, one which will be used in the next chapter when we define the Lebesgue integral.

Theorem 10.20

If the function $f : \Omega \to [0, \infty]$ is Lebesgue measurable, then there is a sequence $(\varphi_n : n \in \mathbb{N})$ of non-negative simple functions such that

(i) $\varphi_{n+1}(x) \geq \varphi_n(x)$ *for all $n \in \mathbb{N}$ and $x \in \Omega$,*
(ii) $\lim_{n \to \infty} \varphi_n(x) = f(x)$ *for all $x \in \Omega$.*

We shall often abbreviate (i) and (ii) by writing $\varphi_n \nearrow f$.

Proof. For every $n \in \mathbb{N}$ we define the sets

$$F_{n,i} = f^{-1}([(i-1)/2^n, i/2^n)), \ i \in \{1, 2, \cdots, n2^n\},$$
$$F_{n,\infty} = f^{-1}([n, \infty]) = f^{-1}([n, \infty)) \cup f^{-1}(\{\infty\}),$$

and the simple function

$$\varphi_n = \sum_{i=1}^{n2^n} \frac{i-1}{2^n} \chi_{F_{n,i}} + n \chi_{F_{n,\infty}},$$

which is measurable since all of the intervals $\left[\dfrac{i-1}{2^n}, \dfrac{i}{2^n}\right]$ and $[n, \infty)$ are Borel sets.

(i) Let $x \in F_{n,i}$ where $1 \leq i \leq n2^n$, that is

$$\frac{i-1}{2^n} \leq f(x) < \frac{i}{2^n}.$$

There are two possibilities:

(a) $\dfrac{i-1}{2^n} \leq f(x) < \dfrac{2i-1}{2^{n+1}}$,

(b) $\dfrac{2i-1}{2^{n+1}} \leq f(x) < \dfrac{i}{2^n}$.

In case (a), $x \in F_{n+1, 2i-1}$ and therefore

$$\varphi_{n+1}(x) = \frac{2i-2}{2^{n+1}} = \frac{i-1}{2^n} = \varphi_n(x).$$

In case (b), $x \in F_{n+1, 2i}$ and therefore

$$\varphi_{n+1}(x) = \frac{2i-1}{2^{n+1}} > \frac{i-1}{2^n} = \varphi_n(x).$$

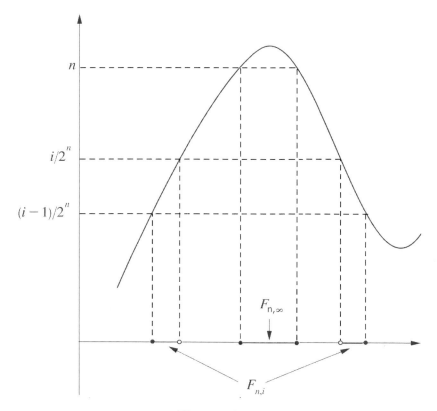

Figure 10.1

If $x \in F_{n,\infty}$ then either $n \leq f(x) < n+1$, or $f(x) \geq n+1$. In the first case,

$$\frac{i-1}{2^{n+1}} \leq f(x) < \frac{i}{2^{n+1}}, \quad n2^{n+1}+1 \leq i < (n+1)2^{n+1}$$
$$\Rightarrow \varphi_{n+1}(x) = \frac{i-1}{2^{n+1}} \geq n = \varphi_n(x).$$

In the second case, $\varphi_{n+1}(x) = n+1 > n = \varphi_n(x)$.

(ii) If $f(x) = \infty$, then $\varphi_n(x) = n \to \infty$. Suppose, therefore, that $f(x) < N$ for some positive integer N. For all $n \geq N$, there is a positive integer $i \leq 2^n N$ such that

$$\frac{i-1}{2^n} \leq f(x) < \frac{i}{2^n},$$

which implies

$$0 \le f(x) - \frac{i-1}{2^n} = f(x) - \varphi_n(x) < \frac{1}{2^n}.$$

Hence $\varphi_n(x) \to f(x)$. □

EXERCISES 10.4

1. Give an example of a Lebesgue non-measurable function, and one of a Borel non-measurable function.

2. Suppose the set E is non-measurable and the function $f : \mathbb{R} \to \mathbb{R}$ is defined by
$$f(x) = \begin{cases} e^x, & x \in E \\ -e^x, & x \notin E. \end{cases}$$
Prove that $\{x : f(x) = c\}$ is measurable for every $c \in \mathbb{R}$, but that f is not measurable.

3. According to Theorem 10.17, $|f|$ is measurable whenever f is measurable. Prove that the converse is false by exhibiting a non-measurable function whose absolute value is measurable.

4. Prove the following equalities:

   (a) $\chi_{A \cap B} = \chi_A \cdot \chi_B$.
   (b) $\chi_{A \cup B} = \chi_A + \chi_B - \chi_{A \cap B}$.
   (c) $\chi_{A \triangle B} = |\chi_A - \chi_B|$.
   (d) $\chi_{A^c} = 1 - \chi_A$.

5. Given a sequence of sets (E_n) with $E^* = \limsup E_n$ and $E_* = \liminf E_n$, prove that $\chi_{E^*} = \limsup \chi_{E_n}$ and $\chi_{E_*} = \liminf \chi_{E_n}$.

6. (a) If $\Omega_1, \Omega_2 \in \mathcal{M}$ and f is a function defined on $\Omega = \Omega_1 \cup \Omega_2$, prove that f is measurable if, and only if, $f|_{\Omega_1}$ and $f|_{\Omega_2}$ are both measurable.

   (b) If $f : \Omega \to \mathbb{R}$ is measurable and $g : \mathbb{R} \to \mathbb{R}$ is defined by
   $$g(x) = \begin{cases} f(x), & x \in \Omega \\ 0, & x \notin \Omega, \end{cases}$$
   prove that g is measurable. This implies that, for the purpose of studying the properties of measurable functions, we need only consider functions defined on all of $\mathbb{R}$.

7. Let $x = \sum_{i=1}^{\infty} s_i/2^i$ be the binary expansion of $x \in [0,1]$ which does not end in a string of 1's if $x > 0$. Define $\Psi : [0,1] \to \mathbb{R}$ by $\Psi(x) = \sum_{i=1}^{\infty} 2s_i/3^i$, so that the range of Ψ is included in the Cantor set (see Example 3.19).

   (a) Prove that Ψ is injective and measurable.
   (b) Prove that $\Psi(E)$ is measurable for every $E \subseteq [0,1]$.
   (c) Prove that $\mathcal{B} \neq \mathcal{M}$ by choosing $E \notin \mathcal{M}$ and showing that $\Psi(E) \in \mathcal{M}$ but $\Psi(E) \notin \mathcal{B}$.

   The function Ψ is often referred to as *Cantor's function*.

8. Let Ψ be the the Cantor function defined in Exercise 10.4.7. Let E be a non-measurable set, and define $g : \mathbb{R} \to \mathbb{R}$ by

$$g(x) = \begin{cases} 1, & x \in \Psi(E) \\ 0, & \text{otherwise.} \end{cases}$$

   Prove that g is measurable but that $g \circ \Psi$ is not.

9. The result of Exercise 10.4.7 shows that the Cantor set contains a subset which is not in $\mathcal{B}$. Use this to conclude that the Borel measure $m|_{\mathcal{B}}$ is not complete.

10. Prove that the convergence $\varphi_n \nearrow f$ in Theorem 10.20 is uniform if f is bounded.

11

Lebesgue Integration

Until the early part of the twentieth century, the standard theory of integration was that of Riemann, which was presented in Chapter 8. Even today, scientists and engineers rarely need to look beyond the Riemann integral in their work. Its definition is simple and clearly motivated as a measure of area, it is well suited to formulating physical laws and performing computations (both analytically and numerically), and it articulates the relationship between integration and differentiation through the fundamental theorem of calculus. The drawback of the theory is that, from an analytical point of view, the class of Riemann integrable functions is not wide enough. Consider the following observations:

1. In Example 8.2 we saw that the function defined on $\mathbb{R}$ by

$$f(x) = \begin{cases} 1, & x \in \mathbb{Q} \\ 0, & x \in \mathbb{Q}^c \end{cases} \quad (11.1)$$

   is not Riemann integrable on any bounded interval, though the function is almost constant ($\mathbb{Q}$ being countable and hence of measure 0).

2. The set of Riemann integrable functions $\mathcal{R}(a,b)$ is not closed under pointwise convergence, as we saw in several examples in Chapter 9. It is not even closed under monotonic pointwise convergence, as the following example demonstrates: Let $(f_n : n \in \mathbb{N})$ be the sequence of functions defined on $[0,1]$ by

$$f_n(x) = \begin{cases} 1, & x = p/q \in \mathbb{Q} \cap [0,1], q \leq n \\ 0, & \text{otherwise,} \end{cases}$$

   where p and q have no common factors. $f_n \in \mathcal{R}(0,1)$ because f_n is continuous except at a finite number of points. Clearly, $f_n(x)$ increases with n and tends to the function $f(x)$ in equation (11.1) as $n \to \infty$.

3. To ensure the closure of $\mathcal{R}(a,b)$ under a limit operation, we have to restrict the operation to something stronger than pointwise convergence, such as uniform convergence (see Theorem 9.4).

4. The Riemann integral is defined for bounded functions over bounded intervals, and its extension beyond that requires a separate treatment.

The Lebesgue integral, which is the subject of this chapter, is based on a more general definition which enlarges the class of Riemann integrable functions, and thereby avoids some of these shortcomings. In fact, such an extension may be viewed as a sort of "completion" of $\mathcal{R}(a,b)$, under certain limiting operations, which addresses the deficiencies of $\mathcal{R}(a,b)$ mentioned above.

11.1 Definition of The Lebesgue Integral

At the outset we assume that $\Omega \subseteq \mathbb{R}$ is a Lebesgue measurable set. The set of Lebesgue measurable functions $f : \Omega \to \bar{\mathbb{R}}$ will be denoted by $\mathcal{L}^0(\Omega)$, and its subset of simple functions by $\mathcal{S}(\Omega)$. $\mathcal{L}^0_+(\Omega)$ and $\mathcal{S}_+(\Omega)$ will denote the subsets of non-negative functions in $\mathcal{L}^0(\Omega)$ and $\mathcal{S}(\Omega)$, respectively. When Ω coincides with $\mathbb{R}$ or is understood from the context, we shall often write $\mathcal{L}^0, \mathcal{S}, \mathcal{L}^0_+$, or $\mathcal{S}_+$.

The definition of the Lebesgue integral of f on Ω will be given in three stages.

Step 1: $f \in \mathcal{S}_+(\Omega)$

Here f can be represented by the sum

$$f = \sum_{i=1}^{n} c_i \chi_{E_i}, \tag{11.2}$$

where $c_i \geq 0$, $E_i \in \mathcal{M}$, and $\{E_i : 1 \leq i \leq n\}$ is a partition of Ω, so that Ω is a disjoint union of the sets $E_1, \cdots, E_n$. The integral of f on Ω, with respect to the measure m, is defined by

$$\int_\Omega f\, dm = \sum_{i=1}^{n} c_i m(E_i), \tag{11.3}$$

where we take $0 \cdot \infty = 0$. The left-hand side of (11.3) is also written as $\int_\Omega f(x)\, dm(x)$. To show that this definition does not depend on the representation (11.2) of f, let

$$f = \sum_{j=1}^{k} d_j \chi_{F_j},$$

where $\{F_j : 1 \leq j \leq k\}$ is any other partition of Ω by measurable sets. We have to prove that

$$\sum_{j=1}^{k} d_j m(F_j) = \sum_{i=1}^{n} c_i m(E_i).$$

Since $F_j = \bigcup_{i=1}^{n} E_i \cap F_j$, for all $1 \leq j \leq k$, is a disjoint union of the collection $\{E_i \cap F_j : 1 \leq i \leq n\}$, we have

$$m(F_j) = \sum_{i=1}^{n} m(E_i \cap F_j).$$

Therefore

$$\sum_{j=1}^{k} d_j m(F_j) = \sum_{j=1}^{k} \sum_{i=1}^{n} d_j m(E_i \cap F_j) = \sum_{(i,j) \in \Lambda} d_j m(E_i \cap F_j), \quad (11.4)$$

where $\Lambda = \{(i,j) : 1 \leq i \leq n,\ 1 \leq j \leq k,\ E_i \cap F_j \neq \varnothing\}$. Similarly,

$$\sum_{i=1}^{n} c_i m(E_i) = \sum_{(i,j) \in \Lambda} c_i m(E_i \cap F_j). \quad (11.5)$$

Now we note that, whenever $E_i \cap F_j \neq \varnothing$, there is a point x such that $x \in E_i$ and $x \in F_j$, which implies $f(x) = c_i = d_j$, and hence the equality of the sums (11.4) and (11.5).

Remark 11.1 Let

$$f = \sum_{i=1}^{n} c_i \chi_{E_i},$$

where $c_i \geq 0$, $E_i \in \mathcal{M}$, and the collection $\{E_i : 1 \leq i \leq n\}$ is pairwise disjoint and contained in Ω, but the union $\bigcup_{i=1}^{n} E_i$ is a proper subset of Ω. By writing

$$f = \sum_{i=1}^{n} c_i \chi_{E_i} + 0 \cdot \chi_E,$$

where $E = \Omega \setminus \bigcup_{i=1}^{n} E_i$, we see that $f \in \mathcal{S}_+(\Omega)$ and that

$$\int_\Omega f\, dm = \sum_{i=1}^{n} c_i m(E_i).$$

In particular, if A is a measurable subset of Ω, then

$$\int_\Omega \chi_A\, dm = m(A).$$

Lemma 11.1
(i) If $f, g \in \mathcal{S}_+$ and $\alpha, \beta \geq 0$, then
$$\int_\Omega (\alpha f + \beta g) dm = \alpha \int_\Omega f dm + \beta \int_\Omega g dm.$$
(ii) If $f, g \in \mathcal{S}_+$ and $f(x) \geq g(x)$ for all $x \in \Omega$, then
$$\int_\Omega f dm \geq \int_\Omega g dm.$$

The proof is left as an exercise.

Step 2: $f \in \mathcal{L}^0_+(\Omega)$

Definition 11.1 The integral of any function f in $\mathcal{L}^0_+(\Omega)$ is defined as
$$\int_\Omega f dm = \sup \left\{ \int_\Omega \varphi dm : \varphi \in \mathcal{S}_+(\Omega), \varphi \leq f \right\},$$
and it is straightforward to verify that this definition agrees with (11.3) when f is a simple function.

We saw in Theorem 10.20 that, for every $f \in \mathcal{L}^0_+(\Omega)$, there is a sequence (φ_n) in $\mathcal{S}_+(\Omega)$ such that $\varphi_n \nearrow f$. Now we shall use this result to arrive at a formula for evaluating $\int_\Omega f dm$.

Lemma 11.2 Let $\psi \in \mathcal{S}_+(\Omega)$. If (φ_n) is a sequence in $\mathcal{S}_+(\Omega)$ such that $\varphi_n \nearrow \psi$, then
$$\lim_{n \to \infty} \int_\Omega \varphi_n dm = \int_\Omega \psi dm.$$

Proof. It should first be noted that $\lim_{n \to \infty} \int_\Omega \varphi_n dm$ exists in $\bar{\mathbb{R}}$, owing to the monotonicity of the integral on $\mathcal{S}_+$.

Suppose $\psi = \sum_{i=1}^k c_i \chi_{E_i}$. There are two cases: (i) $\int \psi dm = \infty$ and (ii) $\int \psi dm < \infty$.

(i) $\int \psi dm = \infty$

In this case there is a j in $\{1, 2, \cdots, k\}$ such that $c_j > 0$ and $m(E_j) = \infty$. For any $\varepsilon \in (0, c_j)$, define
$$A_n = \{ x \in \Omega : \varphi_n(x) > \psi(x) - \varepsilon \}, \; n \in \mathbb{N}. \tag{11.6}$$

A_n is clearly measurable and, since $\varphi_n \nearrow \psi$, $A_n \subseteq A_{n+1}$ for every n. Furthermore, $\lim A_n = \Omega$. Using the continuity of m from below applied to the sequence $(A_n \cap E_j)$, we obtain
$$\lim_{n \to \infty} m(A_n \cap E_j) = m(E_j) = \infty. \tag{11.7}$$

Since
$$\varphi_n \geq \varphi_n \chi_{A_n \cap E_j} \geq (\psi - \varepsilon)\chi_{A_n \cap E_j} = (c_j - \varepsilon)\chi_{A_n \cap E_j},$$
the monotonicity of the integral on $\mathcal{S}_+$ implies
$$\int_\Omega \varphi_n dm \geq \int_\Omega (c_j - \varepsilon)\chi_{A_n \cap E_j} = (c_j - \varepsilon)m(A_n \cap E_j).$$
Now equation (11.7) completes the proof.

(ii) $\int \psi dm < \infty$

The monotonicity of the integral implies $\int \varphi_n dm \leq \int \psi dm$, from which we conclude that
$$\lim_{n \to \infty} \int \varphi_n dm \leq \int \psi dm,$$
and it remains to show that
$$\lim_{n \to \infty} \int \varphi_n dm \geq \int \psi dm. \tag{11.8}$$

In this case $m(E_i) < \infty$ when $c_i > 0$. Let $I = \{i : c_i > 0\}$ and $\Omega^* = \bigcup_{i \in I} E_i$, so that $m(\Omega^*) < \infty$. With $\varepsilon > 0$ and A_n as defined in (11.6), we have
$$\varphi_n + \varepsilon \chi_{A_n \cap \Omega^*} \geq (\varphi_n + \varepsilon)\chi_{A_n \cap \Omega^*} \geq \psi \chi_{A_n \cap \Omega^*} = \sum_{i \in I} c_i \chi_{A_n \cap E_i}$$
$$\Rightarrow \int_\Omega \varphi_n dm + \varepsilon m(A_n \cap \Omega^*) \geq \sum_{i \in I} c_i m(A_n \cap E_i)$$
$$\Rightarrow \lim_{n \to \infty} \int_\Omega \varphi_n dm + \varepsilon m(\Omega^*) \geq \sum_{i \in I} c_i m(E_i) = \int_\Omega \psi dm.$$

Since $m(\Omega^*)$ is finite and $\varepsilon > 0$ is arbitrary, (11.8) follows. $\square$

Theorem 11.1
Let $f \in \mathcal{L}_+^0(\Omega)$ and suppose (φ_n) is a sequence in $\mathcal{S}_+(\Omega)$ such that $\varphi_n \nearrow f$. Then
$$\lim_{n \to \infty} \int_\Omega \varphi_n dm = \int_\Omega f dm.$$

Proof. The monotonicity property of the integral over $\mathcal{S}_+$ guarantees the existence of the limit. By the definition of $\int_\Omega f dm$ we have
$$\lim_{n \to \infty} \int_\Omega \varphi_n dm \leq \int_\Omega f dm,$$

and we have to prove this inequality in the reverse direction. Suppose φ is any function in $\mathcal{S}_+(\Omega)$ such that $\varphi \leq f$. The sequence $(\varphi \wedge \varphi_n)$ is then monotonically increasing in $\mathcal{S}_+(\Omega)$ and converges to φ. Using Lemma 11.2, we conclude that

$$\int_\Omega \varphi\, dm = \lim_{n\to\infty} \int_\Omega (\varphi \wedge \varphi_n)\, dm.$$

Since $\varphi \wedge \varphi_n \leq \varphi_n$,

$$\int_\Omega (\varphi \wedge \varphi_n)\, dm \leq \int_\Omega \varphi_n\, dm$$
$$\Rightarrow \int_\Omega \varphi\, dm \leq \lim_{n\to\infty} \int_\Omega \varphi_n\, dm. \tag{11.9}$$

But φ is an arbitrary function in $\mathcal{S}_+(\Omega)$ which satisfies $\varphi \leq f$, hence, by Definition 11.1 and inequality (11.9), we have

$$\int_\Omega f\, dm = \sup\left\{\int_\Omega \varphi\, dm : \varphi \in \mathcal{S}_+(\Omega), \varphi \leq f\right\} \leq \lim_{n\to\infty} \int_\Omega \varphi_n\, dm. \quad \square$$

Example 11.1 Given $f : [0,1] \to \mathbb{R}$, $f(x) = x$, evaluate $\int_{[0,1]} f\, dm$.

Solution. The sequence of simple functions

$$\varphi_n = \sum_{i=1}^{2^n} \frac{i-1}{2^n} \chi_{[(i-1)/2^n, i/2^n)}$$

clearly lies in $\mathcal{S}_+$ and satisfies $\varphi_n \nearrow f$ (see Theorem 10.20). By Theorem 11.1,

$$\int_{[0,1]} f\, dm = \lim_{n\to\infty} \int_{[0,1]} \varphi_n\, dm$$
$$= \lim_{n\to\infty} \sum_{i=1}^{2^n} \frac{i-1}{2^n} m\left(\left[\frac{i-1}{2^n}, \frac{i}{2^n}\right)\right)$$
$$= \lim_{n\to\infty} \frac{1}{2^{2n}} \sum_{i=1}^{2^n} (i-1)$$
$$= \lim_{n\to\infty} \frac{1}{2^{2n}} \frac{2^n(2^n-1)}{2} = \frac{1}{2}.$$

Example 11.2 Let (a_n) be a sequence in $(0, \infty)$ and suppose f is defined on $[1, \infty)$ by

$$f(x) = a_{[x]},$$

where $[x]$ is the integer part of x.

To calculate $\int_{[1,\infty)} f\,dm$, set $\varphi_n = \sum_{i=1}^{n} a_i \chi_{[i,i+1)}$, which is in $\mathcal{S}_+$, and note that $\varphi_n \nearrow f$. Therefore

$$\int_{[1,\infty)} f\,dm = \lim_{n\to\infty} \int_{[1,\infty)} \varphi_n\,dm$$
$$= \lim_{n\to\infty} \sum_{i=1}^{n} a_i m([i, i+1))$$
$$= \sum_{i=1}^{\infty} a_i.$$

The following lemma shows that linearity and monotonicity are preserved in the extension of the integral to $\mathcal{L}_+^0$.

Lemma 11.3 *Let $f, g \in \mathcal{L}_+^0(\Omega)$.*
(i) For any $\alpha, \beta \geq 0$,

$$\int_\Omega (\alpha f + \beta g)\,dm = \alpha \int_\Omega f\,dm + \beta \int_\Omega g\,dm.$$

(ii) If $f(x) \geq g(x)$ for all $x \in \Omega$, then

$$\int_\Omega f\,dm \geq \int_\Omega g\,dm.$$

Proof
(i) This is left as an exercise.
(ii) Suppose $\varphi \in \mathcal{S}_+(\Omega)$ is such that $\varphi \leq g$. Then $\varphi \leq f$, and hence

$$\int_\Omega f\,dm \geq \int_\Omega \varphi\,dm.$$

$\int_\Omega f\,dm$ is therefore an upper bound for the set

$$\left\{ \int_\Omega \varphi\,dm : \varphi \in \mathcal{S}_+(\Omega), \varphi \leq g \right\},$$

and we immediately conclude that

$$\int_\Omega f\,dm \geq \sup\left\{ \int_\Omega \varphi\,dm : \varphi \in \mathcal{S}_+(\Omega), \varphi \leq g \right\} = \int_\Omega g\,dm. \quad \square$$

Step 3: $f \in \mathcal{L}^0(\Omega)$

Both functions $f^+ = f \vee 0$ and $f^- = -(f \wedge 0)$ lie in $\mathcal{L}_+^0(\Omega)$, hence the integrals $\int_\Omega f^+\,dm$ and $\int_\Omega f^-\,dm$ are well defined. If $\int_\Omega f^+\,dm < \infty$ or $\int_\Omega f^-\,dm < \infty$, the *Lebesgue integral* of f on Ω with respect to the measure m is defined (in $\bar{\mathbb{R}}$) as

$$\int_\Omega f\,dm = \int_\Omega f^+\,dm - \int_\Omega f^-\,dm. \qquad (11.10)$$

f is said to be *integrable on Ω with respect to m*, or is *Lebesgue integrable on Ω*, if $\int_\Omega f\,dm$ exists in $\mathbb{R}$, that is, if both $\int_\Omega f^+\,dm$ and $\int_\Omega f^-\,dm$ are finite. We shall denote the set of Lebesgue integrable functions on Ω by $\mathcal{L}^1(\Omega)$, which is clearly a subset of $\mathcal{L}^0(\Omega)$. Once again, when there is no likelihood of confusion, we shall drop the *Lebesgue* qualification.

For any measurable subset E of Ω, the function $f : \Omega \to \bar{\mathbb{R}}$ is *integrable on E* if its restriction $f|_E$ is integrable on E, in which case we shall write $f \in \mathcal{L}^1(E)$ instead of $f|_E \in \mathcal{L}^1(E)$, and $\int_E f\,dm$ instead of $\int_E f|_E\,dm$.

Remarks 11.2

1. If f is integrable on Ω, then so is $|f|$ and
$$\left|\int_\Omega f\,dm\right| \leq \int_\Omega |f|\,dm.$$
Indeed, $|f|$ is measurable and, since $|f| = f^+ + f^-$, we have
$$\int_\Omega |f|\,dm = \int_\Omega f^+\,dm + \int_\Omega f^-\,dm < \infty.$$
Furthermore,
$$\left|\int_\Omega f\,dm\right| = \left|\int_\Omega f^+\,dm - \int_\Omega f^-\,dm\right| \leq \int_\Omega f^+\,dm + \int_\Omega f^-\,dm \leq \int_\Omega |f|\,dm.$$

On the other hand, if f is measurable on Ω, then
$$|f| \in \mathcal{L}^1(\Omega) \;\Rightarrow\; f \in \mathcal{L}^1(\Omega). \tag{11.11}$$
This follows from the observation that $f^+ \leq |f|$ and $f^- \leq |f|$, so that Lemma 11.3 implies
$$\int_\Omega f^+\,dm \leq \int_\Omega |f|\,dm < \infty,$$
$$\int_\Omega f^-\,dm \leq \int_\Omega |f|\,dm < \infty.$$
Since, however, $|f|$ may be measurable while f is not (see Exercise 10.4.3), the implication (11.11) is not valid in general.

2. If $f \in \mathcal{L}^0(\Omega)$, $E \subseteq \Omega$, and $m(E) = 0$, then $f \in \mathcal{L}^1(E)$ and
$$\int_E f\,dm = 0.$$
We prove this result by following the stages of the definition of the integral. Suppose, to begin with, that f is the simple function
$$f = \sum_{i=1}^n c_i \chi_{E_i}.$$

Then
$$f|_E = \sum_{i=1}^{n} c_i \chi_{E_i \cap E},$$
and therefore
$$\int_E f\,dm = \sum_{i=1}^{n} c_i m(E_i \cap E) = 0.$$

If $f \in \mathcal{L}_+^0(\Omega)$, we choose a sequence (φ_n) in $\mathcal{S}_+(\Omega)$ such that $\varphi_n \nearrow f$. Then $\varphi_n|_E \in \mathcal{S}_+(E)$ and $\varphi_n|_E \nearrow f|_E$. Since, from the previous step, $\int_E \varphi_n\,dm = 0$ for every n, it follows that
$$\int_E f\,dm = \lim_{n\to\infty} \int_E \varphi_n\,dm = \lim_{n\to\infty} 0 = 0.$$

Finally, if $f \in \mathcal{L}^0(\Omega)$, we observe that $(f|_E)^+ = f^+|_E$ and $(f|_E)^- = f^-|_E$ so that
$$\int_E f\,dm = \int_E f^+\,dm - \int_E f^-\,dm = 0.$$

3. If $f \in \mathcal{L}^1(\Omega)$, then $f \in \mathcal{L}^1(E)$ for every measurable set $E \subseteq \Omega$, and
$$\int_E f\,dm = \int_\Omega f\chi_E\,dm. \tag{11.12}$$

Since $|f\chi_E| \leq |f|$, we see that $\int_\Omega |f\chi_E|\,dm < \infty$, so we need only prove equation (11.12). This can be done in stages by following the procedure in Remark 2 above.

Example 11.3 Given $f(x) = x^3$, prove that
$$\int_{[a,b]} f\,dm = \frac{b^4 - c^4}{4}.$$

Proof. We shall only consider the case where $a < 0 < b$. Here
$$f^+(x) = x^3 \chi_{[0,b]}, \quad f^-(x) = -x^3 \chi_{[a,0]}.$$

Let
$$\varphi_n = \sum_{i=1}^{2^n} \left[\frac{(i-1)b}{2^n}\right]^3 \chi_{[(i-1)b/2^n,\,ib/2^n)},$$
$$\psi_n = \sum_{i=1}^{2^n} \left(\frac{-ia}{2^n}\right)^3 \chi_{[ia/2^n,\,(i-1)a/2^n)}.$$

It is straightforward to show that $\varphi_n \nearrow f^+$ and $\psi_n \nearrow f^-$. Using the equality $\sum_{i=1}^{m} i^3 = [m(m+1)/2]^2$ (see Exercise 2.3.4),

$$\int_{[a,b)} f^+ dm = \lim_{n\to\infty} \int_{[a,b)} \varphi_n dm$$

$$= \lim_{n\to\infty} \sum_{i=1}^{2^n} \frac{b^4}{2^{4n}}(i-1)^3$$

$$= \lim_{n\to\infty} \frac{b^4}{2^{4n}} \left[\frac{(2^n-1)2^n}{2}\right]^2 = \frac{b^4}{4}.$$

Similarly, $\int_{[a,b)} f^- dm = a^4/4$, and hence the desired result.

Exercises 11.1

1. Use Theorem 11.1 to evaluate $\int_\Omega f dm$, where

   (a) $\Omega = [0,1]$, $f(x) = \begin{cases} 1, & x \in \mathbb{Q} \\ 0, & x \notin \mathbb{Q} \end{cases}$

   (b) $\Omega = [1,2]$, $f(x) = 3x - 1$

   (c) $\Omega = [0,1]$, $f(x) = x^2$.

2. If $f(x) = 1/x$, where $x \neq 0$, prove that $\int_{[1,\infty]} f dm = \infty$.

3. If $f = \sum_{i=1}^{n} c_i \chi_{E_i} \in \mathcal{L}^1(\Omega)$, where $\{E_i, \cdots, E_n\}$ is a partition of Ω in $\mathcal{M}$ and $c_i \in \mathbb{R}$, prove that $\int_\Omega f dm = \sum_{i=1}^{n} c_i m(E_i)$.

4. Let $f \in \mathcal{L}^0(\Omega)$ and $A = \{x \in \Omega : f(x) \geq 0\}$. Prove that, for all $c > 0$,

$$m\{x \in A : f(x) > c\} \leq \frac{1}{c} \int_A f dm.$$

5. Verify equation (11.12) for $f \in \mathcal{L}^1(\Omega)$ and any measurable $E \subseteq \Omega$.

6. Suppose that $f \in \mathcal{L}^1([a,b])$ and $c > 0$. If $g : [a/c, b/c] \to \mathbb{R}$ is defined by $g(x) = f(cx)$, prove that $g \in \mathcal{L}^1([a/c, b/c])$ and that

$$\int_{[a/c,b/c]} g dm = \frac{1}{c} \int_{[a,b]} f dm.$$

   Hint: Start with $f = \chi_E$, where E is a measurable subset of $[a,b]$.

7. Let $f \in \mathcal{L}^1([a+h, b+h])$ and define $f_h(x) = f(x+h)$. Show that $\int_{[a,b]} f_h dm = \int_{[a+h,b+h]} f dm$.

11.2 Properties of the Lebesgue Integral

The next theorem presents the most basic properties of the Lebesgue integral. Its proof is particularly significant as it illustrates some of the techniques of Lebesgue integration.

Theorem 11.2
(i) Let E and F be two measurable and disjoint subsets of Ω. If $f \in \mathcal{L}^1(E) \cap \mathcal{L}^1(F)$, then $f \in \mathcal{L}^1(E \cup F)$, and

$$\int_{E \cup F} f \, dm = \int_E f \, dm + \int_F f \, dm \tag{11.13}$$

(ii) If $f \in \mathcal{L}^1(\Omega)$ and $g = f$ (a.e) on Ω, then $g \in \mathcal{L}^1(\Omega)$ and

$$\int_\Omega g \, dm = \int_\Omega f \, dm.$$

(iii) If $f \in \mathcal{L}^1(\Omega)$ then

$$m\{x \in \Omega : |f(x)| = \infty\} = 0.$$

(iv) If $f, g \in \mathcal{L}^1(\Omega)$, then $f + g \in \mathcal{L}^1(\Omega)$ and

$$\int_\Omega (f + g) \, dm = \int_\Omega f \, dm + \int_\Omega g \, dm. \tag{11.14}$$

(v) If $f \in \mathcal{L}^1(\Omega)$ and $c \in \mathbb{R}$, then $cf \in \mathcal{L}^1(\Omega)$ and

$$\int_\Omega cf \, dm = c \int_\Omega f \, dm. \tag{11.15}$$

(vi) If $f, g \in \mathcal{L}^1(\Omega)$ and $f(x) \geq g(x)$ (a.e) on Ω, then

$$\int_\Omega f \, dm \geq \int_\Omega g \, dm.$$

(vii) If $f(x) \geq 0$ (a.e.) on Ω and $\int_\Omega f \, dm = 0$, then

$$f(x) = 0 \quad (a.e.) \text{ on } \Omega.$$

(viii) If $f \in \mathcal{L}^1(\Omega)$ and

$$\int_E f \, dm = 0 \quad \text{for all } E \in \mathcal{M}, \ E \subseteq \Omega,$$

then $f(x) = 0$ (a.e.) on Ω.

Note that property (ii) allows us to assume that any integrable function on Ω does not assume the values $\pm\infty$, while (iv), (v) and (vi) express the linearity and monotonicity of the Lebesgue integral.

Proof

(i) Since $E \cap F = \emptyset$,

$$\chi_{E \cup F} = \chi_E + \chi_F$$
$$(f\chi_{E \cup F})^+ = f^+ \chi_E + f^+ \chi_F$$
$$\int_\Omega (f\chi_{E \cup F})^+ dm = \int_\Omega f^+ \chi_E dm + \int_\Omega f^+ \chi_F dm$$
$$= \int_E f^+ dm + \int_F f^+ dm \qquad (11.16)$$

Similarly,

$$\int_\Omega (f\chi_{E \cup F})^- dm = \int_\Omega f^- \chi_E dm + \int_\Omega f^- \chi_F dm$$
$$= \int_E f^- dm + \int_F f^- dm \qquad (11.17)$$

Consequently, $f\chi_{E \cup F}$ is integrable on Ω, that is, f is integrable on $E \cup F$, if and only if f is integrable on E and F, and (11.13) follows by subtracting (11.17) from (11.16).

(ii) Note first that $g \in \mathcal{L}^0(\Omega)$ by Theorem 10.19. Setting $A = \{x \in \Omega : f(x) \neq g(x)\}$, we see that $m(A) = 0$ and $\Omega \backslash A \in \mathcal{M}$. Using Remarks 11.2.2 and 11.2.3, we conclude that $f \in \mathcal{L}^1(\Omega \backslash A)$ and

$$\int_A f dm = \int_A g dm = 0.$$

Since $g\chi_{\Omega \backslash A} = f\chi_{\Omega \backslash A}$, $g \in \mathcal{L}^1(\Omega \backslash A)$ and

$$\int_{\Omega \backslash A} g dm = \int_{\Omega \backslash A} f dm.$$

Using (i) we therefore have $g \in \mathcal{L}^1(\Omega)$ and

$$\int_\Omega g dm = \int_{\Omega \backslash A} g dm + \int_A g dm$$
$$= \int_{\Omega \backslash A} f dm + \int_A f dm$$
$$= \int_\Omega f dm.$$

(iii) Setting $A = \{x \in \Omega : |f(x)| = \infty\}$ and $\varphi_n = n\chi_A$, we see that $\varphi_n \in \mathcal{L}_+(\Omega)$ and $\varphi_n \nearrow |f|\chi_A$. By Theorem 11.1,

$$\int_A |f| dm = \lim_{n \to \infty} n \cdot m(A).$$

Since $f \in \mathcal{L}^1(A)$, $\int_A |f| dm < \infty$ and therefore $m(A) = 0$.

(iv) Using part (iii), we may assume that both f and g are finite-valued functions, so that their sum is well defined on Ω. Since

$$(f+g)^+ \leq |f| + |g|, \quad (f+g)^- \leq |f| + |g|,$$

we conclude, by Lemma 11.3, that
$$\int_\Omega (f+g)^+ \, dm < \infty, \quad \int_\Omega (f+g)^- \, dm < \infty,$$
and this implies that $f + g \in \mathcal{L}^1(\Omega)$. To prove (11.14), observe that
$$f + g = (f+g)^+ - (f+g)^-$$
and
$$f + g = f^+ - f^- + g^+ - g^-,$$
hence
$$(f+g)^+ + f^- + g^- = (f+g)^- + f^+ + g^+$$
$$\Rightarrow \int_\Omega (f+g)^+ dm + \int_\Omega f^- dm + \int_\Omega g^- dm = \int_\Omega (f+g)^- dm + \int_\Omega f^+ dm + \int_\Omega g^+ dm,$$

and we arrive at (11.14) by rearranging terms.

(v) If $c \geq 0$ then $(cf)^+ = cf^+$ and $(cf)^- = cf^-$, and if $c < 0$ then $(cf)^+ = -cf^-$ and $(cf)^- = -cf^-$. In either case we arrive at Equation (11.15) by applying Lemma 11.3.

(vi) We can write $f = g + f - g$, which implies
$$\int_\Omega f \, dm = \int_\Omega g \, dm + \int_\Omega (f-g) \, dm.$$

Since $(f - g) \in \mathcal{L}_+^0$, it follows that $\int_\Omega (f-g) dm \geq 0$ by definition.

(vii) Let $A = \{x \in \Omega : f(x) > 0\}$ and $A_n = \{x \in \Omega : f(x) > 1/n\}$. (A_n) is an increasing sequence in $\mathcal{M}$, and from the theorem of Archimedes we see that it increases to A. Since m is continuous from below (Theorem 10.9),
$$m(A) = \lim_{n \to \infty} m(A_n).$$

For the purpose of evaluating $\int_\Omega f \, dm$ we may assume that $f \geq 0$ on Ω, since the subset of Ω on which f is negative has measure 0. Thus
$$\int_\Omega f \, dm = \int_{A_n} f \, dm + \int_{\Omega \setminus A_n} f \, dm$$
$$\geq \int_{A_n} f \, dm$$
$$\geq \frac{1}{n} m(A_n).$$

But $\int_\Omega f \, dm = 0$, hence $m(A_n) = 0$ for every n, and therefore $m(A) = 0$.

(viii) Let $E = \{x \in \Omega : f(x) \geq 0\}$. We then have

$$\begin{aligned}\int_\Omega |f|\,dm &= \int_E |f|\,dm + \int_{\Omega\setminus E} |f|\,dm \\ &= \int_E f\,dm + \int_{\Omega\setminus E}(-f)\,dm \\ &= \int_E f\,dm - \int_{\Omega\setminus E} f\,dm \\ &= 0,\end{aligned}$$

since both E and $\Omega\setminus E$ are measurable. From (vii) we now see that $|f|$, and hence f, equals 0 almost everywhere. □

Note that (i) implies that

$$\mathcal{L}^1([a,b]) = \mathcal{L}^1([a,b)) = \mathcal{L}^1((a,b]) = \mathcal{L}^1((a,b)),$$

and we shall use $\mathcal{L}^1(a,b)$ to denote any of these sets.

EXERCISES 11.2

1. Suppose $f, g \in \mathcal{L}^0(\Omega)$ satisfy $|f| \leq |g|$ almost everywhere. If g is integrable on on Ω, show that f is also integrable on Ω.

2. Given $f \in \mathcal{L}^1(\Omega)$, prove that $m(\{x \in \Omega : |f(x)| \geq \varepsilon\}) < \infty$ for all $\varepsilon > 0$.

3. For any $f \in \mathcal{L}^1(\Omega)$, show that there is a sequence (E_i) of measurable sets such that

    (a) $E_i \subseteq \Omega$ for all i
    (b) $m(E_i) < \infty$ for all i
    (c) $\{x \in \Omega : f(x) \neq 0\} = \bigcup_{i=1}^\infty E_i$.

4. If $f^2, g^2 \in \mathcal{L}^1(\Omega)$, prove that $fg \in \mathcal{L}^1(\Omega)$ and that the inequality

    $$\int_\Omega |fg|\,dm \leq \left(\int_\Omega f^2 dm\right)^{1/2} \left(\int_\Omega g^2 dm\right)^{1/2},$$

    known as the *Schwarz inequality*, holds.

5. If $m(\Omega) < \infty$ and

    $$\mathcal{L}^2(\Omega) = \{f \in \mathcal{L}^0(\Omega) : f^2 \in \mathcal{L}^1(\Omega)\}$$

is the set of *square integrable* functions on Ω, prove that $\mathcal{L}^2(\Omega) \subseteq \mathcal{L}^1(\Omega)$. Hint: Use the Schwarz inequality.

A more general result, known as *Hölder's inequality*, states that if $|f|^p, |g|^q \in \mathcal{L}^1(\Omega)$, where p and q are positive and $\frac{1}{p} + \frac{1}{q} = 1$, then $fg \in \mathcal{L}^1(\Omega)$ and

$$\int_\Omega |fg|\,dm \leq \left(\int_\Omega |f|^p\,dm\right)^{1/p} \left(\int_\Omega |g|^q\,dm\right)^{1/q}.$$

If $m(\Omega) < \infty$ and $p > q$, show that $\mathcal{L}^p(\Omega) \subseteq \mathcal{L}^q(\Omega)$, where

$$\mathcal{L}^p(\Omega) = \left\{f \in \mathcal{L}^0(\Omega) : |f|^p \in \mathcal{L}^1(\Omega)\right\}.$$

See [ROY] for a more detailed discussion of $\mathcal{L}^p(\Omega)$ spaces.

6. Let $E, F \in \mathcal{M}$ and $f \in \mathcal{L}^1(E)$. If $m(E\Delta F) = 0$, prove that $f \in \mathcal{L}^1(F)$ and $\int_E f\,dm = \int_F f\,dm$.

7. Suppose the sets $E_1, E_2, \cdots, E_n$ are measurable. For each $i \in \{1, \cdots, n\}$ define the set F_i as follows: $x \in F_i$ if and only if x belongs to exactly i members of the class $\{E_1, \cdots, E_n\}$. Prove that

$$\sum_{i=1}^n m(E_i) = \sum_{i=1}^n im(F_i).$$

Hint: $F_i = \{x : \sum_{j=1}^n \chi_{E_j} = i\}$.

8. Let $f \in \mathcal{L}^0(\Omega)$ and $g \in \mathcal{L}^1(\Omega)$. If $a \leq f \leq b$ (a.e.) for some real numbers a and b, prove that $f|g| \in \mathcal{L}^1(\Omega)$ and that there is a number $c \in [a,b]$ such that

$$\int_\Omega f|g|\,dm = c\int_\Omega |g|\,dm.$$

11.3 Lebesgue Integral and Pointwise Convergence

In this section we investigate the behaviour of the Lebesgue integral under limit operations, and demonstrate its superiority over the Riemann integral in this respect.

Theorem 11.3 (Monotone Convergence Theorem)
If $f_n \in \mathcal{L}^0_+(\Omega)$ for every $n \in \mathbb{N}$ and $f_n \nearrow f$, then $f \in \mathcal{L}^0_+(\Omega)$ and

$$\int_\Omega f\,dm = \lim_{n\to\infty} \int_\Omega f_n\,dm.$$

Proof. It was already shown in Theorem 10.18 that f is measurable, hence $f \in \mathcal{L}_+^0(\Omega)$. For each $n \in \mathbb{N}$ choose a sequence $(\varphi_{nk} : k \in \mathbb{N})$ in $\mathcal{S}_+(\Omega)$ such that $\varphi_{nk} \nearrow f_n$, and let

$$\psi_k = \max\{\varphi_{1k}, \varphi_{2k}, \cdots, \varphi_{kk}\}.$$

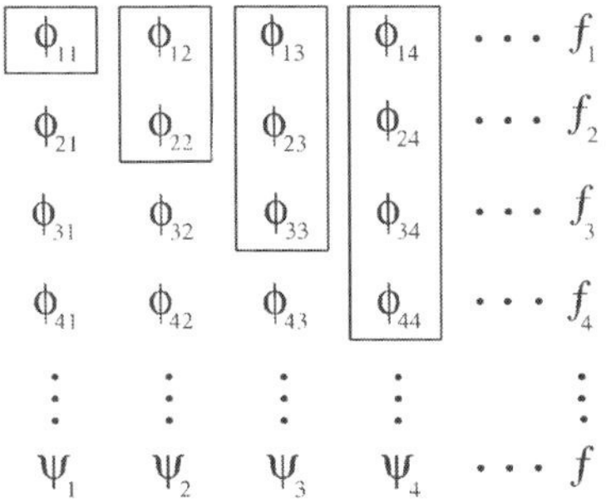

Figure 11.1

Note that
(i) $\psi_k \in \mathcal{S}_+(\Omega)$
(ii) $\psi_1 \leq \psi_2 \leq \psi_3 \leq \cdots$
(iii) For every $x \in \Omega$ and all $k \in \mathbb{N}$, there is a number $m \in \{1, \cdots, k\}$ such that $\psi_k(x) = \varphi_{mk}(x)$, and therefore

$$\psi_k(x) = \varphi_{mk}(x) \leq f_m(x) \leq f_k(x).$$

Thus, for all $k \in \mathbb{N}$ and $n \leq k$,

$$\varphi_{nk} \leq \psi_k \leq f_k. \tag{11.18}$$

By letting $k \to \infty$, we obtain

$$f_n \leq \lim \psi_k \leq f \quad \text{for all } n \in \mathbb{N}.$$

In the limit as $n \to \infty$, this yields

$$\lim \psi_k = f.$$

Consequently, we have $\psi_k \nearrow f$ and, by Theorem 11.1,

$$\int_\Omega f \, dm = \lim_{k \to \infty} \int_\Omega \psi_k \, dm. \tag{11.19}$$

The monotonicity of the integral and equation (11.18) imply

$$\int_\Omega \varphi_{nk} dm \leq \int_\Omega \psi_k dm \leq \int_\Omega f_k dm. \tag{11.20}$$

Now we take the limit in as $k \to \infty$ in (11.20), keeping in mind that $\varphi_{nk} \nearrow f_n$, to obtain

$$\int_\Omega f_n dm \leq \lim_{k\to\infty} \int_\Omega \psi_k dm \leq \lim_{k\to\infty} \int_\Omega f_k dm, \ n \in \mathbb{N}.$$

In view of (11.19), we therefore have

$$\lim_{n\to\infty} \int_\Omega f_n dm = \lim_{k\to\infty} \int_\Omega \psi_k dm = \int_\Omega f dm. \ \square$$

Remarks 11.3

1. We have already encountered two special cases of this theorem: one is Theorem 11.1, and the other is the continuity of m from below, where $E_n \nearrow E$, $f_n = \chi_{E_n}$, $f = \chi_E$.
2. The theorem is not valid if $f_n \searrow f$, i.e., if f_n tends to f from *above*. Consider $E_n = [n, \infty)$. Clearly $E_n \searrow \varnothing$ and $f_n = \chi_{E_n} \searrow f = 0$. Hence $\int_\Omega f dm = 0$ while $\int_\Omega f_n dm = m(E_n) = \infty$ for all n.
3. Theorem 11.3 is not valid for Riemann integrals, as demonstrated at the beginning of this chapter.

Corollary 11.3.1 *Let* (f_n) *be a sequence in* $\mathcal{L}_+^0(\Omega)$ *and*

$$f(x) = \sum_{n=1}^\infty f_n(x), \ x \in \Omega.$$

Then $f \in \mathcal{L}_+^0(\Omega)$ *and*

$$\int_\Omega f dm = \sum_{n=1}^\infty \int_\Omega f_n dm.$$

Compare this result to Theorem 9.4 for Riemann integrals.

Proof. Let $g_n = \sum_{i=1}^n f_i$, in which case $g_n \in \mathcal{L}_+^0(\Omega)$ and $g_n \nearrow f$. The monotone convergence theorem now guarantees that $f \in \mathcal{L}_+^0(\Omega)$ and

$$\int_\Omega f dm = \lim_{n\to\infty} \int_\Omega g_n dm$$
$$= \lim_{n\to\infty} \sum_{i=1}^n \int_\Omega f_i dm$$
$$= \sum_{i=1}^\infty \int_\Omega f_i dm. \ \square$$

Example 11.4 For every $x \in [0,1]$ and $n \in \mathbb{N}$, define $f_n(x)$ to be the distance between x and the nearest point in $[0,1]$ of the form $10^{-n}k$, where k is an integer. If $f(x) = \sum_{n=1}^{\infty} f_n(x)$, evaluate $\int_{[0,1]} f \, dm$.

Solution. Taking $x \in [10^{-n}k, 10^{-n}(k+1))$ with $0 \leq k \leq 10^n$, we have

$$f_n(x) = \begin{cases} x - 10^{-n}k, & x \in I_n^k \\ 10^{-n}(k+1) - x, & x \in J_n^k, \end{cases}$$

where

$$I_n^k = \left[\frac{k}{10^n}, \frac{2k+1}{2 \times 10^n}\right), \quad J_n^k = \left[\frac{2k+1}{2 \times 10^n}, \frac{k+1}{10^n}\right).$$

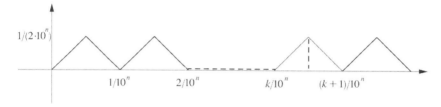

Figure 11.2

Using the result of Example 11.1 and the linearity of the integral, we obtain

$$\int_{I_n^k} f_n \, dm = \int_{J_n^k} f_n \, dm = \frac{1}{8 \times 10^{2n}}$$

$$\Rightarrow \int_{[0,1]} f_n \, dm = \sum_{k=0}^{10^n - 1} \int_{I_n^k \cup J_n^k} f_n \, dm = \frac{1}{4 \times 10^n}.$$

Since $f_n \geq 0$, Corollary 11.3.1 gives

$$\int_{[0,1]} f \, dm = \sum_{n=1}^{\infty} \frac{1}{4 \times 10^n} = \frac{1}{36}.$$

Corollary 11.3.2 (Continuity Theorem) *If $f \in \mathcal{L}^1(\Omega)$, then, for every $\varepsilon > 0$ there is a $\delta > 0$ such that*

$$E \in \mathcal{M}, \ E \subseteq \Omega, \ m(E) < \delta \ \Rightarrow \ \left|\int_E f \, dm\right| < \varepsilon,$$

that is,

$$\lim_{m(E) \to 0} \int_E f \, dm = 0.$$

Lebesgue Integration 411

Proof. For every $n \in \mathbb{N}$ define the function

$$f_n(x) = \begin{cases} |f(x)|, & |f(x)| \le n \\ n, & |f(x)| > n. \end{cases}$$

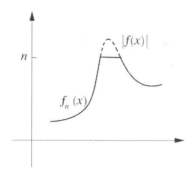

Figure 11.3

Clearly $f_n \in \mathcal{L}_+^0(\Omega)$ and $f_n \nearrow |f|$, hence, by the monotone convergence theorem,

$$\int_\Omega |f|\, dm = \lim_{n \to \infty} \int_\Omega f_n\, dm.$$

This means, given $\varepsilon > 0$, there is an integer N such that

$$n \ge N \implies \int_\Omega (|f| - f_n)\, dm < \varepsilon/2.$$

Let $\delta = \varepsilon/2N$ and suppose $m(E) < \delta$. Since $0 \le f_n(x) \le n$ for all $x \in \Omega$ and $|f| - f_n \ge 0$, we see that

$$\begin{aligned}
\left|\int_E f\, dm\right| &\le \int_E |f|\, dm \\
&= \int_E f_N\, dm + \int_E (|f| - f_N)\, dm \\
&\le N m(E) + \int_\Omega (|f| - f_N)\, dm \\
&< \varepsilon/2 + \varepsilon/2 = \varepsilon. \qquad \square
\end{aligned}$$

Example 11.5 Let $f \in \mathcal{L}^1(a,b)$. Prove that the function $F : [a,b] \to \mathbb{R}$ defined by

$$F(x) = \int_{[a,x]} f\, dm$$

is continuous.

Proof. For any $x \in [a,b]$,

$$|F(x+h) - F(x)| = \left|\int_{[a,x+h)} f\, dm - \int_{[a,x)} f\, dm\right| = \left|\int_{I_h} f\, dm\right|,$$

where $x+h \in [a,b]$ and I_h is the interval with end-points x and $x+h$. Since
$$\lim_{h\to 0} m(I_h) = \lim_{h\to 0} |h| = 0,$$
Corollary 11.3.2 implies
$$\lim_{h\to 0} |F(x+h) - F(x)| = \lim_{h\to 0} \left|\int_{I_h} f\,dm\right| = 0,$$
which means f is continuous at x. $\square$

Theorem 11.4 (Fatou's Lemma)
If (f_n) is a sequence in $\mathcal{L}_+^0(\Omega)$, then $\liminf f_n \in \mathcal{L}_+^0(\Omega)$ and
$$\int_\Omega \liminf f_n\, dm \leq \liminf \int_\Omega f_n\, dm. \tag{11.21}$$

Proof. $\liminf f_n \in \mathcal{L}_+^0(\Omega)$ by Theorem 10.18, and it remains to prove the inequality (11.20). Let
$$g_n = \inf\{f_k : k \geq n\} \text{ for all } n \in \mathbb{N},$$
$$g = \liminf f_n.$$

Then $g_n \in \mathcal{L}_+^0(\Omega)$ and $g_n \nearrow g$, so the monotone convergence theorem gives
$$\int_\Omega g\, dm = \lim_{n\to\infty} \int_\Omega g_n\, dm.$$
But since $g_n \leq f_n$ for all n,
$$\int_\Omega g_n\, dm \leq \int_\Omega f_n\, dm$$
$$\Rightarrow \lim_{n\to\infty} \int_\Omega g_n\, dm \leq \liminf \int_\Omega f_n\, dm. \quad \square$$

The following example shows that (11.21) can be a strict inequality.

Example 11.6 Let $\Omega = [0,2]$ and define the sequence of functions $f_n : [0,2] \to \mathbb{R}$ by
$$f_{2n-1} = \chi_{[0,1)}, \quad f_{2n} = \chi_{[1,2]}.$$
Since $\liminf f_n = 0$, $\int_{[0,2]} \liminf f_n\, dm = 0$, whereas $\int_{[0,2]} f_n\, dm = 1$ for all $n \in \mathbb{N}$, and thereby $\liminf \int_{[0,2]} f_n\, dm = 1$.

Next to the monotone convergence theorem and Fatou's lemma, the following theorem, known as the *dominated convergence theorem*, is another important result of Lebesgue integration with wide applications.

Theorem 11.5 (Dominated Convergence Theorem)

Let (f_n) be a sequence of measurable functions on Ω, and suppose f_n converges to f almost everywhere. If there is a function $g \in \mathcal{L}^1(\Omega)$ such that, for all $n \in \mathbb{N}$,

$$|f_n(x)| \leq g(x), \text{ a.e.},$$

then $f \in \mathcal{L}^1(\Omega)$ and

$$\lim_{n \to \infty} \int_\Omega f_n dm = \int_\Omega f dm. \tag{11.22}$$

Proof. We know from Theorem 10.18 that f is measurable on Ω. Since $|f| \leq g$ (a.e.), we have

$$\int_\Omega |f| \, dm \leq \int_\Omega g \, dm < \infty,$$

which implies $f \in \mathcal{L}^1(\Omega)$.

Let $h_n = 2g - |f_n - f|$, which is clearly in $\mathcal{L}^0_+(\Omega)$, and therefore satisfies

$$\int_\Omega \liminf h_n \, dm \leq \liminf \int_\Omega h_n \, dm,$$

by Fatou's lemma. Since $h_n \to 2g$, we have

$$\int_\Omega 2g \, dm \leq \liminf \left[\int_\Omega 2g \, dm - \int_\Omega |f_n - f| \, dm \right]$$
$$= \int_\Omega 2g \, dm - \limsup \int_\Omega |f_n - f| \, dm,$$

which implies

$$\limsup \int_\Omega |f_n - f| \, dm \leq 0.$$

But since $\int_\Omega |f_n - f| \, dm \geq 0$, it follows that $\liminf \int_\Omega |f_n - f| \, dm \geq 0$, and therefore

$$\lim \int_\Omega |f_n - f| \, dm = 0$$

Now, by taking the limit as $n \to \infty$ in the inequality

$$\left| \int_\Omega f_n dm - \int_\Omega f dm \right| \leq \int_\Omega |f_n - f| \, dm,$$

we obtain (11.22). □

Remarks 11.4

1. In Theorem 11.5 the function g, which is necessarily non-negative, *dominates* the sequence $|f_n|$, in the sense that $|f_n| \leq g$ on Ω, hence the name.

2. Under the hypothesis of the theorem we have actually proved that
$$\lim \int_\Omega |f_n - f|\, dm = 0,$$
which, in general, is stronger than the equality
$$\lim \int_\Omega f_n\, dm = \int_\Omega f\, dm.$$

Corollary 11.5.1 (Bounded Convergence Theorem) *Let (f_n) be a bounded sequence of measurable functions on Ω, where $m(\Omega) < \infty$. If $f_n \to f$ almost everywhere in Ω, then $f \in \mathcal{L}^1(\Omega)$ and*
$$\lim_{n\to\infty} \int_\Omega f_n\, dm = \int_\Omega f\, dm.$$

Proof. Assuming M is an upper bound of $|f_n(x)|$ on Ω for every $n \in \mathbb{N}$, set $g(x) = M$ for all $x \in \Omega$. Since $m(\Omega) < \infty$, $g \in \mathcal{L}^1(\Omega)$, and the dominated convergence theorem gives the desired result. □

Corollary 11.5.2 *Let (f_n) be a sequence in $\mathcal{L}^1(\Omega)$ which satisfies*
$$\sum_{n=1}^\infty \int_\Omega |f_n(x)| < \infty.$$
Then there there is a function $f \in \mathcal{L}^1(\Omega)$ such that
$$f = \sum_{n=1}^\infty f_n \quad (a.e)$$
$$\int_\Omega f\, dm = \sum_{n=1}^\infty \int_\Omega f_n\, dm.$$

Proof. Define the function $g : \Omega \to \bar{\mathbb{R}}$ by
$$g(x) = \sum_{n=1}^\infty |f_n(x)|,$$
which is clearly in $\mathcal{L}^0_+(\Omega)$. From Corollary 11.5.1 we know that
$$\int_\Omega g\, dm = \sum_{n=1}^\infty \int_\Omega |f_n|\, dm < \infty,$$
hence $g \in \mathcal{L}^1(\Omega)$. Let
$$A = \{x \in \Omega : g(x) = \infty\}.$$

By Theorem 11.2, $m(A) = 0$. With $B = \Omega \backslash A$, let

$$f(x) = \chi_B(x) \sum_{n=1}^{\infty} f_n(x), \ x \in \Omega.$$

Now the sum $S_n = \sum_{k=1}^{n} f_k$ is measurable, converges almost everywhere to f, and $|S_n|$ is dominated a.e. by g, hence

$$\int_\Omega f \, dm = \lim_{n \to \infty} \int_\Omega S_n \, dm = \sum_{k=1}^{\infty} \int f_k \, dm. \ \square$$

Example 11.7 Let $f \in \mathcal{L}^1(\Omega)$ and define the function ν on $\mathcal{M}_0 = \{E \subseteq \Omega : E \in \mathcal{M}\}$ by

$$\nu(E) = \int_E f \, dm.$$

Prove that ν is countably additive.

Proof. Suppose (E_n) is a pairwise disjoint sequence in $\mathcal{M}_0$ and $E = \bigcup_{n=1}^{\infty} E_n$. We can then write

$$f\chi_E = \sum_{n=1}^{\infty} f\chi_{E_n}.$$

Applying Corollary 11.3.1, we have

$$\sum_{n=1}^{\infty} \int_\Omega |f\chi_{E_n}| \, dm = \int_\Omega \sum_{n=1}^{\infty} |f| \chi_{E_n} dm$$
$$= \int_\Omega |f| \chi_E dm < \infty.$$

By Corollary 11.3.2,

$$\int_\Omega f\chi_E \, dm = \int_\Omega \sum_{n=1}^{\infty} f\chi_{E_n} dm = \sum_{n=1}^{\infty} \int_\Omega f\chi_{E_n} dm.$$

This implies $\nu(E) = \int_E f \, dm = \sum_{n=1}^{\infty} \int_{E_n} f \, dm = \sum_{n=1}^{\infty} \nu(E_n). \ \square$

Note that, if $f \geq 0$, ν becomes a measure on $\mathcal{M}_0$. Furthermore, if $m(E) = 0$, then $\nu(E) = 0$, in which case we say that the measure ν is *absolutely continuous* with respect to m, and we write

$$\nu \ll m.$$

When Ω is a countable union of measurable sets, each of which has finite measure, this result has a well known converse, called the *Radon-Nykodym theorem* (see [ROY], for example).

Corollary 11.5.3 *Let $f_t \in \mathcal{L}^1(\Omega)$ for every t in the real interval I, and suppose*
$$\lim_{t \to c} f_t(x) = f(x) \quad a.e.,$$
where $c \in \bar{I}$, the closure of I. If there is a function $g \in \mathcal{L}^1(\Omega)$ such that, for all $t \in I$,
$$|f_t(x)| \leq g(x) \quad a.e.,$$
then $f \in \mathcal{L}^1(\Omega)$ and
$$\lim_{t \to c} \int_\Omega f_t \, dm = \int_\Omega f \, dm. \tag{11.23}$$

Proof. Take any sequence $(t_n : n \in \mathbb{N})$ in I such that $t_n \to c$, $t_n \neq c$. By applying the dominated convergence theorem to the sequence (f_{t_n}), we conclude that $f \in \mathcal{L}^1(\Omega)$ and
$$\lim_{n \to \infty} \int_\Omega f_{t_n} \, dm = \int_\Omega f \, dm.$$
Since this equality is true for every sequence (t_n) in I, where $t_n \neq c$, with limit c, Equation (11.23) follows. $\square$

EXERCISES 11.3

1. Suppose (f_n) is sequence in $\mathcal{L}_+^0(\Omega)$ such that $f_n \to f$ and $f_n \leq f$ for all $n \in \mathbb{N}$. Prove that
$$\lim_{n \to \infty} \int_\Omega f_n \, dm = \int_\Omega f \, dm.$$
Note that we do not assume $f_n \nearrow f$.

2. Let $f_n \in \mathcal{L}_+^0(\Omega)$ and $f_n \searrow f$. If there a $k \in \mathbb{N}$ for which $\int_\Omega f_k \, dm < \infty$, prove that
$$\lim_{n \to \infty} \int_\Omega f_n \, dm = \int_\Omega f \, dm.$$
Show the necessity of the condition $\int_\Omega f_k \, dm < \infty$ by a suitable example.

3. Let $f_n \in \mathcal{L}^0(\Omega)$ for all $n \in \mathbb{N}$ and suppose $f \in \mathcal{L}^1(\Omega)$. If
$$|f_n(x)| \leq |f(x)| \quad a.e., \ n \in \mathbb{N},$$

prove that
$$\int_\Omega \limsup f_n \, dm \geq \limsup \int_\Omega f_n \, dm$$
$$\geq \liminf \int_\Omega f_n \, dm$$
$$\geq \int_\Omega \liminf f_n \, dm.$$

Hint: Apply Fatou's lemma to $|f| \pm f_n$.

4. Apply Fatou's lemma to the sequence $f_n : [0,1] \to \mathbb{R}$ defined by
$$f_n(x) = \begin{cases} 0, & x = 0 \\ n, & 0 < x < 1/n \\ 0, & 1/n \leq x \leq 1. \end{cases}$$

5. Given a sequence of measurable sets (E_n), use Fatou's lemma to show that
$$m(\liminf E_n) \leq \liminf m(E_n).$$
Also show that, if $m\left(\bigcup_{i=k}^\infty E_i\right) < \infty$ for some k, then
$$m(\limsup E_n) \geq \limsup m(E_n).$$
(see Exercise 10.3.6)

6. Let (E_n) be a sequence of measurable sets. Prove that
$$\sum_{n=1}^\infty m(E_n) < \infty \Rightarrow m(\limsup E_n) = 0.$$
Hint: Use integration.

7. Let $f \in \mathcal{L}^1(\Omega)$ and $E_n = \{x \in \Omega : f(x) \geq n\}$. Prove that
   (a) $m(E_n) \leq \dfrac{1}{n} \int_{E_n} |f| \, dm$
   (b) $\lim_{n \to \infty} \int_{E_n} |f| \, dm = 0$
   (c) $\lim_{n \to \infty} n m(E_n) = 0$.

8. Suppose $m(\Omega) < \infty$ and $f \in \mathcal{L}^0(\Omega)$. Prove that $f \in \mathcal{L}^1(\Omega)$ if and only if the series $\sum_{n=1}^\infty m\{x \in \Omega : |f(x)| > n\}$ is convergent. Hint: Assume $0 \leq f < \infty$, set $E_n = \{x \in \Omega : f(x) > n\}$, $F_n = E_n \setminus E_{n+1}$, and note that $E_n = \bigcup_{i=n}^\infty F_i$.

9. Let $f \in \mathcal{L}^1(\Omega)$ and suppose $f > 0$. Show that

$$\lim_{n \to \infty} \int_\Omega f^{1/n} dm = m(\Omega).$$

Hint: Consider the integral over the sets $\{x \in \Omega : f(x) \leq 1\}$ and $\{x \in \Omega : f(x) > 1\}$.

10. Suppose the function $f : [a, b] \times [a, b] \to \mathbb{R}$ satisfies the following conditions:

   (a) $f(t, \cdot) \in \mathcal{L}^1(\Omega)$ for all $t \in [a, b]$.

   (b) $\left|\dfrac{\partial}{\partial t} f(t, x)\right| \leq \varphi(x)$ for all $t \in [a, b]$, $x \in \Omega$, where $\varphi \in \mathcal{L}^1(\Omega)$.

   Prove that $\dfrac{d}{dt} \int_\Omega f(t, \cdot) dm = \int_\Omega \dfrac{\partial}{\partial t} f(t, \cdot) dm.$

11. Let $f \in \mathcal{L}^0(0, \infty)$ and define g_t on $(0, \infty)$ by $g_t(x) = x^t f(x)$. If $g_a, g_b \in \mathcal{L}^1(0, \infty)$, where $0 < a < b$, prove that $g_t \in \mathcal{L}^1(0, \infty)$ for all $t \in (a, b)$, and that the function $t \mapsto \int_{(0,\infty)} g_t dm$ is continuous.

12. For any $f \in \mathcal{L}^1(\mathbb{R})$, the *cosine transform* and the *sine transform* of f are defined, respectively, by

$$A(\xi) = \int_\mathbb{R} f(x) \cos(\xi x) dx, \quad B(\xi) = \int_\mathbb{R} f(x) \sin(\xi x) dx, \quad \xi \in \mathbb{R}.$$

Prove that A and B are continuous on $\mathbb{R}$.

11.4 Lebesgue and Riemann Integrals

Here we consider the Lebesgue integral over an interval in order to study its relation to the Riemann integral, and show that it is an extension of the latter. Furthermore, with the tools of measure theory at our disposal, we can now determine the necessary and sufficient condition for a bounded function to be Riemann integrable on a bounded interval.

Theorem 11.6
Let f be a bounded function on the compact interval $I = [a, b]$. If $f \in \mathcal{R}(a, b)$, then $f \in \mathcal{L}^1(I)$ and

$$\int_{[a,b]} f dm = \int_a^b f(x) dx.$$

Proof. Let $P = \{x_0, x_1, \cdots, x_n\}$ be a partition of I, and define the functions φ_P and ψ_P on I as follows:

$$\varphi_P = \sum_{i=0}^{n-1} m_i \chi_{[x_i, x_{i+1})},$$

$$\psi_P = \sum_{i=0}^{n-1} M_i \chi_{[x_i, x_{i+1})},$$

where, as usual,

$$m_i = \inf\{f(x) : x_i \leq x \leq x_{i+1}\},$$
$$M_i = \sup\{f(x) : x_i \leq x \leq x_{i+1}\}.$$

Note that

$$\varphi_P \leq f \leq \psi_P, \quad \varphi_P, \psi_P \in \mathcal{S}(I).$$

Referring to Definitions 8.2 and 8.3,

$$\int_I \varphi_P \, dm = L(f, P),$$
$$\int_I \psi_P \, dm = U(f, P).$$

Furthermore,

$$Q \supseteq P \implies \varphi_Q \geq \varphi_P, \; \psi_Q \leq \psi_P.$$

Choose a sequence of partitions (P_n) of I such that, for each $n \in \mathbb{N}$,

$$P_{n+1} \supseteq P_n, \; \|P_n\| \to 0.$$

For example, P_n could be the uniform partition of I into 2^n equal parts. Setting $\varphi_n = \varphi_{P_n}$ and $\psi_n = \psi_{P_n}$, we conclude that the sequence (φ_n) is increasing while (ψ_n) is decreasing. Their limits

$$\varphi(x) = \lim_{n \to \infty} \varphi_n(x), \; \psi(x) = \lim_{n \to \infty} \psi_n(x)$$

are clearly measurable and satisfy $\varphi \leq f \leq \psi$. Since f is bounded, the sequence $(\psi_n - \varphi_n)$ is also bounded, and the bounded convergence theorem (Corollary 11.5.1) implies

$$\int_I (\psi - \varphi) \, dm = \lim_{n \to \infty} \int_I (\psi_n - \varphi_n) \, dm$$
$$= \lim_{n \to \infty} [U(f, P_n) - L(f, P_n)]$$
$$= U(f) - L(f) = 0,$$

where we use Darboux's theorem and the fact that f is Riemann integrable in the last two equalities. Since $\psi - \varphi \geq 0$ on I, it follows from Theorem 11.2(vii) that $\psi - \varphi = 0$ a.e., hence

$$f = \varphi = \psi \quad \text{a.e.,} \tag{11.24}$$

which means f is measurable. Furthermore,

$$\begin{aligned}
\int_I f\,dm &= \int_I \varphi\,dm \\
&= \lim_{n\to\infty} \int_I \varphi_n\,dm \\
&= \lim_{n\to\infty} L(f, P_n) \\
&= L(f) \\
&= \int_a^b f(x)\,dx. \quad \square
\end{aligned}$$

Remarks 11.5

1. According to Theorem 11.6, we have $\mathcal{R}(a,b) \subseteq \mathcal{L}^1(a,b)$ for any bounded interval (a,b). Since the function

$$f(x) = \begin{cases} 1, & x \in \mathbb{Q} \cap [0,1] \\ 0, & x \in \mathbb{Q}^c \cap [0,1], \end{cases}$$

is Lebesgue (but not Riemann) integrable, we actually have the proper inclusion $\mathcal{R}(a,b) \subset \mathcal{L}^1(a,b)$.

2. Theorem 11.6 allows us to use the methods of Chapter 8 to evaluate $\int_{[a,b]} f\,dm$ when f is Riemann integrable. In particular, if f is continuous on $[a,b]$ and F is a primitive function of f on $[a,b]$, then

$$\int_{[a,b]} f\,dm = F(b) - F(a).$$

Refer to Examples 11.1 and 11.3 in this connection. Conversely, where limits are involved, we also use the more powerful results of Lebesgue integration to evaluate a Riemann integral (see Examples 11.8 and 11.9 below).

3. If f is continuous on $[a,b]$ and $F(x) = \int_{[a,x]} f\,dm$, then F is a primitive function of f. In fact, if $f \in \mathcal{L}^1(a,b)$, we have the following significant result which generalizes Theorem 8.11(ii), and whose proof may be found in [ROY],

$$F'(x) = f(x) \quad \text{a.e.}$$

The tools used in the proof of Theorem 11.6 allow us now to completely characterize the set $\mathcal{R}(a,b)$ of Riemann integrable functions, as promised at the beginning of this section.

Theorem 11.7
Let f be a bounded function on a compact interval $I = [a,b]$, and suppose $C \subseteq I$ is it domain of continuity. Then

$$f \in \mathcal{R}(a,b) \Leftrightarrow m(I\backslash C) = 0,$$

that is, f is Riemann integrable on $[a,b]$ if, and only if, f is continuous almost everywhere.

Proof. Let P_n, φ_n, ψ_n, φ, ψ be as defined in the proof of Theorem 11.6.

(i) Suppose $f \in \mathcal{R}(a,b)$. It suffices to show that

$$I\backslash C \subseteq E \cup \{x \in I : \varphi(x) \neq \psi(x)\}, \tag{11.25}$$

where $E = \bigcup_{n=1}^{\infty} P_n$ is the set of all points of the partitions P_n. That is because, on the one hand, E is countable (hence $m(E) = 0$) and, on the other, $m(\{x \in I : \varphi(x) \neq \psi(x)\}) = 0$ as we saw in equation (11.24). To prove (11.25), let $x \in I\backslash C$. Then there is a positive number ε and a sequence (x_k) in I such that $x_k \to x$ and $|f(x_k) - f(x)| \geq \varepsilon$ for all k. If $x \notin E$ then, for every $n \in \mathbb{N}$, there is a k such that x_k and x both lie in the same subinterval of P_n. But since $|f(x_k) - f(x)| \geq \varepsilon$, we obtain

$$\psi_n(x) - \varphi_n(x) \geq \varepsilon$$
$$\Rightarrow \psi(x) - \varphi(x) \geq \varepsilon$$
$$\Rightarrow x \in \{x \in I : \varphi(x) \neq \psi(x)\}.$$

(ii) Assume now that $m(I\backslash C) = 0$. If $x \in C$ then, for every $\varepsilon > 0$, there is a $\delta > 0$ such that

$$x' \in (x - \delta, x + \delta) \Rightarrow |f(x) - f(x')| < \varepsilon. \tag{11.26}$$

Since $\|P_n\| \to 0$ we can choose N so that the subinterval of P_N which contains x is contained in $(x - \delta, x + \delta)$. In view of (11.26) we therefore have

$$\psi_N(x) - \varphi_N(x) \leq 2\varepsilon$$
$$\Rightarrow \psi(x) - \varphi(x) \leq 2\varepsilon.$$

Since $\varepsilon > 0$ is arbitrary, $\psi(x) = \varphi(x)$, and since $m(I \setminus C) = 0$, $\psi = \varphi$ a.e. Referring back to equation (11.23), this implies that $U(f) = L(f)$ and hence $f \in \mathcal{R}(a,b)$. □

Theorem 11.6 does not address the relationship between the Lebesgue integral and the improper integral. We consider here the case of the semi-infinite interval $[a, \infty)$, and relegate the other types of improper integrals to the exercises.

To begin with, we should note that the existence of the improper integral $\int_a^\infty f(x)dx$ entails the measurability of f on $[a, \infty)$. Indeed, when $f \in \mathcal{R}(a,b)$ for every $b > a$, then $f \in \mathcal{L}^1(a,b)$ by Theorem 11.6. Since $f = \lim_{n \to \infty} f\chi_{[a,n]}$, we conclude that $f \in \mathcal{L}^0(a, \infty)$.

(i) Suppose $f \geq 0$ on $[a, \infty)$ and $\int_a^\infty f(x)dx$ is convergent. Then, since $f\chi_{[a,n]} \nearrow f$, Theorems 11.3 and 11.6 imply

$$\int_a^\infty f(x)dx = \lim_{n \to \infty} \int_a^n f(x)dx = \lim_{n \to \infty} \int_{[a,n]} f\,dm = \int_{[a,\infty)} f\,dm. \quad (11.27)$$

(ii) Suppose a is fixed and $f \in \mathcal{R}(a,b)$ for all $b > a$. If $f \in \mathcal{L}^1(a, \infty)$, then f has an improper Riemann integral over $[a, \infty)$ and

$$\int_a^\infty f(x)dx = \int_{[a,\infty)} f\,dm.$$

To see this, let $b_n > a$ and $b_n \to \infty$. Using the dominated convergence theorem, with $g = |f|$ and $f_n = f\chi_{[a,b_n]}$, we obtain

$$\int_{[a,\infty)} f\,dm = \lim_{n \to \infty} \int_{[a,b_n)} f(x)dx.$$

But, from Theorem 11.6, $\int_{[a,b_n)} f\,dm = \int_a^{b_n} f(x)dx$. Therefore

$$\int_a^\infty f(x)dx = \lim_{n \to \infty} \int_a^{b_n} f(x)dx = \lim_{n \to \infty} \int_{[a,b_n)} f(x)dx = \int_{[a,\infty)} f\,dm.$$

(iii) Suppose $f \in \mathcal{R}(a,b)$ for all $b > a$. If $\int_a^\infty |f(x)|\,dx$ is convergent, then both the Lebesgue integral $\int_{[a,\infty)} f\,dm$ and the improper integral $\int_a^\infty f(x)dx$ exist and are equal. Indeed, from (i), we see that $|f| \in \mathcal{L}^1(a, \infty)$. Since we have already observed that $f \in \mathcal{L}^0(a, \infty)$, it follows that $f \in \mathcal{L}^1(a, \infty)$. From (ii) we therefore conclude the existence of the improper integral $\int_a^\infty f(x)dx$ and its equality with $\int_{[a,\infty)} f\,dm$. Thus if both $\int_a^\infty f(x)dx$ and $\int_a^\infty |f(x)|\,dx$ exist, then so does $\int_{[a,\infty)} f\,dm$. In fact, the existence of $\int_a^\infty f(x)dx$ ensures that $f \in \mathcal{R}(a,b)$ for all $b > a$ and, by (iii), the existence of $\int_a^\infty |f(x)|\,dx$ implies $f \in \mathcal{L}^1(a, \infty)$.

An example where the improper integral of f exists and that of $|f|$ does not is provided by the function $\sin x/x$ over $(0,\infty)$, which is not in $\mathcal{L}^1(a,\infty)$ (see Exercise 9.3.12). Therefore not every function which is integrable in the improper sense is Lebesgue integrable.

(iv) Similar results can be obtained for other types of improper integrals. In general, the existence of the improper integral of $|f|$ ensures that f is Lebesgue integrable. If, furthermore, the improper integral of f exists, then it equals the Lebesgue integral.

Example 11.8 Evaluate

$$\lim_{n\to\infty} \int_0^1 \frac{nx}{1+n^2x^2} dx. \tag{11.28}$$

Solution. For every $n \in \mathbb{N}$, define the sequence of functions

$$f_n(x) = \frac{nx}{1+n^2x^2}, \quad x \in [0,1],$$

which clearly converges to 0. Since f_n attains its maximum value at $x = 1/n$, we have

$$\sup_{x\in[0,1]} |f_n(x)| = \frac{1}{2},$$

so the convergence $f_n \to 0$ is not uniform, and we cannot justify taking the limit inside the integral (11.28) on the basis of the properties of the Riemann integral. But we can also consider (11.28) as a Lebesgue integral, in which case we can use the bounded convergence theorem to write

$$\lim_{n\to\infty} \int_0^1 f_n(x)dx = \int_0^1 \lim_{n\to\infty} f_n(x)dx = 0.$$

Example 11.9 Evaluate

$$\lim_{n\to\infty} \int_0^n \left(1-\frac{x}{n}\right)^n e^{-2x} dx. \tag{11.29}$$

Solution. To express (11.29) as a Lebesgue integral, we define

$$f_n(x) = \chi_{[0,n]}(x) \cdot \left(1-\frac{x}{n}\right)^n e^{-2x},$$

so that

$$\int_0^n \left(1-\frac{x}{n}\right)^n e^{-2x} dx = \int_{[0,\infty)} f_n dm.$$

Noting that $|f_n(x)| \le e^{-2x}$ for all $n \in \mathbb{N}$, $x \ge 0$, and that $\int_{[0,\infty)} e^{-2x} dm$ can be evaluated as an improper Riemann integral,

$$\int_{[0,\infty)} e^{-2x} dm = \lim_{n \to \infty} \int_0^n e^{-2x} dx = \frac{1}{2},$$

we can apply the dominated convergence theorem to obtain

$$\lim_{n \to \infty} \int_0^n \left(1 - \frac{x}{n}\right)^n e^{-2x} dx = \int_{[0,\infty)} \lim_{n \to \infty} f_n dm$$

$$= \int_{[0,\infty)} e^{-3x} dx$$

$$= \frac{1}{3}.$$

Example 11.10 Prove that

$$\int_0^1 \left(\frac{\log x}{1-x}\right)^2 dx = 2 \sum_{n=1}^{\infty} \frac{1}{n^2}.$$

Solution. For every $x \in (-1, 1)$ we have the power series representation

$$\frac{1}{1-x} = \sum_{n=0}^{\infty} x^n.$$

Differentiating both sides, we obtain

$$\frac{1}{(1-x)^2} = \sum_{n=1}^{\infty} n x^{n-1}.$$

Using Corollary 11.3.1, we can write

$$\int_0^1 \left(\frac{\log x}{1-x}\right)^2 dx = \sum_{n=1}^{\infty} n \int_0^1 x^{n-1} (\log x)^2 dx. \tag{11.30}$$

When $n = 1$ the integral $\int_0^1 (\log x)^2 dx$ on the right-hand side is improper, whereas the other integrals are Riemann integrals. In either case we use integration by parts (and L'Hôpital's rule) to obtain

$$\int_0^1 x^{n-1} (\log x)^2 dx = \frac{x^n}{n} (\log x)^2 \Big|_0^1 - \int_0^1 \frac{2 x^{n-1} \log x}{n} dx$$

$$= -\frac{2}{n} \left(\frac{x^n \log x}{n} \Big|_0^1 - \int_0^1 \frac{x^{n-1}}{n} dx \right)$$

$$= \frac{2}{n^3}.$$

Substituting into (11.30) yields the desired equality.

The final theorem of this chapter allows us to approximate the Lebesgue integral of an $\mathcal{L}^1$ function over an interval by the integral of a continuous function over a compact interval, which can therefore be treated as a Riemann integral. A simple function $\varphi = \sum_{i=1}^{n} c_i \chi_{E_i}$ is called a *step function* if E_i is a real interval for each $i \in \{1, \cdots, n\}$.

Theorem 11.8
If $f \in \mathcal{L}^1(\Omega)$ then, for every $\varepsilon > 0$, there exists
(i) a step function ψ on Ω and a compact interval I such that $\psi(x) = 0$ for all $x \notin I$, and
$$\int_\Omega |f - \psi| \, dm < \varepsilon,$$
(ii) a continuous function g on Ω and a compact interval I such that $g(x) = 0$ for all $x \notin I$, and
$$\int_\Omega |f - g| \, dm < \varepsilon.$$

Proof
(i) we shall prove this part in stages:
(a) To begin with, assume $f = \chi_E$ where E is a measurable subset of Ω. Since $f \in \mathcal{L}^1(\Omega)$, $m(E) < \infty$. By Theorem 10.15 there is a set $F \in \mathcal{E}$ such that $m(E \Delta F) < \varepsilon$. Let $\psi = \chi_F$. Since $F = \bigcup_{i=1}^{n} [a_i, b_i)$, where the intervals $[a_i, b_i)$ are pairwise disjoint, ψ is a step function. Clearly, $\psi = 0$ outside $[a, b]$, where $a = \min\{a_1, \cdots, a_n\}$ and $b = \max\{b_1, \cdots, b_n\}$. Furthermore,
$$\begin{aligned} \int_\Omega |f - \psi| \, dm &= \int_\Omega |\chi_E - \chi_F| \, dm \\ &= \int_\Omega \chi_{E \Delta F} \, dm \\ &= m((E \Delta F) \cap \Omega) < \varepsilon. \end{aligned}$$

(b) Suppose $f = \sum_{i=1}^{n} c_i \chi_{E_i} \in \mathcal{S}_+(\Omega)$. Let $c = \max\{c_1, \cdots, c_n\}$, and, for each $1 \leq i \leq n$, choose a step function ψ_i as in (a) such that
$$\int_\Omega |\chi_{E_i} - \psi_i| \, dm < \frac{\varepsilon}{nc}.$$
Now $\psi = \sum_{i=1}^{n} c_i \psi_i$ is a step function which vanishes (i.e., equals 0) outside a compact interval and satisfies
$$\int_\Omega |f - \psi| \, dm \leq \sum_{i=1}^{n} c_i \int_\Omega |\chi_{E_i} - \psi_i| \, dm < \varepsilon.$$

426   Elements of Real Analysis

(c) Suppose now that $f \in \mathcal{L}_+^0(\Omega)$. By the definition of the integral, we can choose $\varphi \in \mathcal{S}_+(\Omega)$ such that $\varphi \leq f$ and $0 \leq \int_\Omega f \, dm - \int_\Omega \varphi \, dm < \varepsilon/2$. Since $\varphi \in \mathcal{L}^1(\Omega)$, by part (b) there is a step function ψ which vanishes outside a compact interval and satisfies

$$\int_\Omega |\varphi - \psi| \, dm < \frac{\varepsilon}{2},$$

and therefore

$$\int_\Omega |f - \psi| \, dm \leq \int_\Omega |f - \varphi| \, dm + \int_\Omega |\varphi - \psi| \, dm < \varepsilon.$$

(d) Finally, let $f \in \mathcal{L}^1(\Omega)$. According to part (c), there are two step functions ψ_1 and ψ_2, both of which vanish outside a compact interval, such that

$$\int_\Omega |f^+ - \psi_1| \, dm < \frac{\varepsilon}{2}, \quad \int_\Omega |f^- - \psi_2| \, dm < \frac{\varepsilon}{2}.$$

Consequently, $\psi = \psi_1 - \psi_2$ is a step function which vanishes outside a compact interval and satisfies

$$\int_\Omega |f - \psi| \, dm \leq \int_\Omega |f^+ - \psi_1| \, dm + \int_\Omega |f^- - \psi_2| \, dm < \varepsilon.$$

(ii) To prove the second part of the theorem, choose a step function ψ which vanishes outside a compact interval and satisfies

$$\int_\Omega |f - \psi| \, dm < \frac{\varepsilon}{2}.$$

Suppose $\psi = \sum_{i=1}^n c_i \chi_{[a_i, b_i)}$, where the intervals are pairwise disjoint. Let $0 < \varepsilon' < \varepsilon/n \max |c_i|$, and define $g_i : [a_i, b_i) \to \mathbb{R}$ by

$$g_i(x) = \begin{cases} 2(x - a_i)/\varepsilon', & x \in [a_i, a_i + \varepsilon'/2) \\ 1, & x \in [a_i + \varepsilon'/2, b_i - \varepsilon'/2) \\ -2(x - b_i)/\varepsilon', & x \in [b_i - \varepsilon'/2, b_i) \\ 0, & x \notin [a_i, b_i). \end{cases}$$

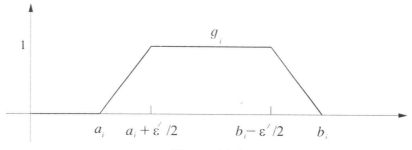

Figure 11.4

Each g_i, and hence $g = \sum_{i=1}^{n} c_i g_i$, is a continuous function which vanishes outside $\bigcup_{i=1}^{n}[a_i, b_i]$. Furthermore,

$$\int_\Omega |\psi - g|\, dm \leq \sum_{i=1}^{n} |c_i| \int_\Omega \left|\chi_{[a_i,b_i)} - g_i\right| dm$$

$$= \sum_{i=1}^{n} |c_i| \frac{\varepsilon'}{2} < \frac{\varepsilon}{2}.$$

Hence $\int_\Omega |f - g|\, dm \leq \int_\Omega |f - \psi|\, dm + \int_\Omega |\psi - g|\, dm < \varepsilon$. □

Example 11.11 If f and $|f|$ have improper integrals on $\mathbb{R}$, prove that

$$\lim_{n \to \infty} \int_{-\infty}^{\infty} f(x) \cos nx\, dx = 0.$$

Proof. The function $f(x) \cos nx$ is clearly Lebesgue integrable on $\mathbb{R}$ and we use Theorem 11.8 to assert that, for any $\varepsilon > 0$, there is a step function $\psi = \sum_{i=1}^{k} c_i \chi_{[a_i, b_i)}$, which vanishes outside a compact interval, and which satisfies

$$\int_\mathbb{R} |f - \psi|\, dm < \frac{\varepsilon}{2}.$$

Thus

$$\left|\int_{-\infty}^{\infty} f(x) \cos nx\, dx\right| \leq \left|\int_{-\infty}^{\infty} [f(x) - \psi(x)] \cos nx\, dx\right| + \left|\int_{-\infty}^{\infty} \psi(x) \cos nx\, dx\right|$$

$$< \frac{\varepsilon}{2} + \left|\sum_{i=1}^{k} c_i \int_{a_i}^{b_i} \cos nx\, dx\right|$$

$$= \frac{\varepsilon}{2} + \left|\sum_{i=1}^{k} \frac{c_i}{n} (\sin nb_i - \sin na_i)\right|$$

$$\leq \frac{\varepsilon}{2} + \frac{2}{n} \sum_{i=1}^{k} |c_i| < \varepsilon \text{ for all } n \geq \frac{4}{\varepsilon} \sum_{i=1}^{k} |c_i|.$$

Hence $\lim_{n \to \infty} \int_{-\infty}^{\infty} f(x) \cos nx\, dx = 0$. This result is sometimes referred to as the *Riemann-Lebesgue lemma*. □

Exercises 11.3

1. Evaluate $\lim_{n\to\infty} \int_0^n \left(1+\frac{x}{n}\right)^n e^{-2x} dx.$

2. Prove that $\lim_{n\to\infty} \int_0^1 f_n(x) dx = 0$, where

   (a) $f_n(x) = \dfrac{n\sqrt{x}}{1+n^2 x^2},$

   (b) $f_n(x) = \dfrac{nx \log x}{1+n^2 x^2}.$

3. Evaluate $\lim_{n\to\infty} \int_0^\infty \left(1+\frac{x}{n}\right)^{-n} \sin\frac{x}{n} dx.$

4. Prove that, for all $r > 0$,
$$\lim_{n\to\infty} \int_0^n \left(1-\frac{x}{n}\right)^n x^{r-1} dx = \int_0^\infty e^{-x} x^{r-1} dx.$$

5. Prove that $\int_0^1 \sin x \log x\, dx = \sum_{n=1}^\infty \dfrac{(-1)^n}{2n(2n)!}.$

6. Prove that $\int_0^1 (e^x - 1)\left(\log x + \dfrac{1}{x}\right) dx = \sum_{n=1}^\infty \dfrac{n^2+n+1}{(n-1)!(n^2+n)^2}.$

7. If $p, q > 0$ prove that $\int_0^1 \dfrac{x^{p-1}}{1+x^q} dx = \sum_{n=0}^\infty (-1)^n \dfrac{1}{p+nq}$, then conclude that $\log 2 = \sum_{n=0}^\infty \dfrac{(-1)^n}{1+n}.$

8. Prove that $\int_0^\infty e^{-x} \cos\sqrt{x}\, dx = \sum_{n=0}^\infty \dfrac{(-1)^n n!}{(2n)!}.$

9. Prove that $\int_0^1 \dfrac{(x\log x)^2}{1+x^2} dx = 2\sum_{n=1}^\infty \dfrac{(-1)^{n-1}}{(2n+1)^3}.$

10. Prove that $\lim_{t\to 0} \int_a^b f(x) \sin tx\, dx = 0$ for every $f \in \mathcal{L}^1(a,b).$

11. If $f \in \mathcal{L}^1(\mathbb{R})$ prove that $\lim_{h\to 0} \int_\mathbb{R} |f(x+h) - f(x)|\, dm = 0.$ Hint: A continuous function on $\mathbb{R}$ which vanishes outside a compact interval is uniformly continuous.

References

[APO] T. M. Apostol, *Mathematical Analysis*, 2nd ed., Addison-Wesley, Reading, MA, 1974.

[B&S] R. G. Bartle and D. R. Sherbert, *Introduction to Real Analysis*, 3rd ed., Wiley, New York, 2000.

[B&M] Garrett Birkhoff and Saunders Mac Lane, *A Survey of Modern Algebra*, 3rd ed., Macmillan, New York, 1965.

[BUC] R. C. Buck, *Advanced Calculus*, 3rd ed., McGraw-Hill, Tokyo, 1978.

[BUR] J. C. Burkill, *A First Course in Mathematical Analysis*, Cambridge Univ. Press, Cambridge, 1962.

[CAN] Georg Cantor, *Contributions to the Founding of the Theory of Transfinite Numbers*, Dover, New York, 1915.

[COH] P. J. Cohen, *Set Theory and the Continuum Hypothesis*, Benjamin, Reading, MA, 1966.

[HAL] P. R. Halmos, *Measure Theory*, Van Nostrand, Princeton, NJ, 1950.

[HAR] G. H. Hardy, *A Course of Pure Mathematics*, 10th ed., Cambridge Univ. Press, Cambridge, 1955.

[KAM] E. Kamke, *Theory of Sets*, Dover, New York, 1950.

[KRA] Steven G. Krantz, *Real Analysis and Foundations*, 2nd ed., Chapman and Hall/CRC, Boca Raton, FL, 2005.

[ROY] H. L. Royden, *Real Analysis*, 3rd ed., Macmillan, New York, 1988.

[RUD] Walter Rudin, *Principles of Mathematical Analysis*, McGraw-Hill, New York, 1964.

[SPI] Michael Spivak, *Calculus*, Benjamin, New York, 1967.

[TAR] Alfred Tarski, *Introduction to Logic*, Oxford Univ. Press, New York, 1965.

[TAY] S. J. Taylor, *Introduction to Measure and Integration*, Cambridge Univ. Press, Cambridge, 1973.

Notation

$\mathbb{N} = \{1, 2, 3, \cdots\}$
$\mathbb{N}_0 = \{0, 1, 2, 3, \cdots\}$
$\mathbb{Z} = \{\cdots, -2, -1, 0, 1, 2, \cdots\}$
$\mathbb{Q}$ set of rational numbers
$\mathbb{R}$ set of real numbers
$\bar{\mathbb{R}} = \mathbb{R} \cup \{-\infty, \infty\}$ extended real numbers, 83
$B \backslash A$ complement of A in B, 2
A^c complement of A in $\mathbb{R}$, 2
A° interior of A, 111
$\bar{A}$ closure of A, 111
$\hat{A}$ cluster points of A, 89
$A \Delta B = (A/B) \cup (B/A)$ symmetric difference between A and B, 6
$A \times B$ cartesian product of A and B, 7
$\mathcal{P}(A)$ power set (set of subsets) of A, 6
$|A|$ cardinal number of A, 60
$\aleph_0$ cardinal number of $\mathbb{N}$
$\aleph_1$ cardinal number of $\mathcal{P}(\mathbb{N})$
c cardinal number of $\mathbb{R}$
$\sim$ equivalence relation between sets, 55
$f \circ g$ composition of the functions f and g, 14
D_f, R_f domain and range of f, 9-10
$\mathcal{P}(a, b)$ set of partitions of $[a, b]$, 258
$U(f, P)$, $L(f, P)$ upper and lower Riemann sums of f on P, 258
$S(f, P, \boldsymbol{\alpha})$ Riemann sum of f on P with mark $\boldsymbol{\alpha}$, 273
$U(f) = \inf\{U(f, P) : P \in \mathcal{P}(a, b)\}$, 262
$L(f) = \sup\{U(f, P) : P \in \mathcal{P}(a, b)\}$, 262
$\|P\|$ norm of the partition P, 268
$\mathcal{R}(a, b)$ Riemann integrable functions on $[a, b]$, 262
$\mathcal{I} = \{[a, b) : a, b \in \mathbb{R}, a < b\}$, 351
$\mathcal{I}_0 = \{(a, b) : a, b \in \mathbb{R}\}$, 355
$\mathcal{I}_1 = \{[a, b] : a, b \in \mathbb{R}\}$, 355
$\mathcal{I}_2 = \{(a, \infty) : a \in \mathbb{R}\}$, 355
$\mathcal{I}_3 = \{[a, \infty) : a \in \mathbb{R}\}$, 355

Notation

$\mathcal{E}$ ring of finite, disjoint unions of intervals in $\mathcal{I}$, 351

$a \vee b = \max\{a,b\}, \ a \wedge b = \min\{a,b\}$, 352

$(f \vee g)(x) = f(x) \vee g(x), \ (f \wedge g)(x) = f(x) \wedge g(x)$, 385

$f^+ = f \vee 0, \ f^- = -(f \wedge 0)$, 385

$\mathcal{A}(\mathcal{D})$ σ-algebra generated by the class $\mathcal{D}$, 354

$\mathcal{B} = \mathcal{A}(\mathcal{I})$ Borel sets, 354

m^* Lebesgue outer measure, 361

m Lebesgue measure, 372

$\mathcal{M}$ Lebesgue measurable sets, 369

μ general measure function, page 365

χ_E characteristic function of E, 388

$\mathcal{S}(\Omega)$ simple functions defined on $\Omega \subseteq \mathbb{R}$, 388

$\mathcal{S}_+(\Omega)$ non-negative simple functions on Ω, 394

$\mathcal{L}^0(\Omega)$ Lebesgue measurable functions on Ω, 394

$\mathcal{L}^0_+(\Omega)$ non-negative Lebesgue measurable functions on Ω, 394

$\mathcal{L}^1(\Omega)$ Lebesgue integrable functions on Ω, 400

$\mathcal{L}^2(\Omega)$ Lebesgue square-integrable functions on Ω, 406

$f_n \to f$ the sequence of functions f_n converges pointwise to f, 303

$f_n \xrightarrow{u} f$ f_n converges uniformly to f, 307

$f_n \nearrow f$ f_n converges to f from below, 389.

Index

Abel's partial summation formula, 327
Abel's test for uniform convergence, 329
Abel's continuity theorem, 335
algebra, σ-algebra, 353
absolute value, 25
accumulation point, 89
Archimedes theorem, 43

Bernoulli's inequality, 230
binary operation, 17
binary relation, 8
    inverse relation, 9
    reflexive relation, 16
    symmetric relation, 16
    transitive relation, 16
    equivalence relation, 16
binomial theorem, 38
Bolzano-Weierstrass theorem, 93
bounded convergence theorem, 414

Cantor theorem, 92
Cantor set, 108
Cantor function, 392
Carathéodory condition, 369
cardinal number, 60
cartesian product, 7
Cauchy criterion, 89, 310
Cauchy's mean value theorem, 238
Cauchy principal value, 297, 298
Cauchy-Hadamard theorem, 333

chain rule, 215
cluster point, 89
completeness axiom, 41
completeness of measure, 374
continuum hypothesis, 61
convergence of sequence of functions,
    pointwise, 304
    uniform, 307
convergence of series of functions, 323
countable additivity, 350
countable subadditivity, 362, 365
critical point, 224

Darboux's theorem, 233, 272
De Morgan's laws, 3
density, 43
derivative, 208
Dini's theorem, 312
Dirichlet function, 263
Dirichlet test for uniform convergence, 327
discontinuity, 163
    removable, 164
    non-removable, 164
    jump, 165
    oscillatory, 165
    infinite, 166
disjoint union, 351
dominated convergence theorem, 413

equivalence,

of statements, 5
of sets, 55
relation, 16

Fatou's lemma, 412
field axioms, 19
finite additivity, 358
finite subadditivity, 360
function, 9
    domain, 9
    co-domain, 9
    range, 10
    natural domain, 12
    graph, 11
    injection, 12
    surjection, 13
    bijection, 13
    inverse, 12
    restriction, 14
    extension, 14
    composition, 14
    exponential, 45, 341
    logarithmic, 46, 342
    trigonometric, 143, 343
    monotonic, 27, 155
    continuous, 160
    uniformly continuous, 188
    domain of continuity, 161
    differentiable, 208
    analytic, 338
    measurable, 381
    characteristic, 388
    simple, 388
fundamental theorem of calculus, 288

Heine-Borel theorem, 196
Hölder's inequality, 407

implication (logical), 4

improper integral, 295
indeterminate form, 241
infimum, 39
intermediate value property, 176
intersection of sets, 2
irrational numbers, 42
integers, 34
integration by parts, 292
intersection (of sets), 2
inverse function theorem, 232
isolated point, 89

Lebesgue integral, 399
Leibnitz' rule, 222
L'Hôpital's rule, 238
Lipschitz condition, 193
limit,
    of a sequence, 67
    superior, 101' 155
    inferior, 101, 155
    of a function, 132
    right-hand, 149
    left-hand, 149
local (relative) extremum, 223
lower bound, greatest lower bound, 39
lower integral, 262
lower sum, 258

mark (on a partition), 273
mathematical induction, 30, 32
maximum, 23, 174
mean value theorem,
    of differentiation, 227
    of integration, 280
measure, 350
    outer, 361
    general, 365
    Dirac, 366
    Lebesgue, 372

Index   435

Borel, 374
minimum, 23, 174
monotone convergence theorem, 407

Natural numbers, 30
neighborhood, 67
norm (of a partition), 268

ordered pair, 7
order axioms, 23

partition, 258
prime number, 33
prime factorization theorem, 33
power series, 332
proof,
    direct, 4
    contrapositive, 4
    by contradiction, 5
property,
    well ordering, 31
    intermediate value, 176

radius of convergence, 333
rational numbers, 34
real interval, 26, 27
real line, 26
real numbers, 19
    decimal expansion, 49
    binary expansion, 52
    ternary expansion, 52
    extended, 83
Riemann integral, 262
    linearity property, 277
    positivity property, 277
    monotonicity property, 277
Riemann criterion, 264
Riemann-Lebesgue lemma, 427
Riemann sum, 273
ring, σ-ring, 351, 353

Rolle's theorem, 226

Schwarz inequality, 29, 406
sequence, 65
    convergent, 67
    divergent, 67
    shuffled, 71
    Fibonacci, 72
    bounded, 73
    monotonic, 82
    Cauchy, 88
series, 113
    convergence, 113
    absolute convergence, 116
    conditional, 117
    divergence, 113
    geometric, 114
    harmonic, 115
    rearrangement, 117
    alternating, 127
series convergence tests, 121
    comparison test, 121
    limit comparison test, 121
    root test, 124
    ratio test, 126
    alternating series test, 127
    integral test, 299
set, 1
    empty, 2
    universal, 2
    complement of, 2
    power, 6
    inductive, 30
    countable, 56
    uncountable, 60
    denumerable, 56
    open, 105
    closed, 107
    interior of, 111
    closure of, 111

boundary of, 111
　　　compact, 194
　　　Borel, 354
　　　measurable, 372
singular point, 166
subsequence, 97
subset, 2
supremum, 39
symmetric difference, 6

Taylor's remainder,
　　　Lagrange form, 249
　　　Cauchy form, 249
　　　integral form, 294
Taylor's theorem, 247
Taylor series, 338
topology, 106

uniform continuity, 186
union (of sets), 2
upper bound, least upper bound, 39
upper integral, 262
upper sum, 258

Weierstrass approximation theorem, 318
Weierstrass M-test, 325

Young's theorem, 252

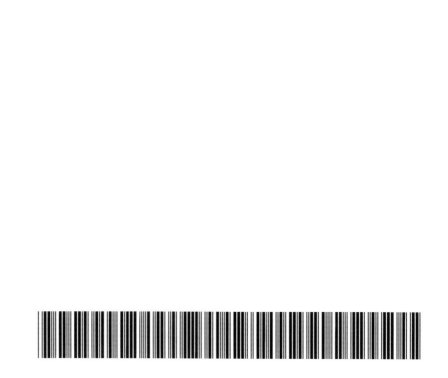